Toward Modern Humans

The Yabrudian and Micoquian

400–50 k-years ago

Proceedings of a Congress held at the University of Haifa
November 3-9, 1996

Edited by

Avraham Ronen
Mina Weinstein-Evron

BAR International Series 850
2000

Published in 2016 by
BAR Publishing, Oxford

BAR International Series 850

Toward Modern Humans

ISBN 978 1 84171 056 3

BAR Publishing is the trading name of British Archaeological Reports (Oxford) Ltd. British Archaeological Reports was first incorporated in 1974 to publish the BAR Series, International and British. In 1992 Hadrian Books Ltd became part of the BAR group. This volume was originally published by Archaeopress in conjunction with British Archaeological Reports (Oxford) Ltd / Hadrian Books Ltd, the Series principal publisher, in 2000. This present volume is published by BAR Publishing, 2016.

Printed in England

BAR
PUBLISHING

BAR titles are available from:

BAR Publishing
122 Banbury Rd, Oxford, OX2 7BP, UK
EMAIL info@barpublishing.com
PHONE +44 (0)1865 310431
FAX +44 (0)1865 316916
www.barpublishing.com

CONTENTS

INTRODUCTION

Avraham Ronen and Mina Weinstein-Evron

In the 1930s and 1940s, before the term 'Yabrudian' was coined (Rust, 1950), the lithic industries of Tabun E (Garrod and Bate, 1937) and Umm Qatafa D1 (Neuville, 1931) were labeled ' Micoquian'. Subsequently, Tabun E was renamed Yabrudian. Unlike the Upper Acheulean entity, which occurs throughout the Levant and in neighbouring areas, the Yabrudian is limited to a restricted territory in the central part of the Levant.

The Micoquian and Yabrudian have in common some special forms of bifaces, and neither adopted the Levallois technology. This explains why these Levantine occurrences were initially called Micoquian. With advanced research, the term Micoquian has taken on different meanings (Bosinski, 1967), and now appears to have existed mainly in central and eastern Europe. In south-central Europe (Richter, 1997) and western Europe (Farizy, 1995), Late Micoquian layers may sometimes contain Levallois products. In present-day terminology, it is mainly the 'bifacial backed-knife' which is shared by the Yabrudian and the Micoquian.

Key-traits of the Yabrudian, besides the bifacial backed knives, are Quina-type flake production, Quina tool forms and Quina retouch technique. It is therefore all the more surprising that during the subsequent stages of the Palaeolithic in the Levant, Quina-type Mousterian is entirely absent. The only Mousterian ever found in the Levant broadly fits the definition of 'Typical Mousterian'.

Surprising, too, is the fact that blade industries were encountered within the Yabrudian entity. Termed Pre-Aurignacian (Rust, 1950) and Amudian (Garrod, 1970), their distribution in the Levant was limited to the same territory - in fact, to the same very sites - as the Yabrudian. While different in details, common to all these Pre-Upper Palaeolithic blade assemblages (Ronen, 1992) is that, technologically speaking, they are closer to Upper Palaeolithic blade production than anything produced earlier, in the Early Palaeolithic, or later, in the Middle Palaeolithic. To these Levantine blade occurrences were added, more recently, other early blade assemblages in Europe, Africa and central Asia. The early blade assemblages now seem to have had a very wide geographical distribution and all seem to have occurred towards the end of the Lower Palaeolithic or the onset of the Middle Palaeolithic. The Intra-Yabrudian blade occurrences in the Levant seem to be the oldest.

Within Palaeolithic time scale, the phenomena mentioned above took place at the end of the eons-long Oldowan-Acheulean tradition, related to *Homo habilis* - *H. erectus*. Although their makers are either unknown (blade assemblages, Micoquian) or only poorly known (Yabrudian), it seems safe to say that from an evolutionary perspective, the phenomena concerned here occurred somewhat before the global appearance of modern humans.

We gathered in November 1996 to discuss these issues in the light of recent research. Our purpose was to evaluate the possible relationships between the Yabrudian, Micoquian, Quina and the early blade assemblages, considering the recently established chronological framework and environmental constraints.

The congress was organized by the Zinman Institute of Archaeology, under the auspices of the Faculty of Humanities, University of Haifa. The papers delivered at the meeting and the discussions which followed are presented here. The contributions by Copeland, Foltyn et al. and Otte were written especially for this volume, though their authors were unable to attend the meeting.

Bibliography

Bosinski, G., 1967. *Die mittelpalaolithische Funde im westlichen Mitteleuropa*. Fundamenta A/4. Köln.

Farizy, C., 1995. Industries Charentiennes a influences micoquiennes, l'example de l'Est de la France. *Paléo*, Suppl. 1, 173-178.

Garrod, D. A. E., 1970. Pre-Aurignacian and Amudian: a comparative study of the earliest blade industries of the Near East. *Fundamenta* A/2, 224-229. Köln.

Garrod, D. A. E. and Bate, D. M. A., 1937. *The Stone Age of Mount Carmel*, Vol. I. Oxford.

Neuville, R., 1931. L'Acheuleen superieur de la grotte d'Oumm Qatafa. *L'Anthropologie* 41, 13-51 and 249-263.

Richter, J., 1997. *Der G-Schichten-Komplex der Sesselfelsgrotte. Sesselfelsgrotte III.* Saarbrucker Druckerei und Verlag.

Ronen, A., 1992. The emergence of blade technology: cultural affinities. In: Akazawa, T., Aoki, A. and Kimura, T. (eds.), *The Evolution and Dispersal of Modern Humans in Asia*, pp. 217-228. Hokusen-Sha, Tokyo.

Rust, A., 1950. *Die Hohlenfunde von Jabrud (Syrien)*. Karel Wacholtz, Neumunster.

Acknowledgments

Ms. Ruth Maos conducted with efficiency and a smile the secretarial work of preparing the congress, the logistics during the congress and the tedious post-congress coordination of authors, discussions, manuscripts and proofs. Ruth's help in the publication of this volume has been paramount. Ms. Ann Miller and Dr. Daniel Kaufman helped to improve the English texts and Ms. Danielle Friedlander helped with the French texts. Ms. Hana Katz and Dr. Dani Nadel transcribed the discussions' recordings with great devotion. A large part of editing the proceedings was done when the senior editor was on a sabbatical leave at the Department of Prehistory, the University of Basel. Cordial thanks are due to Prof. Dr. Jean-Marie Le Tensorer, Head of the Department, and to faculty and students, especially Reto Jagher, for helpful discussions, and to the Fonds zur Förderung der Studien auf dem Gebiete der agyptologischen, orientalischen und klassischen Altertumskunde der Universität Basel for financial help.

Generous contributions made by the following sponsors have rendered the congress possible. We express our thanks and deep gratitude to:
Mrs. Betty Zinman and the Zinman Foundation
The Department of Public Relations, University of Haifa
The Rector, University of Haifa
The Matla and Feival Coastal and Underwater Archaeological Foundation
The Dean, Faculty of Humanities, University of Haifa
The Israel National Academy of Sciences
The Zinman Institute of Archaeology

THE ORGANIZING COMMITTEE

Prof. Dr. Gerhard BOSINSKI
Prof. Marcel OTTE
Dr. Árpád RINGER
Prof. Avraham RONEN
Prof. Mina WEINSTEIN-EVRON

CONTRIBUTORS (* not present in the meeting)

Bluszcz, Andrzej *
Bratlund, Bodil *
Burdukiewicz, Jan M.
Burke, Ariane M. *
Butrym, Jan *
Conard, Nicholas J.
Copeland, Lorraine *
Dobosi, Viola T.
Fischer, Birgit *
Foltyn, Edelgarda M. *
Foltyn, Eugeniusz *
Gautier, Achilles *
Jensen, Helle J. *
Kozłowski, Janusz K. *
Królik, Halina *
Mania, Dietrich
Maruszczak, Henryk *
Mojski, Józef E. *
Nadachowski, Adam *
Oliva, Martin
Otte, Marcel *
Ranov, Vadim
Ringer, Árpád
Rolland, Nicolas
Ronen, Avraham
Schäfer, Joachim
Schild, Romuald
Shifroni, Anat
Stworzewicz, Ewa *
Sulgostowska, Zofia *
Tomaszewski, Andrzej, J. *
Tsatskin, Alexander
Valoch, Karel
Vishnyatsky, Leonid B.

DISSCUSSANTS

Banesz, Ladislav
Belfer-Cohen, Anna
Bosinski, Gerhard
Callow, Paul
Goring-Morris, Nigel
Goren-Inbar Naama
Horowitz, Aharon
Laukhin, Stanislav A.
Mania, Ursula
Marder, Ofer
Porat, Naomi
Praslow, Nikolai D.
Sivan, Dorit

PART I

THE SETTING

TOWARD MODERN HUMANS: YABRUDIAN AND MICOQUIAN, 400 – 50 kyears ago.
Edited by A. Ronen and M. Weinstein-Evron. BAR S850, 2000.

L'ACHEULEEN COMME POPULATION

Marcel Otte

Introduction

L'aire de répartition des techniques primitives définie à travers l'Eurasie par H. L. Movius (1948) reste bien connue et diversement appréciée (Ikasawa-Smith, 1978; Vishnyatsky, 1996; Bosinski, 1992; Rolland, 1992). Cependant, elle semble bien correspondre à l'existence de phénomènes profonds qui ont affecté les premières formes d'installation humaine, dans les zones septentrionales. Plusieurs découvertes récentes forcent à reconsidérer cette théorie objectivement, car elle touche à la notion même d'humanité à ses aptitudes primordiales et à sa relation à la Nature lors de sa genèse (Lorenz, 1970).

Techniquement, les industries de l'Extrême-0rient ne correspondent en rien à la conception africaine contemporaine. En Chine (Lan Po et Chuhnkum, 1990; Seonbok and Clark, 1983) comme à Java (Swischer et al., 1994), les documents 'datés' entre 2 et 1 millions d'années possèdent des critères associés aux travaux de pièces en bois: petits éclats tranchants irréguliers et blocs massifs rendus tranchants ou denticulés. Quelle que soit la signification (fonctionnelle, traditionnelle ou conceptuelle) de ces ensembles, ils échappent totalement aux conceptions acheuléennes occidentales. Leurs hautes datations (accompagnées de restes de fossiles humains) prouvent une nappe de population autonome et puissante dans le Sud-Est asiatique, très ancienne (Ranov et al., 1995; Raposo and Santonja, 1995). Quelle qu'en soit l'origine, la nature ou le sens, cette masse de population aux techniques particulières doit être considérée dans les tableaux de reconstitution ultérieur de l'Ancien Monde, en dehors des influences africaines.

Aux origines donc (apparemment dès 2 millions d'années), une aire d'adaptation humaine importante, non acheuléenne, s'est constituée dans la ceinture forestière asiatique. Ce noyau essentiel de population a pu se constituer sur un contexte environnemental analogue à l'Afrique originelle. Quoiqu'il en fut, leur destin fut désormais distinct des populations africaines (Merrick et al., 1973).

Caucase

La découverte, l'analyse, puis la datation des restes anthropiques à Dmanisi en Géorgie éclairent elles aussi ce débat sous un jour nouveau. Accompagnée d'une faune africaine mais aussi d'une industrie non acheuléenne, cette mandibule de Dmanisi, datée vers 1,8 millions d'années (Gabunia, 1992; Swischer et al., 1994), démontre l'extension hors Afrique vers les steppes, encore avant l'étape acheuléenne (Henke, 1997). La limite semble se percevoir en ces zones montagneuses intermédiaires entre la plaque afro-arabique et l'aire s'ouvrant sur les steppes d'Asie centrale. Toutefois, la distinction n'est pas seulement géographique mais, surtout, chronologique et culturelle. Dès le Caucase du Nord en effet, on voit apparaître les traditions acheuléennes très élaborées (Liubine, 1984), mais dans un stade bien ultérieur aux prémisses africaines ou aux sites levantins.

Le Levant

Grâce à des études spécialement précises, actives et documentées, on connaît mieux l'information du 'couloir levantin' que celles de bien d'autres régions cruciales. Il faut toutefois prendre garde de ne pas confondre l'intensité même de cette activité avec l'importance de la région où elle s'applique. D'autres voies de passage ont pu être empruntées (Yemen, Sicile, Gibraltar) et, par ailleurs, les foyers d'évolution principaux furent probablement concentrés aux centres des continents plutôt qu'en périphérie.

Néanmoins, on y reconnaît diverses phases, mieux attestées qu'ailleurs, quant à leur définition technique et chronologique (Ronen, 1991; Bar-Yosef, 1994; Tchernov, 1987). Des traces d'industries primitives, non-acheuléennes, s'y sont accumulées récemment sous une forme toujours plus convaincante: sites d'Evron, par exemple. Mais aussi, les ensembles 'Oldowayens' bien caractéristiques se retrouvent à Oubeidya (Tchernov, 1987), tandis que l'Acheuléen, classiquement africain, se retrouve à Gesher Benot Ya'aqov (Goren-Inbar et al., 1992) et à Latamné sur l'Oronte en Syrie (Clark, 1965). Ce tableau général semble reproduire fidèlement les successions observées en Afrique de l'Est.

Le Sens de cette Divergence

Si l'on compare l'Asie du Sud-Est avec le Proche-Orient au Paléolithique ancien, le sens de la différence apparaît clairement: des gestes techniques stéréotypés se trouvent liés à des traditions elles-mêmes forgées par des populations autonomes, dans leur histoire, leur genèse, leurs modes de vie et leur environnement. L'idée de fonds communs distincts et anciens peut difficilement être écartée, bien qu'aucune preuve biologique ne puisse jusqu'ici y être apportée (Biagi et al., 1996). Par ailleurs, le reste de l''histoire' semble également illustrer le maintien de cette distinction jusqu'aux phases les plus récentes du Paléolithique (Lan Po et Chuhnkum, 1990). Par contre, le continent africain manifeste, en succession chronologique rigoureuse, ces deux modes adaptatifs distincts. Ils ont donc pu se séparer géographiquement, portés par des populations migrantes dès les phases les plus anciennes de l'évolution africaine.

Enfin et surtout, lorsque l'on compare les restes de pierre laissés par le hasard de la préservation des outils ou des armes selon les conditions particulières de présentation, on y constate des enchaînements logiques totalement différents. Dans un cas (l'Acheuléen), le bloc est 'sculpté', telle une statue de marbre, au sein de la roche, progressivement dégagée vers la forme. L'outil y est donc conçu pour soi et en soi, tel un objectif final à atteindre et à manipuler. L'autre conception technique va finalement presque à contre-sens : l'outil de pierre y est quelconque, à peine ébauché, afin d'être utilisé tel quel, rapidement, provisoirement, dans une tâche dont nous ne savons rien; la conception était ici *au-delà* de l'objet retrouvé. Cet 'enchaînement' de transfert d'un matériau vers l'autre rend opaque l'intentionnalité des procédures techniques, toutes situées en aval de notre emprise, par les 'gestes retrouvés'.

Nous tenons cette distinction, relative à l'emprise sur la matière lithique, comme fondamentale et non anodine. Elle restitue, en d'autres termes, des façons de penser différentes, des manières distinctes d'agir sur le monde et, finalement, des valeurs différentes quant à la place prise par un acte intentionnel sur son environnement naturel.

L'Europe

Quoiqu'il en fut des sens donnés aux gestes techniques aussi anciens, une évidence s'impose à tous: ces gestuels codifiés s'instituent en véritables traditions que l'on peut suivre à travers les âges et les espaces successivement conquis par l'humanité originelle.

Ainsi, par sa marginalité, notre continent européen a-t-il 'subi' des courants migratoires appartenant à l'une et à l'autre traditions, voire populations (Otte , 1996). Le fonds essentiel de notre continent s'assimile assez aisément aux vagues 'asiatiques', soit réellement orientales, soit appartenant encore aux pratiques antérieures à l'Acheuléen en Afrique de l'Est. La continentalité, très marquée vers l'Orient où s'ouvre la zone des steppes, plaide plutôt en faveur de la première hypothèse, c'est-à-dire, issue d'une migration asiatique. En l'absence de Mer Méditerranée en Asie centrale, l'adaptation et la migration aux aires steppiques septentrionales a pu s'y faire plus précocement qu'en Europe. Cette adaptation permit alors la migration occidentale. Quoiqu'il en soit, toutes les industries connues du continent européen dans les stades archaïques sont à la fois très 'récentes' (après 1 million d'années) et non-acheuléennes, exactement comme en Asie dès les stades antérieurs. Autant en Europe de l'Est (Korolevo; Gladilin et Sitlivi, 1990), centrale (Bilzingsleben; Mania et al., 1980-86), qu'occidentale (Atapuerca; Carbonell et al., 1995) et méridionale (Isernia La Pineta; Peretto, 1991), la même tendance se vérifie. L'hypothèse d'une vague migratoire asiatique originelle semble donc la plus 'économique' pour expliquer de si profondes divergences avec le continent africain contemporain. La signification des restes osseux qui y sont associés restent d'un autre domaine de compétence (Mauer, Atapuerca). Mais leur interprétation devrait tenir compte des observations archéologiques (donc finalement comportementales) aussi élémentaires.

Vers 500 mille ans, le 'tableau' européen se modifie encore sensiblement. Toute l'aire sud-occidentale connaît alors l'extension des traditions techniques, connues depuis longtemps en Afrique du Nord. Les importantes séries acheuléennes d'Espagne, du Portugal et du Sud de la France contiennent à la fois les blocs sculptés 'bifaces' et les outils à tranchants terminaux ('hachereaux') sur le mode africain. Cette extension s'étend, en perdant de sa puissance et de sa netteté, jusqu'aux îles britanniques (Roe, 1981), la Belgique (Ulrix-Closset, 1975) et la Rhénanie (Bosinski, 1976). L'ensemble du mouvement, relativement récent (vers 450.000 ans à Boxgrove; Roberts, 1986) provient de l'ouest et, plus probablement, de Gibraltar. Le détroit, parfois plus étroit qu'aujourd'hui, possède en outre une 'île sous-marine', exondée temporairement lors des baisses du niveau marin. Le passage y était donc facilité épisodiquement. L'extrême contraction vers l'ouest et la relative lenteur de ce phénomène peuvent être expliquées par ce passage marin, retardant mais non empêchant les migrations africaines. Il s'agit donc d'une situation inverse de l'aire orientale où l'expansion à travers les steppes fut très précoce.

L'Espace et les Hommes

Si l'on cherche à assimiler ces techniques distinctes aux informations anthropologiques, on constate la carence générale de données pour la tradition acheuléenne. Néanmoins, les quelques restes dont on dispose à cet égard forment des 'isolats' anthropologiques classiques: Steinheim, Swanscombe furent diversement interprétés (Piveteau, 1973), mais toujours situés à l'extérieur ou en parallèle aux lignées classiques des *Homo erectus* européens (Mauer, Tautavel, Atapuerca, Petralona) dont l'origine ultime semble bien être asiatique.

La phase 'médiane' du Paléolithique européen voit à la fois la mixité des caractères anatomiques (aboutissant aux Néandertaliens) et l'uniformisation des techniques (éclats préparés et persistances locales d'outils bifaciaux). Ceci explique à la fois les traits annonçant les Néandertaliens présents chez les *erectus* classiques (Petralona, Mauer) et chez les 'Acheuléens' (Steinheim) (Leroi-Gourhan, 1983; Wolpoff, 1996). Ils sont en effet tous considérés rétrospectivement *par rapport à* leur future évolution commune. Les 'persistances' de traditions acheuléennes occidentales, sur un fond par ailleurs moustérien, forment un autre indice de cette 'évolution combinée' entre deux courants, de pensées et d'ethnies.

Le continent européen se présente alors simplement comme un cas rendu particulier par sa position marginale, en Extrême-Occident. Les phénomènes évolutifs de très grande ampleur, profonds, puissants et complexes semblent s'être déroulés essentiellement en dehors de ce territoire: l'immense Asie et l'insondable Afrique. Ces deux aires nucléaires ont vu s'amorcer deux formes d'adaptations convergentes et analogues d'humanité, tendant toutes deux vers les espaces septentrionaux, l'alimentation carnée, en développant des modes d'équilibre distincts sur un monde semblable. Le retard relatif du mouvement africain en Europe est dû à la Méditerranée, formant un filtre davantage qu'une barrière. L'Asie centrale a constitué une autre catégorie de 'filtre' dans l'océan terrestre de ses steppes infinies. Les deux mouvements se sont finalement rejoints, soit en bordure du Caucase au Proche-Orient (Denell, 1997), soit à travers l'Anatolie (Otte et al., 1995), soit enfin, à l'ouest européen où les deux traditions sont superposées.

Le cas de l'Italie reste ouvert à interprétation, car si le site d'Isernia appartient au monde oriental (Peretto, 1991), les découvertes de Ceprano (Ascenzi et al., 1996) semblent montrer une alternance d'influences africaines et continentales, d'interprétation encore délicate à nos yeux.

Conclusion

La division jadis pressentie par H.L. Movius se confirme donc globalement aujourd'hui, mais sur des bases toutes différentes. Davantage qu'un espace, il s'agit d'un temps qui sépare ces deux traditions dans l'Ancien Monde. Et plutôt qu'une simple pratique technique, il s'agit de deux conceptions des rapports à l'outillage, à la Nature et au monde, qui s'opposent dans ces traditions. Cette dichotomie reflète en fait une division originelle des populations humaines aux tout premiers temps de leur formation. Que celle-ci soit africaine nous paraît rester l'hypothèse jusqu'ici la plus acceptable. Cependant, ce scénario, s'il explique bien la profonde spécificité acheuléenne, implique par ailleurs l'unité biologique de l'espèce humaine dès cette haute époque. L'évolution ultérieure n'est alors plus qu'affaire d'adaptation, de convergences et de déplacements, c'est-à-dire un phénomène 'historique' et non plus biologique pur. Ceci ne rend toutefois pas l'explication plus accessible, mais permet de s'écarter de toute emprise dogmatique par d'autres disciplines que la préhistoire de l'humanité, infiniment plus complexe que beaucoup d'autres.

Adresse de l'auteur: Service de Préhistoire, Université de Liège, 4000 Liège, Belgique. prehist@ulg.ac.be

Bibliography

Ascenzi, A., Bidittu, I., Cassoli, P. F., Segre, A.-G. and Segre-Naldini, E., 1996. A calvarium of late *Homo erectus* from Ceprano, Italy. *Journal of Human Evolution*, 31, 409-423.

Bar-Yosef, O., 1994. The Lower Palaeolithic of the Near East. *Journal of World Prehistory*, 8/3, 211-265.

Biagi, P., Mukhtiar, K. and Negrino, F., 1996. An Acheulean workshop at Ziarat Pir Schaban on the Rohri Hills, Sindh, Pakistan. *South Asian Studies*, 12, 49-62.

Bosinski, G., 1976. L'Acheuléen en Europe centrale du Nord. Dans Combier, J., *L'évolution de l'Acheuléen en Europe*, Colloque X, p. 52-64. Congrès UISPP (Nice).

Bosinski, G., 1992. Die Ersten Mensche in Eurasian. *Jahrbuch des Römisch-Germanischen Zentral-museums Mainz*, 39, 131-181.

Carbonell, E., Bermudez, J.-M., Arsuaga, I.-L. and Diez, T. C., 1995. Lower Pleistocene Hominids and Artifacts from Atapuerca (Spain). *Science*, 269, 826-829.

Clark, D., 1965. *Further Excavations (1965) at the Middle Acheulian Occupations Site of Latamne, Northern Syria: General Results, Definitions and Interpretation,* Berkeley, 35p.

Denell, R., 1997. The Pre-Acheulean of Southern Asia and its Implications for the Earliest Humans in Europe. *Abstracts of the 3rd Meeting of the European Association of Archaeologists*, p. 14. Ravenne (24-28 septembre 1997).

Gabunia, L. K., 1992. Der Menschliche Unterkiefer von Dmanisi (Georgien, Kaukasus). *Jarhburch des Römisch-Germanischen Zentralmuseums Mainz,* 39 (1), 185-208.

Gladinin, V. et Sitlivi, V., 1990. *Achel tchentralnai. Europi,* Kiev, Naykova Doumka.

Goren-Inbar, N., Belitzky, S., Goren, Y., Rabinovich, R. and Saragusti, I., 1992. Gesher Benot Ya'aqov, the 'Bar': An Acheulian Assemblage. *Geoarchaeology,* 7 (1), 27-40.

Henke, W., 1997. The Dmanisi mandibula. Early Eurasian dispersal of the genus *Homo* reconsidered. *Colloque de Bordeaux, Aspects biologiques et culturels des changements en Europe* (pré-tirages, non paginés).

Ikawa-Smith, 1978. *Early Paleolithic in South and East Asia,*. p. 97-128. The Hague, Mouton,

Lan, Po Jia et Chuhnkum, Ho, 1990. Lumière nouvelle sur l'archéologie paléolithique chinoise. *L'Anthropologie,* 94 (4), 851-860.

Leroi-Gourhan, A., 1983. *Mécanique vivante*, Paris, Fayard.

Liubine, V., 1984. *Romii paleolit Kavkaza* (Le Paléolithique ancien du Caucase), Saint Petersbourg, Académie des Sciences.

Lorenz, K., 1970. *Essais sur le comportement animal et humain*, Paris, Seuil.

Mania, D. *et al.*, 1980-86. *Bilzingsleben. Homo erectus, seine Kultur und seine Um toelt,* vols I-II-III, Berlin, V. E. B. Deutscher Verlag der Wissenschaften, Veröffentlichen des Landesmuseums für Vorgeschichte in Halle, 32, 36, 39.

Merrick, H. V., De Heinzelin, T., Haesaerts, P. and Howell, F. C., 1973. Archaeological occurence of Early Pleistocene age from the Shungura Formation, Lower Omo Valley, Ethiopa. *Nature*, 242, 572-575.

Movius, H., 1948. The Lower Palaeolithic cultures of Southern and Eastern Asia. *Transactions of the American Philosophical Society*, 38 (4), 329-420.

Otte, M., 1996. *Le Paléolithique inférieur et moyen en Europe,* Paris, A. Collin.

Otte, M., Yalcinkaya, I., Taskiran, H., Kozlowski, J., Bar-Yosef, O. and Noiret, P., 1995. The Anatolian Paleolithic: New research in Karain Cave. *Journal of Anthrop. Research,* 51 (4), 287-299.

Peretto, C. (ed.), 1991. *Isernia La Pineta. Nuovi contributi scientifici,* Isernia.

Piveteau, J., 1973. *Origine et destinée de l'homme,* Paris, Masson.

Ranov, V., Carbonell, E. and Rodriguez, X.P., 1995. Kuldara: earliest human occupation in Central Asia in its Afro-Asian context. *Current Anthropology*, 36, 337-346.

Raposo, L. et Santonja, M., 1995. The Earliest occupation of Europe: The Iberian peninsula. In Roebroeks, W. and Van Kolfschoten, Th., *The Earliest Occupation of Europe,* p. 7-26. Leiden.

Roe, D., 1981. *The Lower and Middle Palaeolithic Periods in Britain,* Routledge and Kezan, London.

Roberts, M. B., 1986. Excavations at the Lower Palaeolithic Site at Amey's Eartham Pit, Boxgrove, West Sussex : a Preliminary Report. *Proceedings of the Prehistoric Society*, 52, 215-245.

Rolland, N., 1992. The Palaeolithic colonization of Europe: an archaeological and biogeographic perspective. *Trabajos de Prehistoria*, 49, 69-111.

Ronen, A., 1991. The Yiron-Gravel lithic assemblage, artifacts older than 2.4 my in Israel. *Archägologisches Korrespondenzblatt*, 21, 159-164.

Seonbok, Y. and Clark, G. A., 1983. Observations on the Lower Palaeolithic of Northeast Asia. *Current Anthropology,* 24, (2), 181-202.

Swischer, C. C., Curtis, G. H., Jacvob, T., Getty, A.-G., Surriyo, A. and Widiasmoro, 1994. Age of the earliest known hominids in Java, Indonesia. *Science*, vol. 263, 1118-1121.

Tchernov, E., 1987. The age of the Ubeidiya formation an Early Pleistocen hominid site in the Jordan Valley, Israel *Israel Journal of Earth Sciences*, 36, 3-30.

Ulrix-Closset, M., 1975. *Le Paléolithique moyen dans le bassin mosan en Belgique,* Wetteren, Universa.

Vishnyatsky, L. B., 1996. *Paléolithique d'Asie septentrionale et du Caucase,* Saint-Petersbourg, Maison européenne (en russe).

Wolpoff, M.-H., 1996. *Human Evolution*, MC. Graw-Hill. New-York.

TOWARD MODERN HUMANS: YABRUDIAN AND MICOQUIAN, 400 – 50 kyears ago.
Edited by A. Ronen and M. Weinstein-Evron. BAR S850, 2000.

ARE THERE RECOGNIZABLE CULTURAL ENTITIES IN THE GERMAN MIDDLE PALAEOLITHIC?

Nicholas J. Conard and Birgit Fischer

Introduction

This paper addresses the problems associated with defining cultural entities within the German Middle Palaeolithic. In examining this question we are confronted with the problem of linking the human actions that lead to the production, use and discard of artefacts with the usually coarse, temporal resolution of the archaeological and geological data that constitute the Palaeolithic record. In this short paper we cannot fully address the complex problems connected with placing people and not just artefacts into our interpretations of the Palaeolithic. Instead, we will focus on the artefacts themselves and first ask the more basic question of whether Middle Palaeolithic artefacts occur in recognizable spatio-temporal groups. This being done, some conclusions about the cultural and ethnic implications that follow from the existence of such groups can be considered.

The first difficulty one encounters when examining the German Palaeolithic record regards the general terminology of the period. As in other parts of Europe, the use and meaning of the term Middle Palaeolithic is not self-evident (e.g., Ronen, 1982; Gamble, 1986). Here we use the term to refer to lithic assemblages from the last three glacial cycles that contain prepared cores and diverse flake tools. We do not envision a sharp archaeological break between the Lower and Middle Palaeolithic; however, as the Middle Palaeolithic progresses, a far greater diversity of assemblages is apparent than in the Lower Palaeolithic.

While there appears to be general agreement that Lower Palaeolithic assemblages cannot be placed into meaningful cultural entities beyond the level of the industrial complex (e.g., Clark et al., 1966), there is a broad consensus that the European Upper Palaeolithic can be divided into archaeological cultures that may or may not correspond to ethnic groups. Ethnic interpretations of Middle Palaeolithic variability (e.g., Bordes, 1961a) have received extensive criticism in recent decades (e.g., Binford and Binford, 1966; Mellars, 1996), such that researchers today rarely suggest that assemblages dating to this period can be associated with specific ethnic or cultural groups. At present there is little agreement about how to describe and understand the temporal and spatial distribution of archaeological remains from the Middle Palaeolithic.

Before beginning a discussion of the taxonomy of the Middle Palaeolithic, we turn to the history of Palaeolithic research in Germany. Several decades after the first planned Palaeolithic excavations took place in the 1860s and 1870s, attempts were made to place the German Palaeolithic finds in a general chronostratigraphic and taxonomic system. The work of Moritz Hoernes (1903), Hugo Obermaier (1911/12) and Robert Rudolf Schmidt (1912) are important early attempts to organize the available data. Subsequent publications by Fritz Wiegers (1928), Julius Andree (1928) and others tried to document the chronostratigraphy and cultural history of the region more fully. Much of this early research applied information gained in France as a guide to the German record. In this work differences of opinion often exist between scholars seeking to place primacy with typological arguments and researchers who emphasize the geological and chronological arguments over typological links.

In more recent literature, some authors chose to use the term Middle Palaeolithic (e.g., Narr 1951), while others did not. Parallel to the use of the term Middle Palaeolithic, a tendency exists, especially in southern Germany, of subsuming the Middle Palaeolithic within the Lower Palaeolithic. This was the case in Hansjürgen Müller-Beck's (1956) *Das obere Altpaläolithikum in Süddeutschland.* In subsequent publications he often refers to the Middle Palaeolithic as the Late Lower Palaeolithic (*oberes Altpaläolithikum*) (e.g., Müller-Beck, 1988). The majority of studies from the middle of the century focused on only a portion of the country rather than attempting a synthesis of the finds from all of Germany. This coverage was far from homogeneous, with the important sites of southern Germany forming the focus of the work by archaeologists including Riek (e.g., 1934; 1938), Wetzel (e.g., 1958; Wetzel and Bosinski, 1969), and Zotz (e.g., 1955; 1959), while various researchers (cf. Bosinski, 1967) addressed developments in other regions. The political division of the country after World War II blocked scientific contacts between scholars in East and West Germany, thereby accentuating the problems associated with publishing on the entirety of the German Middle Palaeolithic.

Gerhard Bosinski's ambitious work from 1967, *Das Mittelpaläolithikum im westlichen Mitteleuropa,* represented a major attempt to move away from provincial research within the Palaeolithic. This monograph addressed the entirety of available information from West Germany and the immediately surrounding regions. In all, Bosinski examined 318 artefact assemblages and isolated finds. His book along with several articles (e.g., Bosinski 1963; 1982; 1986) did much to determine how the Middle Palaeolithic was defined and organized in the following three decades. These publications succeeded in pushing back the accepted date for the use of prepared cores - Bosinski's hallmark for the start of the Middle Palaeolithic - and readied the way for the acceptance of an early Middle Palaeolithic in northwestern Europe and Central Europe.

Bosinski divided his Middle Palaeolithic into a series of typological and technological groups which he called *Formengruppen*, a term that had been used earlier by archaeologists including Narr (1954) and Müller-Beck (1956). He specifically avoided using the term *Kultur* for these groups, out of a hesitance to equate these often long-lived entities directly with the archaeologically or ethnographically documented cultures of modern humans. He did, however, state that his *Formengruppen,* as represented by artefact assemblages, document the existence of groups of hominids that transferred their stone knapping practices via learned behavior from one generation to the next and recognized themselves as belonging to a social group (Bosinski, 1967, 84).

Bosinski's scheme for the Middle Palaeolithic in west central Europe is complicated, and can perhaps be best summarized graphically (Fig. 1). Noteworthy in his classification are the following *Formengruppen*: Lebenstedt or the *Jungacheuléen*; the Micoquian which is divided into the Bockstein, Klausennische, Schambach and Rörshain groups; the Mousterian which is divided into Rheindahlen, Kartstein and Balve IV groups; and finally the Altmühl or *Blattspitzen* group. Bosinski places his *Jungacheuléen*, which he equates with the *Acheuléen Supérieur* of the French literature, in the Saale/Riss glaciation. Here the stratigraphic arguments placing Markkleeberg, Hundisburg and other northern German sites in the first phase of the Saale play a key role. Then follows the Micoquian which, mainly on the basis of faunal and stratigraphic arguments from Bockstein, he places in the early Weichsel/Würm. Using *Leitformen* (*fossils directeurs*) and stratigraphic evidence, he divides the Micoquian into the four assemblage types mentioned above. The Mousterian is often characterized by more generalized assemblages and is difficult to pinpoint in time. Early assemblages such as Rheindahlen B3 date to the second half of the Saalian (Bosinski, 1982), Kartstein assemblages generally appear to postdate the Micoquian, while Balve IV assemblages date to very late in the Middle Palaeolithic sequence. Throughout his monograph, Bosinski consistently refers to the Mousterian in quotation marks and alludes to the difficulty of organizing this heterogeneous group of assemblages. Finally, the *Altmühlgruppe,* with its characteristic *Blattspitzen* and close resemblance to the Szeletian, correlates with the end of the Middle Palaeolithic.

Current Research

Most researchers familiar with the German data acknowledge the importance of Bosinski's attempt to create a structure for studying such a vast quantity of material; however, even the first reviews of the book pointed out weaknesses and contradictions in his framework (Freund, 1969). Since the publication of the monograph, some attempts to refine the system have been made (Bosinski, 1969, 1972, 1982), but Bosinski has never totally revised it. In recent years it has become increasingly clear that, despite its importance over three decades, this scheme contains many problems and must be reevaluated. New interpretations of Middle Palaeolithic variability have been published in recent years (e.g., Richter, 1997; Veil et al., 1994; Weismüller, 1995), but aside from several attempts to use attribute analysis to separate archaeological entities (Kind, 1992; Schäfer, 1993; 1997; Weber, 1986; 1996), no one has advanced a general framework to rival Bosinski's system.

Most current research in German does not focus on cultural history. Instead, issues of subsistence, settlement and technology are emphasized. In recent decades materials from earlier excavations have been reexamined and much field work has been conducted. Important new information comes from many localities including: Rheindahlen (Klostermann and Thissen, 1995; Thieme, 1983), Buhlen (Fiedler, 1997; Jöris, 1994), Königsaue (Mania and Toepfer, 1973), Sesselfelsgrotte (Richter, 1997; Weismüller, 1995), Salzgitter-Lebenstedt (Busch and Schwabedissen, 1991; Tode, 1982), Balve (Jöris, 1992), the East Eifel volcanoes (Bosinski et al., 1986; Conard, 1992; Schäfer, 1990), Lichtenberg (Veil et al., 1994), Wallertheim (Conard et al., 1995) and Zwochau (Pasda, 1996; Weber, 1996). The new influx of archaeological and geological data combined with new methodological and theoretical approaches is creating renewed interest in the variation within the Middle Palaeolithic. Much of this research is available only in the form of unpublished master's and doctoral theses and is not yet readily available inside the country, let alone abroad. In this short paper we touch on some issues emerging from the new research.

In the context of the Haifa meeting focusing on the Jabrudian and Micoquian, Richter's (1997) recent monograph on the G Layers of Sesselfelsgrotte is of particular importance. Richter rejects Bosinski's interpretation of the Micoquian as well as Kind's (1992) and Schäfer's (1993) results based on attribute analysis. He proposes a new system in which he argues that the Micoquian of the Sesselfelsgrotte can best be described as a Mousterian with Micoquian Option (MMO). Following Weismüller (1995), he correlates the sequence from the G Layers with the Oerel Interstadial at the start of OIS 3 and dates the assemblages between 57 and 53 kyears B.P. Richter bases his results on the study of 13 find horizons from Sesselfelsgrotte, which he defines with the help of his *Rohmaterialnachbarschaften* using statistical studies of raw materials and stratigraphic associations. He then separates these assemblages into two main phases, the first being characterized by non-Levallois (MMO-A) and the latter by Levallois technology (MMO-B). Each of these is further divided into three subphases. Perhaps the most radical aspect of Richter's approach is his claim that the Mousterian and Micoquian are one in the same, and that typical aspects of the Micoquian, such as relatively high percentages of bifacial forms including various kinds of *Keilmesser* (backed bifacial knives), handaxes and *Faustkeilblätter*, are determined primarily by the extent to which assemblages have been reduced. This approach emphasizes the biography of artefacts and rejects the notion that most tools are manufactured during a single episode of production. Richter, in another controversial step, uses the Sesselfels sequence as the basis for organizing and dating the Micoquian assemblages of Europe. Like Bosinski, he emphasizes the importance of comparative material from western and eastern Europe.

In addition to site-focused studies like Richter's, prehistorians including Weber (1986; 1996), Schäfer (1993; 1997) and Kind (1992) have turned to attribute analysis combined with various statistical approaches as a means of classifying early Palaeolithic assemblages. In recent years, Schäfer (1993; 1997) has pursued this approach enthusiastically. Aside from various methodological problems (see Steguweit, 1997), the assumption that morphological and technological characteristics of lithic assemblages will form clusters in time and space is not universally valid. Functional considerations, raw material availability and other factors can complicate the picture and raise questions about this approach. Despite the appeal of the seemingly more rigorous methods used by advocates of attribute analysis, the results have not been widely accepted.

Perhaps the most useful application of attribute analysis has been made by Dieter Schäfer (1997). Based on the study of unmodified flakes from several dozen pre-Upper Palaeolithic assemblages, Schäfer constructs three main groups: *unteres, mittleres and oberes Altpaläolithikum*. In the terminology we offer here, these groups correspond to the Lower Palaeolithic, Early Middle Palaeolithic and Late Middle Palaeolithic. Schäfer mentions some possible contradictions within his system, such as Rheindahlen B3 and Weimar-Ehringsdorf, but he ultimately argues that these divisions are real, reproducible and more objective than those based on the *Leitformen* method. He also acknowledges that other methods besides attribute analysis may well be needed to construct a system of classification at a more detailed level than his three divisions of the pre-Upper Palaeolithic (Schäfer, 1997,17).

At present, researchers are confronted with a broad array of Middle Palaeolithic data that appears to defy any coherent system of organization. Here we address some of the strengths and weaknesses of the archaeological record and offer a general scheme to help organize the German Middle Palaeolithic.

The Problem of Defining Cultural Entities

In considering the Middle Palaeolithic evidence from Germany, it is evident that, while many localities are known, relatively few have provided large, well dated assemblages from primary contexts. Bosinski (1967) encountered this problem and attempted to get around it by studying as much material as possible. He operated under the assumption that lithic artefacts of a given form must be from the same period, and that they can be used as chronological markers. Given the small size of many of these assemblages, Bosinski was unable to use statistical methods similar to those developed by Bordes (1961b) and instead focused on *Leitformen*. This approach led to a lively exchange with Freund (1969), in which Bosinski (1970a) reiterated the futility of a statistical analysis of most of these assemblages.

The best illustration of Bosinski's faith in *Leitformen* is his willingness to assume that all northern German *Jungacheuléen* assemblages from his *Lebenstedtergruppe* are roughly the same age and that all must date to the Saalian glaciation. Bosinski thus gave priority to typology and argued that Salzgitter-Lebenstedt, contrary to several lines of geological evidence (Tode et al., 1953; Tode, 1982; Busch and Schwabedissen, 1991), must date to the Saalian. Following this approach we arrive at the unusual circumstance that the type locality of the *Lebenstedtergruppe* is widely considered to be two glacial cycles younger than other sites like Markkleeberg and Hundisburg that belong to the main phase of the *Jungacheuléen* and date to the early Saalian.

Here we take a more restrictive approach and generally focus on assemblages that are dated, and are large enough to characterize from a technological and typological point of view. When possible, we emphasize assemblages from primary contexts. In this way, one arrives at a much smaller number of assemblages than Bosinski studied, but gains better chronological control. Nonetheless, with few exceptions, we still lack the temporal control needed to date most archaeological assemblages accurately. This shortcoming is nowhere more acute than in the last glacial cycle, during which an increase in lithic variability is well documented and interpreting this variability depends to a large extent on establishing better temporal resolution between the types of assemblages.

Another observation that indicates we should be prudent in advocating new taxonomic systems for the Middle Palaeolithic is the realization in recent years that new kinds of assemblages are still being found. Finds from Tönchesberg 2B (Conard, 1992), Wallertheim D (Conard and Adler, 1997) and the current excavations at Rheindahlen B1 (Thissen and Schmitz, 1997; Thissen, 1998) demonstrate that new assemblage types are still being identified. Thus, we must conclude that the Middle Palaeolithic record still provides highly selective excerpts from what is probably a much more complicated and regionally varied pattern of cultural and technological change. At present some general trends and taxonomic units can be identified, but it would be naive to think that we are on the verge of producing the definitive synthesis of the Middle Palaeolithic. We still lack basic information about the organization of settlement and subsistence, and the interplay between hominid behavior and environmental change remains poorly understood. Thus, depending on the questions being posed, more rather than fewer taxonomic schemes are needed. We assume that an ideal taxonomic system is neither attainable nor desirable (cf. Brew, 1946), and with these reservations in mind we present an outline for the Middle Palaeolithic (Fig.2).

Outline for the Classification of the German Middle Palaeolithic

Following Bosinski's (1963; 1967) suggestion that prepared cores and a high proportion of flake tools distinguish the Middle Palaeolithic from the Lower Palaeolithic, we classify Saalian/Rissian assemblages within the Early Middle Palaeolithic (EMP), and Eemian and Weichselian/Würmian assemblages within the Late Middle Palaeolithic (LMP). This suggestion is by no means new, nor should it be too controversial. The more serious problems concern how the EMP and LMP should be subdivided and to what extent the Eemian interglacial actually marks a significant boundary.

With the last glaciation one sees a larger number of sites and a greater diversity among the assemblages. This diversity, which is in many respects the hallmark of the Middle Palaeolithic, may simply result from the larger number of better documented sites, and at present it seems most likely that the increasing differentiation of assemblages does not take a quantum leap after the Eemian but builds in a gradual, perhaps punctuated, manner throughout the EMP. This question, however, cannot be answered until the density of high quality information on the EMP approaches that of the LMP.

The Early Middle Palaeolithic

The Early Middle Palaeolithic has two main assemblage types: the *Jungacheuléen* and the Early Mousterian, and three major geological and chronological phases: the Early Saalian/Rissian (OIS 8), the penultimate interglacial (OIS 7), and the Late Saalian/Rissian (OIS 6). The chronostratigraphy of the Middle Pleistocene in northern Germany, however, is controversial. Mania (1997) favors a long chronology for the Saalian glaciation, and Eissmann (1994) a short chronology without an Intersaalian interglacial. Bibus and Kösel's (1996) study of the Rissian deposits in Baden-Württemberg also suggests that the chronostratigraphy of the EMP may be more complex than the highly simplified presentation here would indicate. Given the rarity of pre-Eemian cave deposits, it is not surprising that all the EMP sites mentioned here are from open-air settings. These groups are largely unknown in southern Germany where for over a century research has focused on caves.

If we limit ourselves to relatively well dated assemblages, the *Jungacheuléen* is known from Early Saalian contexts in Markkleeberg, Hundisburg, Zwochau, Eythra, Lübbow and other localities (e.g., Baumann and Mania, 1983; Eissmann et al., 1991; Grahmann, 1955; Pasda, 1996; Steguweit, 1997; Toepfer, 1978; 1981; Weber, 1996; Figs. 2 and 3). These localities rarely have yielded finds from primary contexts. It has long been known that handaxes at sites including Markkleeberg are by no means common and that Levallois technology is well documented. The *Jungacheuléen* also appears to be present in the Late Saalian at the poorly dated sites of Herne and Bottrop (Bosinski, 1963; Schmitz, 1988). The securely dated site of Schweinskopf has been associated with the *Jungacheuléen*, but the assemblage provides relatively few diagnostic artefacts (J. Schäfer, 1990). There are no well dated *Jungacheuléen* assemblages belonging to the intersaalian interglacial, but the absence of such assemblages cannot be taken as proof that they do not exist, especially given that interglacial sediments are rarely preserved. Although there are also Weichselian/Würmian assemblages with handaxes, these have typically not been classified as *Jungacheuléen*, and we follow this convention here.

The other main group within the Saalian is the Early Mousterian. This assemblage type is known from find horizons within the Lower Travertine in Ehringsdorf (Behm-Blancke, 1960; Blackwell and Schwarcz, 1986; Feustel, 1983; Steiner, 1979), which appears to date to the penultimate interglacial, although a correlation with the Eemian interglacial cannot be entirely ruled out (Schäfer, 1991). Early Mousterian assemblages from Rheindahlen B3 (Bosinski, 1967; 1986; Narr, 1951; Thieme, 1983) appear to be securely dated to the Late Saalian. In northern France (e.g., Tuffreau, 1989; Tuffreau and Sommé 1988) and the Netherlands (Roebroeks, 1988) similar assemblages have been the focus of recent research. Based on these finds, the Early Mousterian is present during the intersaalian interglacial and in the Late Saalian. Technologically and typologically, these assemblages are indistinguishable from Mousterian assemblages dominated by various scraper and point types that characterize the Mousterian of the last glacial cycle.

The Late Middle Palaeolithic

There are several well-dated Eemian sites in Germany including Weimar (Belvederer Allee), Taubach, Wallertheim A, and the small assemblages of Lehringen and Gröbern (Behm-Blancke, 1960; Mania, et al. 1990; Wenzel, 1996). These assemblages, however, are difficult to classify. The term 'Taubachian', which Valoch (1977, 1984; 1988), following Collins (1969), has suggested to describe the assemblage from Kůlna 11, is misleading because Taubach (Behm-Blancke, 1960; Eichhorn, 1909) is not the best example of an irregular, microlithic assemblage. D. Schäfer (1990) has discussed this point at length and concludes that the assemblage from Taubach does not fit the definition of the Taubachian offered by Valoch (1977). This being the case, we suggest that the better described assemblage from Kůlna 11 (Valoch, 1988) rather than Taubach be used as the type site for such irregular, microlithic, or nearly microlithic, assemblages of Eemian age. Furthermore, the use of the term 'Taubachian' (Valoch, 1977; 1984) to describe a single tradition of lithic technology spanning much of the Middle Pleistocene and Eemian interglacial is of questionable utility.

At present it remains difficult to generalize about the nature of the Eemian assemblages. Several small assemblages from sites including Lehringen and Gröbern (Mania et al., 1990; Thieme and Veil, 1985) appear to correlate with butchering activities. The same can be said for Neumark-Nord (Mania et al., 1990), although it is not certain whether the locality dates to the last or penultimate interglacial. It is also uncertain to what extent environmental factors or constraints related to raw material availability might be responsible for the tendency toward irregular, and at times microlithic assemblages. One relevant factor could be the ubiquitous presence of wood and other plant fibers during interglacials. The presence of these resources would open a wide range of technological options to hominids that were not available in the open landscapes of cool and cold periods. These materials could in some cases replace lithic and faunal resources as starting materials for making tools. Additionally, stone tool assemblages could be expected to change in response to the need of working more organic materials. The diminution of artefacts during Middle and Late Pleistocene interglacials (Valoch, 1988), the *Federmesser* period and the Mesolithic may be related to increased availability of these materials. Alternatively, thicker interglacial vegetation may have hindered access to lithic raw materials and necessitated heavier use and reduction of the available lithic resources. Finally, interglacial conditions affected patterns of subsistence and settlement and placed new constraints on lithic technology that in many settings seem to have favored a reduction in the size of artefacts. At present, we need better data to test hypotheses focusing on the changing availability of raw materials and changing mobility and subsistence strategies as key variables for explaining the diminution of lithic artefacts during interglacial periods.

In the Rhineland, Belgium and Northern France, Middle Palaeolithic sites rich in blades and blade tools date to the beginning of the last glaciation (Révillion, 1994; Révillion and Tuffreau, 1994). These laminar Middle Palaeolithic assemblages are by no means homogeneous, but may be placed tentatively in a group (Conard, 1990). The key find horizons in Germany are Tönchesberg 2B (Conard, 1992) and Wallertheim D (Conard and Adler, 1997). The status of Rheindahlen B1 (Bosinski, 1966; Klostermann and Thissen, 1995; Thieme, 1983; Thissen, 1998) is more complex and can only be addressed after the results of the current excavation by Thissen and Schmitz have been reported.

In Germany these assemblages appear to be limited to open-air contexts of the Rhineland. In France and Belgium, laminar Middle Palaeolithic sites are also typically from open-air settings. Blade-rich assemblages with backed tools similar to those of the northwest European Plain are unknown in the caves of central and southern Germany. Tönchesberg 2B and Wallertheim D are the only find horizons of this group that include faunal remains. Both find horizons appear to represent brief occupations where portions of hunted bovids, cervids and equids were transported for further processing and consumption. Both horizons include a relative abundance of points and provide evidence for primary lithic production and curation of existing tools (Conard and Adler, 1997; Conard et al., 1998).

A controversial group of assemblages belongs to the Micoquian, which, following Veil et al. (1994), we prefer to call the *Keilmessergruppe*. Numerous find horizons from both open-air and cave settings belong to this group. These include: Lichtenberg, Königsaue A and C, Bockstein III, Sesselfelsgrotte G, Abri Schulerloch, Klausennische, Buhlen IIIb, Hohler Stein-Schambach, and Balve II-III (Günter, 1964; Bosinski, 1967; Wetzel and Bosinski, 1969; Jöris, 1992; Richter, 1997; Veil et al., 1994). In most cases, these assemblages can be placed with certainty in the last glacial, but their relative ages are difficult to determine. The results from Königsaue, Bockstein, Buhlen and Balve suggest an early glacial age, while Lichtenberg and Sesselfelsgrotte appear to be younger, possibly belonging to OIS 3. Richter's (1997) suggestion that all these horizons date to the beginning of OIS 3 requires disregarding evidence for earlier correlations. A key problem remains the dating of the many cave deposits in Baden-Württemberg and Bavaria. At present, it appears that the *Keilmessergruppe* can be documented in both the early glacial, as has been argued by Bosinski (e.g., 1967; 1969), Mania (Mania and Toepfer, 1973) and Jöris (1992), and in the middle part of the last glaciation, rather than exclusively to the Oerel Interstadial as Richter argues. The Polish sites belonging to this group appear to date primarily to the Early Weichselian (J. Kozlowski and S. Kozlowski, 1996), but the presence of Pre-eemian *Keilmesser* assemblages cannot be ruled out (J. Kozlowski, personal communication, 1997).

At many sites, including Buhlen, Lichtenberg, Königsaue and others, *Keilmesser* are far more numerous than Micoquian handaxes and *Faustkeilblätter*. Thus we favor the name *Keilmessergruppe* over the Micoquian. Furthermore, renaming this group eliminates the necessity to correlate the Central European material directly with La Micoque as often has been considered imperative (Bosinski, 1967; Richter, 1997; Thissen, 1998). The vast majority of the La Micoque sequence is from much earlier fluviatile deposits, and the degree to which similarities exist between these assemblages and the later German material is debatable (see Bosinski, 1970b; Delpech et al. 1995; Pratte, 1971; Rolland, 1986; Schwarcz and Grün, 1988). Basing the connection on the uppermost Layer 6 at La Micoque is also problematic since these finds stem from early excavations and are of unknown age (Delpech et al., 1995). The most recent work at La Micoque has not been fully published, but promises to resolve some of the controversial issues surrounding the site. Before crossing whole continents and looking far to the east and west for answers to questions concerning the *Keilmessergruppe*, researchers need better data from, and better chronostratigraphic control of the find horizons from Central Europe.

We agree with Jöris (1994) and Richter (1997) that Bosinki's four-part division of this group should be rejected, and the suggestion that Bosinski's different types of *Keilmesser* represent stages along a continuum of reduction seems plausible. Still we are unconvinced by Richter's hypothesis that the Micoquian and Mousterian are one and the same. We do not see how the typical bifacial forms of the *Keilmessergruppe* can be explained as being the result of further reduction of unifacial Mousterian tools. Nor is it clear how this hypothesis would explain the temporal and spatial distribution of the Mousterian and *Keilmessergruppe*. Certainly, the Mousterian is documented long before the appearance of *Keilmesser* assemblages, and, for example, the rich *Keilmesser* assemblage of Buhlen IIIb shows little technological affinity with the Mousterian. Nonetheless, Richter is correct in stressing the apparent close relationship between the two groups. Particularly in cases where *Keilmesser* and other bifacial forms are rare, such as in the G layers of Sesselfelsgrotte, it becomes difficult to distinguish between Mousterian and *Keilmesser* assemblages. A certain level of ambiguity seems to be nearly inherent in systems of classifying the Middle Palaeolithic, and it would be unjustified to claim that the assemblage types discussed here represent discrete and fully independent cultural entities.

Another important issue is the question of how narrowly to restrict the *Keilmessergruppe*. In assemblages such as Buhlen IIIb, Lichtenberg, and Königsaue A and C, the term fits well. In other cases the term *Keilmessergruppe* or Micoquian can become a wild mix of vaguely bifacial assemblages from the last glaciation. The current work at Rheindahlen B1 is an interesting example. Here, Thissen (Klostermann and Thissen, 1995; Thissen, 1998) places the most impressive German Micoquian handaxe in find horizon B1 and suggests that the highly diverse and often laminar assemblages from this horizon belong in the 'Micoquian.' He, furthermore, suggests, despite the absence of skeletal remains to support his point of view, that 'Taubachian', 'Micoquian' and *Blattspitzen* assemblages are all the product of *Homo sapiens sapiens,* while Mousterian assemblages are the product of Neanderthals. This hypothesis is highly speculative and unconvincing in light of the lack of supporting skeletal evidence. Given the overall nature of the lithic assemblages from horizon B1, Thissen's assessment requires a very liberal definition of the 'Micoquian.'

The role of para-burin technology within the *Keilmessergruppe* also deserves consideration. Find horizons such as Balve III and Buhlen IIIb contain many *Keilmesser* with resharpening blows along the cutting edge (Bosinski, 1969; Jöris, 1992; 1994). This resharpening technique is by no means limited to backed bifacial knives and is well documented on scrapers and other forms in find horizons including Buhlen IIIb and Rheindahlen B1 (Jöris, 1994; Thissen, 1998). Such sharpening and resharpening blows are also well documented on pradniks (*Keilmesser*) in Poland at sites including Ciemna, Okiennik and Wylotne (Chmielewski, 1969; Krukowski, 1939; Kowalski, 1967; Kozlowski, 1972; J. Kozlowski and S. Kozlowski, 1996). Polish colleagues including Chmielewski (1969) and J. Kozlowski (1972) refer to negatives from these sharpening blows as para-burin scars, as do Schild and Wendorf (1977) in their analyses of Polish and Egyptian material. Since the term pradnik in the Polish literature is equivalent to the German term *Keilmesser*, it should not be confused with a method of sharpening or resharpening a cutting edge. At present the term pradnik is often used to refer to any tool that has been sharpened using the para-burin technique. This usage creates a contradiction because these so-called 'pradniks' are often not *Keilmesser* at all. Use of the term para-burin technique instead of 'pradnik technique' (Bosinski, 1969; Jöris, 1992) would eliminate this problem. Finally, it is entirely reasonable to consider the term Pradnik Group as synonymous with the term *Keilmessergruppe.*

Another assemblage type is the Mousterian of the Late Middle Palaeolithic. This is a mixed group of find horizons from both cave and open-air sites. Mousterian assemblages are known from above, below and between horizons of the *Keilmessergruppe*, and there are no obvious characteristics that distinguish these assemblages from the Mousterian of the EMP. Richter (1997) addresses the connections between the *Keilmessergruppe* (Micoquian) and the Mousterian and concludes that they are stratigraphically and technologically inseparable. For him, this pattern can be explained on the basis of lithic reduction. He finds support for his model in Kind's (1992) identification of many assemblages that appear to be mixtures of the Micoquian and Mousterian.

While it is clear that the distinctions between Late Mousterian and *Keilmesser* assemblages are not always sharp, it seems unlikely that the observed variation can be explained solely on the basis of reduction sequences. On the other hand, Richter is correct in stressing the life histories of tools and the important role that reduction cycles play in shaping assemblages. Some examples of the Late Mousterian are Sirgenstein I and II, Große Grotte VIII-II, Königsaue B, Buhlen II, Kartstein III, Großes Schulerloch and Balve IV (e.g., Bosinski, 1967; Mania and Toepfer, 1973; Schmidt, 1912; Wagner, 1983). As was the case with Bosinski's 'Mousterian', our Late Mousterian also tends to serve as a catch-all for assemblages dominated by flake tools with or without the use of the Levallois concept.

One of the most enigmatic aspects of the LMP is the presence of a variety of handaxe-bearing assemblages that do not fit into the *Keilmessergruppe*. Examples include Salzgitter-Lebenstedt (Busch and Schwabedissen, 1991; Tode, 1982; Tode et al., 1953), Große Grotte Layer IX (Wagner, 1983) and Beixenstein Höhle (Rieder, 1989), although the ages of these sites are far from certain. These assemblages cannot be equated with the *Moustérien de Tradition Acheuléen*. The thin, cordiform and triangular handaxes of the MTA are rare east of the Rhine and generally occur as single finds. The Late Middle Palaeolithic with Handaxes is a group of assemblages with bifacial components that are characterized by diverse forms of handaxes and a lack of large numbers of *Keilmesser,* Micoquian handaxes and *Blattspitzen.* No find horizons belonging to the Late Middle Palaeolithic with Handaxes are well-dated, and without new chronostratigraphic studies the Late Pleistocene handaxe assemblages cannot be localized within the archaeological sequence. Technological and typological similarities between these and earlier handaxe assemblages exist, as is illustrated by Salzgitter-Lebenstedt, which was long accepted as the type site for the German *Jungacheuléen* (Bosinski, 1963; 1967).

The Middle Palaeolithic sequence ends with the horizons belonging to the *Blattspitzengruppe* (e.g., Bosinski, 1967; Freericks, 1995; Freund, 1952; Zotz, 1955). These assemblages date to the middle of the Middle Pleniglacial and presumably have ages of roughly 40,000 B.P. Finds of *Blattspitzen* from open-air context are known especially from surface collections including Rörshain (Hahn, 1990), Kösten (Zotz, 1959), Zeitlarn (Schönweiß and Werner, 1996), Lenderscheid, Wahlen (Fiedler, 1994; 1997) and Mundelsheim (Wagner, 1996). Stratified Blattspitzen assemblages have been documented at caves including Weinberghöhlen (von Königswald et al., 1974; Zotz, 1955), Haldensteinhöhle, Obere Klause III, Kleine Ofnet (Freund, 1952), Ilsenhöhle (Hülle, 1977), and the Oberneder Höhle (Freund, 1987). Of all the taxonomic subdivisions of the Middle Palaeolithic, the *Blattspitzengruppe* is the most restricted in time, though combined with the Szeletian it covers much of Central Europe (Allsworth-Jones, 1986). While J. Kozlowski and S. Kozlowski (1996) and others place this group in the Upper Palaeolithic, we prefer to keep it within the Middle Palaeolithic and begin the Upper Palaeolithic with the appearance of the Aurignacian and the first documented presence of modern humans in the region around 40,000 years ago (Hahn, 1996; Housley et al., 1997; Richter, 1998).

Discussion

The taxonomic approach to the Middle Palaeolithic proposed here has been kept simple (Figs. 2, 3, and 4). We take a conservative approach and do not assume that similar assemblages are of similar ages. The research of the last several decades has repeatedly shown that there is no necessary correlation between artefact types, hominid forms and chronology (Conard, 1990). This point is illustrated by the abundant evidence for Early Mousterian sites from across Europe as well as by the controversy surrounding Bosinski's attempt to date the *Lebenstedtergruppe* using typological and technological arguments (Richter, 1997; Tode, 1982; Veil, 1983/84).

In fact, the Haifa meeting can only serve to warn us against constructing cultural links between typologically similar assemblages without adequate chronological control. During the Haifa meeting participants examined Jabrudian assemblages from Mount Carmel that even in the short chronology of Tabun (Bar-Yosef, 1995; Klein, 1996) far predate the last interglacial. These assemblages are in many respects similar to finds from the European *Keilmesser* group, but are perhaps 200,000 years older. To our knowledge, no one at the meeting wished to see a direct cultural link between the Levant and Europe in this case. While obviously stone tools provide our best hope of establishing the cultural history of regions, they must be viewed with caution. Certainly in the Middle Palaeolithic they are the result of learned behavior and cultural transmission, but they are above all else utilitarian objects that may or may not allow long-distance cultural connections to be established.

To conclude, we turn with mixed results to the question posed in the title of this paper: Are there recognizable cultural entities within the German Middle Palaeolithic? The data from the Early Middle Palaeolithic remain too sparse to go beyond the *Jungacheuléen* and Early Mousterian groupings offered here. We are obviously dealing with phenomena that are not narrowly restricted in time and space and that overlap in both dimensions. Without more and better chronologically controlled data, there will be little chance for isolating narrowly defined cultural entities. Based on the evidence in hand, there is no indication that broadly defined cultural groups like those that form the basis of the Upper Palaeolithic sequence existed during the EMP.

In the Late Middle Palaeolithic there is a greater variation in assemblages and far more data to consider. As a result of the higher quality data, we are in a better position to address this question. Most of the Eemian assemblages are too small to classify on the basis of typology or technology. In the early and middle parts of the last glaciation, we continue to encounter major chronological problems, but here assemblage types can be recognized.

Across northwestern Europe many Middle Palaeolithic assemblages with strong laminar components dating to the early glacial have been documented (Conard, 1992; Conard and Adler, 1997; Révillion, 1994). These find horizons are highly diverse, at times with Upper Palaeolithic blade production and at times with less systematic blade production. The assemblages appear to be restricted to open-air settings and often contain small or larger backed points, backed blades and bladelets. Most assemblages are from contexts lacking fauna, but at Tönchesberg 2B and Wallertheim D, the relative abundance of points in association with what appears to be hunted fauna suggests that these were sites where hunted game was processed and stone tools were produced and maintained. These occupations appear to be linked to specific seasons and activities and offer particularly good prospects for placing the observed variation in the assemblages within the context of a system of subsistence and settlement (Conard, 1998a; Conard et al., 1998). In the most general sense, these sites seem to be confined in space and time, but the high degree of variation between the assemblages warns against arguing that they are part of a single cultural phenomenon.

The *Keilmesser* assemblages form a diverse group characterized by numerous bifacial tools, especially *Keilmesser* with and without accompanying para-burin technology. Unfortunately, many *Keilmesser* sites remain poorly dated. Jöris (1992) has argued that para-burin technology occurs only within a brief period of the early glacial, and Richter (1997) suggests that *Keilmesser* assemblages are temporally restricted to the Oerel Interstadial and undergo a technological shift from Quina to Levallois concepts of reduction during this period. While these authors are correct in rejecting Bosinski's model with four subdivisions for the Micoquian using data from the *chaines operatoires* and artefact life history approaches, neither Jöris's nor Richter's chronological hypotheses can be confirmed with the data now available. Still, the *Keilmessergruppe* remains a recognizable part of the LMP that is restricted to several tens of thousands of years. Rich *Keilmesser* assemblages are known from caves and open-air settings and do not appear to be associated with a specific set of topographic or environmental conditions. Better data on subsistence and settlement practices are needed to help improve our understanding of this period.

Diverse Late Mousterian assemblages are present throughout the early and middle parts of the last glaciation. Similarly, the LMP with Handaxes includes diverse assemblages that do not appear to be narrowly restricted in time or space. At present one cannot argue that either group forms a tightly defined archaeological entity.

The *Blattspitzengruppe*, however, does seem to form an archaeological entity with a temporal resolution and distinctiveness comparable to the major cultural units of the European Upper Palaeolithic. This is not to say that such archaeological 'cultures' based on artefact forms that persist for thousands of years should be interpreted as units reflecting the ethnic affiliation of a social group. Ethnic affiliation based on the archaeological record is a question of scale. At the most general level it seems likely that the makers of *Blattspitzen* would recognize themselves as distinct from other Middle Palaeolithic groups and from the makers of Aurignacian artefacts if encounters between these groups occurred (Conard, 1998b). It is, however, extremely unlikely, given the nature of ever shifting ethnic identity with all its dynamic aspects (e.g Jones, 1997; Shennan, 1989; Steele and Shennan, 1996), that a single ethnic group could have persisted over many thousand years over large portions of continents.

Finally, it is probable that stone artefacts provide a very conservative record of the Palaeolithic past, and that many kinds of social and cultural change occurred within a context of a stable lithic technology. On the other hand, technological innovation could take place within a stable social setting. The perpetual questions concerning the link between material cultural and social identity are bewildering within the archaeological context of modern humans in the Holocene, and even in contemporary ethnographic settings. Thus we should not expect easy answers to this problem in the far more distant past in the socio-cultural contexts of archaic hominids. Furthermore, while individual find horizons can provide high temporal resolution and glimpses into specific episodes of Palaeolithic behavior, and perhaps even social interaction (Conard, 1994), the groups of assemblages that can be defined in the Middle Palaeolithic remain difficult to interpret within an ethnographic time scale.

Acknowledgments

In preparing this manuscript we have profited from conversations and suggestions from many people. We especially wish to thank Daniel Adler, Michael Bolus, Gerhard Bosinski, Jan Burdukiewicz, Janus Kozlowski, Hansjürgen Müller-Beck, Romuald Schild, Leif Steguweit and Karel Valoch. These colleagues bear no responsibility for our interpretations and oversights. We are grateful to Tim Prindiville for preparing the figures and to Helga van den Boom for providing translations of Polish literature.

Authors' address: Institut für Ur- und Frühgeschichte, Schloss Hohentübingen, 72070 Tübingen, Germany. nicholas.conard@uni-tuebingen.de

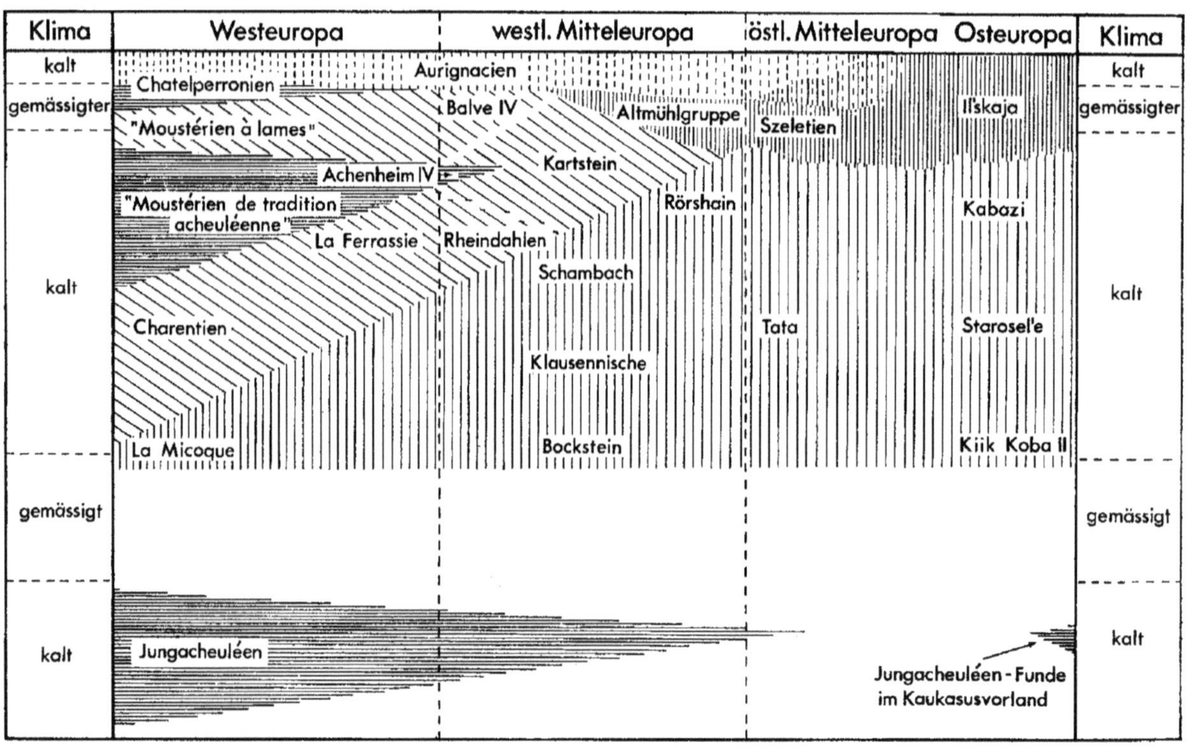

Fig. 1. Bosinski's graphic summary of his classification for the Middle Palaeolithic (after Bosinski 1967).

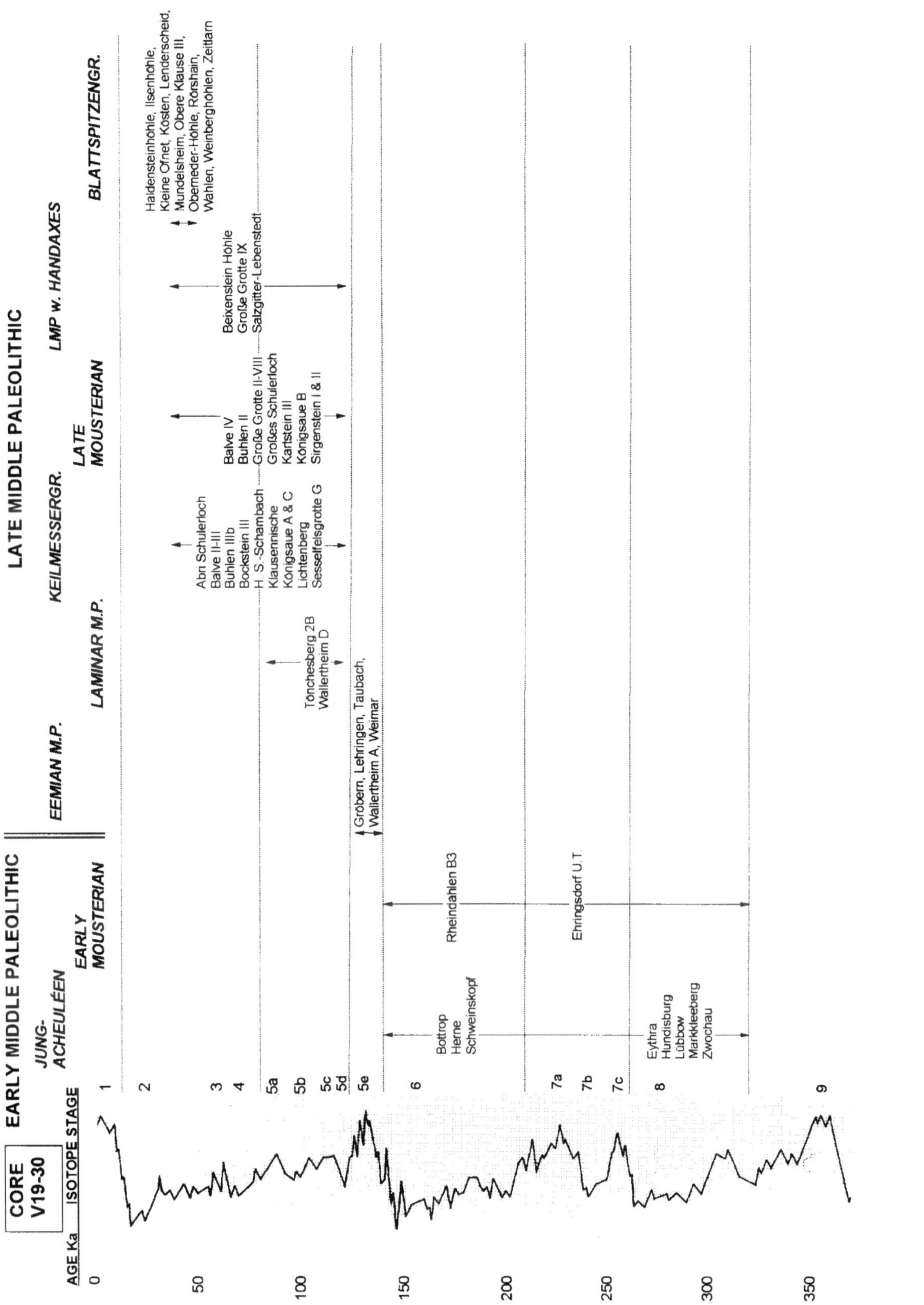

Fig. 2. Schematic representation of the proposed system for classifying the German Middle Palaeolithic including preliminary correlations with the oxygen isotope record. Sites are listed in alphabetical, not chronological order (oxygen isotope curve after Shackleton and Pisias (1985) and Roebroeks (1988)).

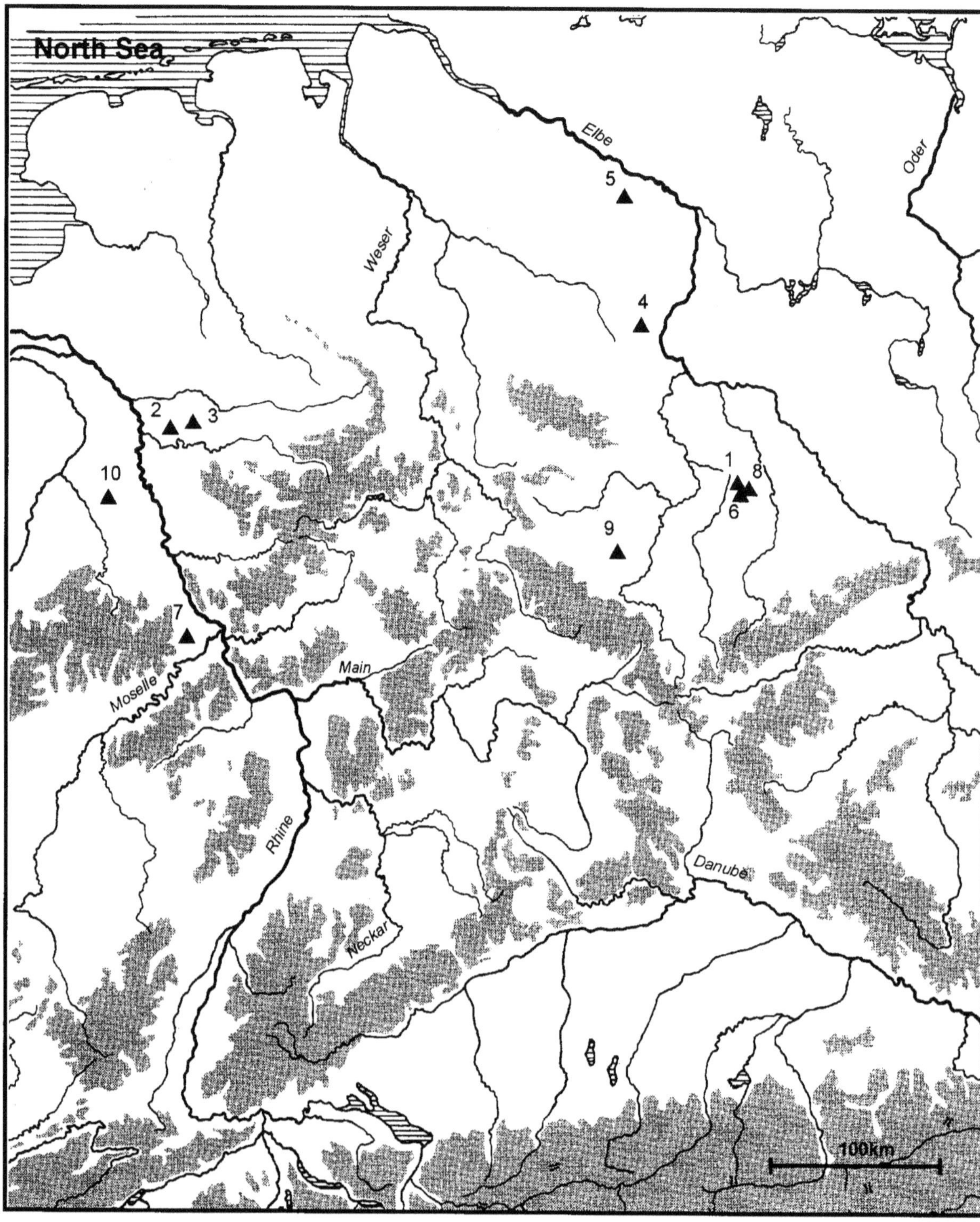

Fig. 3. Early Middle Palaeolithic sites mentioned in text. *Jungacheuléen*: Eythra (1), Bottrop (2), Herne (3), Hundisburg (4), Lübbow (5), Markkleeberg (6), Schweinskopf (7), Zwochau (8). Early Mousterian: Ehringsdorf U.T. (9), Rheindahlen B3 (10).

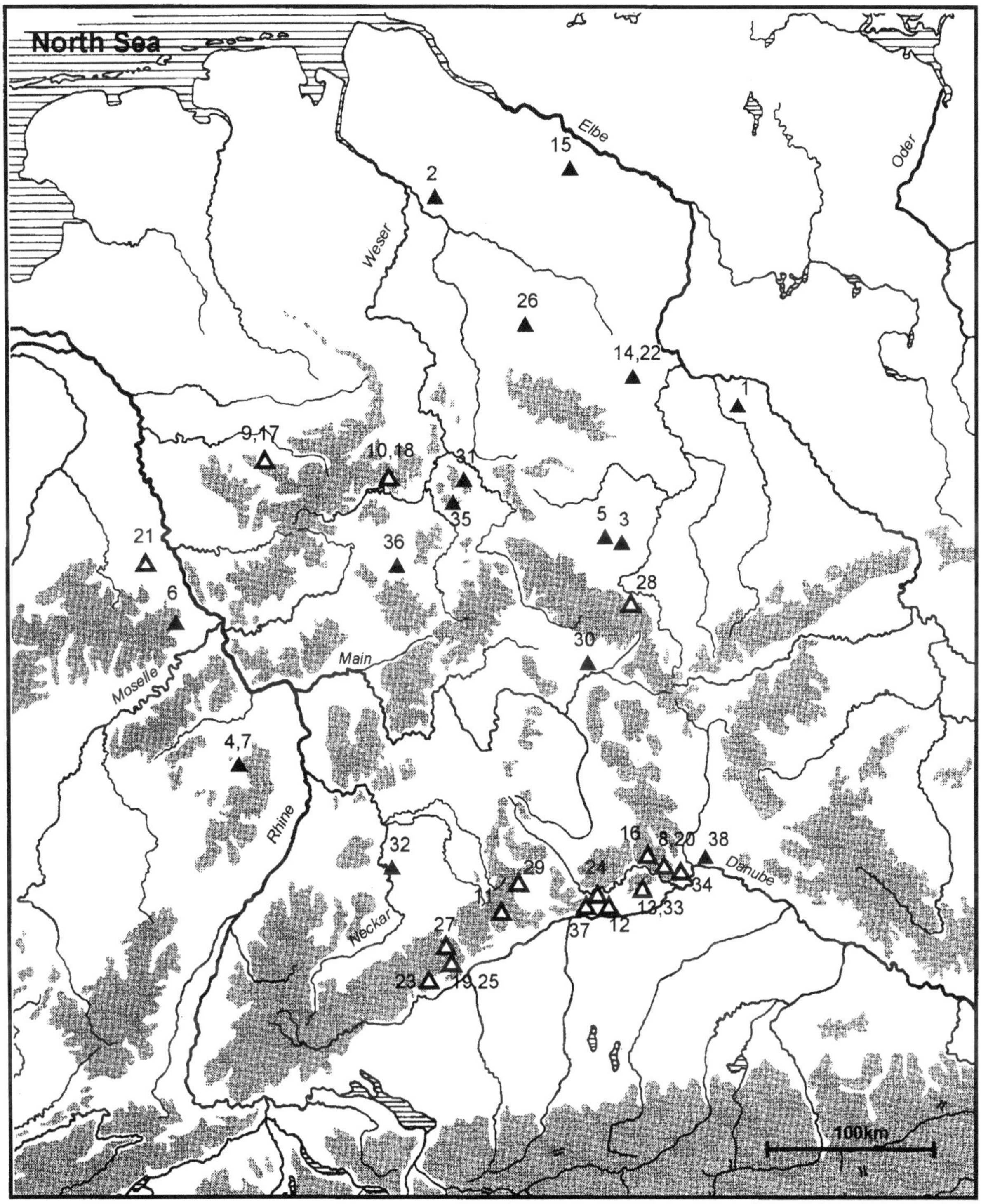

Fig. 4. Late Middle Palaeolithic sites mentioned in text. Solid triangles depict open-air sites. Open triangles depict caves and rock shelters. Eemian Middle Palaeolithic: Gröbern (1), Lehringen (2), Taubach (3), Wallertheim A (4), Weimar (5). Laminar Middle Palaeolithic: Tönchesberg 2B (6), Wallertheim D (7). *Keilmessergruppe*: Abri Schulerloch (8), Balve II-III (9), Buhlen IIIb (10), Bockstein III (11), Hohler Stein-Schambach (12), Klausennische (13), Königsaue A & C (14), Lichtenberg (15), Sesselfelsgrotte G (16). Late Mousterian: Balve IV (17), Buhlen II (18), Große Grotte VII-II (19), Großes Schulerloch (20), Kartstein III (21), Königsaue B (22), Sirgenstein I & II (23). Late Middle Palaeolithic with Handaxes: Beixenstein Höhle (24), Große Grotte IX (25), Salzgitter-Lebenstedt (26). *Blattspitzengruppe*: Haldensteinhöhle (27), Ilsenhöhle (28), Kleine Ofnet (29), Kösten (30), Lenderscheid (31), Mundelsheim (32), Obere Klause III (33), Oberneder Höhle (34), Rörshain (35), Wahlen (36), Weinberghöhlen (37), Zeitlarn (38). *The cultural association of Rheindahlen B1 is unclear at present.

Bibliography

Allsworth-Jones, P. 1986. *The Szeletian and the Transition from Middle to Upper Palaeolithic in Central Europe*. Oxford: Clarendon Press.

Andree, J. 1928. *Das Paläolithikum der Höhlen des Hönnetals in Westfalen*. Mannus Bibliothek 42. Leibzig: Kalitzsch Verlag

Bar-Yosef, O. 1995. The Lower and Middle Paleolithic in the Mediterranean Levant: chronology and cultural entities, in *Man and Environments in the Paleolithic*. Edited by H. Ullrich, pp. 246-263. Liège: Université de Liège.

Baumann, W. and D. Mania 1983. *Die paläolithischen Neufunde von Markkleeberg bei Leipzig*. Veröffentlichungen des Landesmuseums für Vorgeschichte Dresden 16. Berlin: Deutscher Verlag der Wissenschaften.

Behm-Blancke, G. 1960. *Altsteinzeitliche Rastplätze im Travertingebiet von Taubach, Weimar, Ehringsdorf.* Alt-Thüringen 4. Weimar: Hermann Böhlaus Nachfolger.

Bibus, E. and M. Kösel, 1996. Paläopedologische Klimazeugen zur Untergliederung des Rißeiszeiten-Komplexes im Rheingletschergebiet. *Eiszeitalter und Gegenwart* 46, 65-90.

Binford, L. R. and S. R. Binford 1966. A preliminary analysis of functional variability in the Mousterian of Levallois facies, in Recent studies in paleoanthropology. Edited by J. D. Clark and F. C. Howell. *American Anthropologist* 68 (2), 238-295.

Blackwell, B. and H. P. Schwarcz. 1986. U-series analysis of the lower travertine at Ehringsdorf, D.D.R., *Quaternary Research* 25, 215-222.

Bordes, F. 1961a. Mousterian cultures in France. *Science* 134, 803-810.

Bordes, F. 1961b. *Typologie du Paléolithique Ancien et Moyen*. Publications de l'Institut de Préhistoire de l'Université de Bordeaux, Mémoire 1. Bordeaux.

Bosinski, G. 1963. Eine mittelpaläolithische Formengruppe und das Problem ihrer geochronologischen Einordnung. *Eiszeitalter und Gegenwart* 14, 124-140.

Bosinski, G. 1966. Der paläolithische Fundplatz Rheindahlen, Ziegelei Dresen-Westwand. *Bonner Jahrbücher* 166, 318-343.

Bosinski, G. 1967. *Die mittelpaläolithischen Funde im westlichen Mitteleuropa*. Fundamenta A4. Köln: Böhlau Verlag.

Bosinski, G. 1969. Eine Variante der Micoque-Technik am Fundplatz Buhlen, Kreis Waldeck. *Jahresschrift der mitteldeutschen Vorgeschichte* 53, 59-74.

Bosinski, G. 1970a. Erwiderung zu der Besprechung meiner Arbeit 'Die mittelpaläolithischen Funde im westlichen Mitteleuropa' durch G. Freund. *Quartär* 21, 99-101.

Bosinski, G. 1970b. Bemerkungen zu der Grabung D. Peyronys in La Micoque, in *Frühe Menschheit und Umwelt*. Festschrift Alfred Rust. Edited by K. Gripp, R. Schütrumpf and H. Schwabedissen, pp. 52-56. Köln: Böhlau Verlag.

Bosinski, G. 1972. Late middle paleolithic groups in northwestern Germany and their relations to early upper paleolithic industries, in *The Origin of Homo sapiens*. Edited by F. Bordes, pp. 153-160. Paris: Unesco.

Bosinski, G. 1982. The transition Lower/Middle Palaeolithic in northwestern Germany, in *The Transition from Lower to Middle Palaeolithic and the Origin of Modern Man*. Edited by A. Ronen, pp. 165-175. British Archaeological Reports International Series 151. Oxford: BAR.

Bosinski, G. 1986. Chronostratigraphie du Paléolithique Inférieur et Moyen en Rhénanie, in *Chronostratigraphie et faciès culturels du Paléolithique Inférieur et Moyen dans l'Europe du Nord-Ouest*. Supplément au Bulletin de l'Association Française pour L'Étude du Quaternaire. Edited by A. Tuffreau and J. Sommé, pp. 15-34. Paris.

Bosinski, G., K. Kröger, J. Schäfer, and E. Turner. 1986. Altsteinzeitliche Siedlungsplätze auf den Osteifel-Vulkanen. *Jahrbuch des Römisch-Germanischen Zentralmuseums* 33, 97-130.

Brew, J. O. 1946. The use and abuse of taxonomy, in *The Archaeology of Alkali Ridge Utah*. Papers of the Peabody Museum of Archaeology and Ethnology 21, 44-46.

Busch, R., and H. Schwabedissen (ed.) 1991. *Der altsteinzeitliche Fundplatz Salzgitter-Lebenstedt, Teil II*. Fundamenta A 2/2. Köln: Böhlau Verlag.

Chmielewski, W. 1969. Ensembles micoquo-prondnikiens en Europe Centrale. *Geographia Polonica* 17, 371-386.

Clark, J. D., G. H. Cole, G. L. Isaac and M. R. Kleindienst 1966. Precision and definition in African archaeology. *South African Archaeological Bulletin* 21, 114-121.

Collins, D. M. 1969. Cultural traditions and environment of early man. *Current Anthropology* 10, 267-316.

Conard, N. J. 1990. Laminar lithic assemblages from the last interglacial complex in northwestern Europe. *Journal of Anthropological Research* 46, 243-262.

Conard, N. J. 1992. *Tönchesberg and its Position in the Paleolithic Prehistory of Northern Europe*. Römisch-Germanisches Zentralmuseum Mainz, Monographien 20. Bonn: Habelt.

Conard N. J. 1994. On the prospects for an ethnography of extinct hominids. *Current Anthropology* 35, 281-282.

Conard, N. J., D. S. Adler, D. T. Forrest and P. J. Kaszas 1995. Preliminary archaeological results from the 1991-1993 excavations in Wallertheim. *Archäologisches Korrespondenzblatt* 25:13-27.

Conard, N. J. and D. S. Adler. 1997. Lithic reduction and hominid behavior in the Middle Paleolithic of the Rhineland. *Journal of Anthropological Research* 53, 147-176.

Conard, N. J. 1998a. Middle Paleolithic settlement in the Rhineland. in *Middle Paleolithic and Middle Stone Age Settlement Systems*. Edited by N. J. Conard and F. Wendorf. Proceedings of the XIII Congress of the UISPP Vol. 6. pp. 255-268. Forli:

A.B.A.C.O. Editioni.

Conard, N. J., 1998b Comments on d'Errico et al. Neanderthal acculturation in Western Europe? A critical review of the evidence and its interpretation. *Current Anthropology* (in press).

Conard, N. J., T. J. Prindiville and D. S. Adler 1998. Refitting bones and stones as a means of reconstructing Middle Paleolithic subsistence in the Rhineland. in *Économies préhistorique: les comportements de susistance au Paléolithique.* Edited by J.-P. Brugal, L. Meignen and M. Patou-Mathis. pp. 273-290. Sophia Antipolis: Éditions APDCA.

Delpech, F., J.-M. Geneste, J.-P. Rigaud and J.-P. Texier 1995. Les industries anterieures a la derniere glaciation en Aquitaine Septentrionale: chronologie, paléoenvironnements, technologie, typologie et économie de susistance. *Actes du colloque de Miskolc. Paleo Supplément* 1, 133-163.

Eichhorn, G. 1909. *Die paläolithischen Funde von Taubach in den Museen zu Jena und Weimar. Festschrift zum 350-jährigen Jubiläum der Universität Jena.* Jena: Fischer Verlag.

Eissmann, L. 1994. Grundzüge der Quartärgeologie Mitteldeutschlands (Sachsen, Sachsen-Anhalt, Südbrandenburg, Thüringen) in *Das Quartär Mitteldeutschlands: Ein Leitfaden und Exkursionsführer mit einer Übersicht über das Präquartär des Saale-Elbe-Gebietes.* Edited by L. Eissmann and T. Litt., Altenburger Naturwissenschaftliche Forschungen 7, pp. 55-135.

Eissmann, L., W. Bernhardt and A. Rudolph. 1991. Eine mittelpaläolithische Fundstelle von Eythra bei Leipzig. *Veröffentlichungen des Naturkundemuseums Leipzig* 9, 33-54.

Feustel, R. 1983. Zur zeitlichen und kulturellen Stellung des Paläolithikums von Weimar-Ehringsdorf. *Alt-Thüringen* 19, 16-42.

Fiedler, L. 1994. *Alt- und mittelsteinzeitliche Funde in Hessen.* Führer zur hessischen Vor- und Frühgeschichte 2. Stuttgart: Konrad Theiss Verlag.

Fiedler, L. 1997. *Jäger und Sammler der Frühzeit: Alt- und Mittelsteinzeit in Nordhessen.* Vor- und Frühgeschichte im Hessischen Landesmseum in Kassel 1. Kassel: Staatliche Museen.

Freericks, M. 1995. Transition du Paléolithique moyen au Paléolithique supérieur en Allemagne: quelques exemples. *Actes du colloque de Miskolc. Paleo Supplément* 1, 117-122.

Freund, G. 1952. *Die Blattspitzen des Paläolithikums in Europa.* Quartär Bibliothek 1. Bonn: Röhrscheid Verlag.

Freund, G. 1969. Review of G. Bosinski's Die Mittelpaläolithischen Funde im westlichen Mitteleuropa. *Quartär* 20, 238-242.

Freund, G. 1987. Das Paläolithikum der Oberneder-Höhle (Landkreis Kehleim/Donau). Quartär-Bibliothek 5. Bonn: Röhrschied Verlag

Gamble, C. 1986. *The palaeolithic settlement of Europe.* Cambridge: Cambridge University Press.

Grahmann, R. 1955. The lower paleolithic site of Markkleeberg and other comparable localities near Leipzig. *Transactions of the American Philosophical Society, N. S.* 45, 509-686.

Günter, K. 1964. *Die altsteinzeitlichen Funde der Balver Höhle.* Bodenaltertümer Westfalens 8. Münster: Aschendorff.

Hahn, J. 1990. La technologie des pointes bifaciales de Rörshain et leurs relations avec l'Allemagne du Sud, in *Feuilles de pierre. Les industries à pointes foliacées du Paléolithique Supérieur européen.* Actes du colloque de Cracovie 1989. Études et recherches archéologiques de l'Université de Liège 42. Edited by J. K. Kozlowski, pp. 79-93. Liège: ERAUL.

Hahn, J. 1996. Le Paléolithique supérieur en Allemagne méridionale (1991-1995). in *Le Paléolithique supérieur européen bilan quinquennal 1991-1996.* Actes du UISPP Congrés de Forli. Études et recherches archéologiques de l'Université de Liège 76. Edited by M. Otte. pp. 181-186. Liège: ERAUL.

Housley, R. A., C. S. Gamble, M. Street and P. Pettitt 1997. Radiocarbon evidence for the Lateglacial human recolonisation of Northern Europe. *Proceedings of the Prehistoric Society* 63, 25-54.

Hoernes, M. 1903. *Der diluviale Mensch in Europa.* Braunschweig: Verlag Vieweg und Sohn.

Hülle, W. W. 1977. *Die Ilsenhöhle unter Burg Ranis/Thüringen: Eine paläolithische Jägerstation.* Stuttgart: Fischer.

Jones, S. 1997. *The Archaeology of Ethnicity: Constructing Identities in the Past and Present.* New York: Routledge.

Jöris, O. 1992. Pradniktechnik im Micoquien der Balver Höhle. *Archäologisches Korrespondenzblatt* 22, 1-12.

Jöris, O. 1994. Neue Untersuchungen zum Mittelpaläolithikum von Buhlen, Hessen: Technologische Studien zur Pradniktechnik in Horizont III b des Oberen Fundplatzes. *Ethnographisch-Archäologische Zeitschrift* 35, 88-97.

Kind, C.-J. 1992. Bemerkungen zur Differenzierung des süddeutschen Mittelpaläolithikums. *Archäologisches Korrespondezblatt* 22, 151-159

Klein, R. G. 1996. Neanderthals and modern humans in West Asia: A conference summary. *Evolutionary Anthropology* 5, 187-193

Klostermann, J. and J. Thissen 1995. Die stratigraphische Stellung des Lößprofils von Mönchengladbach-Rheindahlen. *Eiszeitalter und Gegenwart* 45, 42-85.

Königswald, W. v., H. Müller-Beck, E. Pressmar 1974. *Die Archäologie und Paläontologie in den Weinberghöhlen bei Mauern (Bayern). Grabungen 1937-1967.* Archaeologica Venatoria 3. Tübingen: Archaeologica Venatoria.

Kowalski, S. 1967. Ciekawsze zabytki paleolityczne z najnowoszych badan archeologicznych (1963-1965) w Jaskini Ciemnej w Ojcowie, pow. Olkusz. *Materialy Archeologiczne* 8, 39-46.

Kozlowski, J. K. 1972. On the typological classification of stone age artefacts. *Sprawozdania Archeologiczne*

24, 494-502.

Kozlowski, J. K. and S. K. Kozlowski 1996. *Le Paléolithique en Pologne*. Grenoble: Jérome Millon.

Krukowski, S. 1939. Paleolit, in *Prehistoria ziem polskich*, Zeszyt 1. Edited by S. Krukowski and J. Kostrzewski, pp. 1-117. Kraków: Polska Akademia Umiejetnosci.

Mania, D. 1997. Das Quartär des Saalegebiets und des Harzvorlandes unter besonderer Berücksichtigung der Travertine von Bilzingsleben - Ein Beitrag zur zyklischen Gliederung des eurasischen Quartärs. in *Bilzingsleben V: Homo erectus - seine Kultur und Umwelt*. Bad Homburg: Verlag Ausbildung und Wissen. pp. 23-103.

Mania, D., M. Thomae, T. Litt and T. Weber 1990. *Neumark-Gröbern. Beiträge zur Jagd des mittelpaäolithischen Menschen*. Veröffentlichungen des Landesmuseums für Vorgeschichte in Halle 43. Berlin: Deutscher Verlag der Wissenschaften

Mania, D. and V. Toepfer, 1973. *Königsaue: Gliederung, Ökologie und mittelpaläolithische Funde der letzten Eiszeit*. Veröffentlichungen des Landesmuseums für Vorgeschichte in Halle 26. Berlin: Deutscher Verlag der Wissenschaften.

Mellars, P. 1996. *The Neanderthal Legacy: An Archaeological Perspective from Western Europe*. Princeton: Princeton University Press.

Müller-Beck, H. 1956. *Das obere Altpaläolithikum in Süddeutschland*. Hamburg: Auerdruck.

Müller-Beck, H. 1988. The ecosystem of the 'Middle Paleolithic' (late Lower Paleolithic) in the Upper Danube Region: A stepping-stone to the Upper Paleolithic, in *Upper Pleistocene Prehistory of Western Eurasia*. University Museum Monograph 54. Edited by H. L. Dibble and A. Montet White, pp. 233-254. Philadelphia: University of Pennsylvania Press.

Narr, K. 1951. Alt- und mittelpaläolithische Funde aus rheinischen Freilandstationen. *Bonner Jahrbücher* 151, 5-51.

Narr, K. 1954. Formengruppen und Kulturkreise im europäischen Paläolithikum.. *Bericht der Römisch-Germanischen Kommission* 34, 1-40.

Obermaier, H. 1911/12. *Der Mensch der Vorzeit*. Berlin-Wien: Allgemeine Verlagsgesellschaft.

Pasda, C. 1996. Silexverarbeitung am Rohmaterialvorkommen im Mittelpleistozän: Ergebnisse einer Rettungsgrabung in Zwochau (Lkr. Delitzsch). *Arbeits- und Forschungsberichte zur Sächsischen Bodendenkmalpflege* 38, 13-56.

Pratte, E. 1971. L'industrie de La Micoque. *L'Anthropologie* 75, 369-396.

Révillion, S. 1994. *Les industries laminaire du Paléolithique moyen en Europe septentrionale: L'example des gisements de Saint-Germain-des-Vaux/Port-Racine (Manche), de Seclin (Nord), et de Riencourt-les-Bapaume (Pas-de-Calais)*. Publications du Centre d'Etudes et de Recherches Préhistoriques 5. Villeneuve d'Ascq: CERP.

Révillion, S., and A. Tuffreau (eds.) 1994. *Les industries laminaire au Paléolithique Moyen*. Dossier de Documentation Archéologique 18. Paris: Éditions CNRS.

Richter, D. 1998. *Thermolumineszenzdatierungen erhitzter Silices aus paläolithischen Fundstellen*. Doctoral dissertation, University of Tübingen.

Richter, J. 1997. *Der G-Schichten-Komplex der Sesselfelsgrotte: Zum Verständnis des Micoquien*. Forschungsprojekt 'Das Paläolithikum und Mesolithikum des Unteren Altmühltals II', Teil III. Quartär-Bibliothek 7. Saarbrücken: Saarbrücker Verlag.

Rieder, K-H. 1989. Die mitteleuropäische Station in der Höhlenruine Beixenstein im Wellheimer Trockental, in *Steinzeitliche Kulturen an Donau und Altmühl*. Edited by K. H. Rieder, A. Tillmann and J. Weinig, pp. 43-50. Ingolstadt: Courier Druckhaus.

Riek, G. 1934. *Die Eiszeitjägerstation Vogelherd*. Akademische Verlagsbuchhandel F. F. Heine: Tübingen.

Riek. G. 1938. Ein Beitrag zur Kenntnis des süddeutschen Solutréen. *Germania* 22: 147-150.

Ronen, A. (ed.) 1982. *The transition from Lower to Middle Palaeolithic and the origin of modern man*. British Archaeological Reports International Series 151. Oxford: BAR.

Rolland, N. 1986. Recent findings from La Micoque and other sites in south-western and Mediterranean France: their bearing on the 'Tayacian' problem and Middle-Palaeolithic emergence, in *Stone Age Prehistory. Studies in Memory of Charles McBurney*. Edited by G. N. Bailey and P. Callow, pp. 121-151. Cambridge: Cambridge University Press.

Schäfer, D. 1990. Merkmalanalyse mittelpaläolithischer Steinartefakte. *Ethnographisch-Archäologische Zeitschrift* 31, 54-64.

Schäfer, D. 1991. Weimar-Ehringsdorf: Diskussionsstand zur geochronologischen und archäologischen Einordnung sowie aktuelle Aufschlußsituation. *Quartär* 41/42, 19-43.

Schäfer, D. 1993. Grundzüge der technologischen Entwicklung und Klassifikation vor-jungpaläolithischer Steinartefakte in Mitteleuropa. *Bericht der Römisch-Germanischen Komission* 74, 49-193.

Schäfer, D. 1997. Überlegungen zur Älterpaläolithischen Technologieentwicklung und Nomenklatur. *Germania* 75, 1-18.

Schäfer, J. 1990. *Der altsteinzeitliche Fundplatz auf dem Vulkan Schweinskopf-Karmelenberg*. Doctoral dissertation, University of Käln.

Schild, R. and F. Wendorf. 1977. *The Prehistory of the Dakhla Oasis and Adjacent Desert*. Polska Akademia Nauk Instytut Historii Kultury Materialnej. Wroclaw: Ossolineum.

Schmidt, R.R. 1912. *Die diluviale Vorzeit Deutschlands*. Stuttgart: Schweizerbartsche Verlagsbuchhandlung.

Schmitz, R.-W. 1988. Die mittelpaläolithischen Fundplätze Herne und Bottrop im Emschertal. *Archäologisches Korrespondenzblatt* 18, 311-321.

Schönweiß, W. and H.-J. Werner. 1986. Ein Fundplatz des

Szeletien in Zeitlarn bei Regensburg. *Archäologisches Korrespondenzblatt* 16, 7-12.

Schwarcz, H. and R. Grün 1988. ESR dating of level 2/3 at La Micoque (Dordogne), France: Excavations of Debénath and Rigaud. *Geoarchaeology* 3, 293-296.

Shackleton, N J. and N.G. Pisias 1985. Atmospheric carbonic dioxide, orbital forcing and climate. *in The Carbon Cycle and Atmospheric CO2: Natural Variations Archean to Present.* Edited by E. T. Sundquist and W. S. Broecker. Washington, D.C.: American Geophysical Union. pp. 449-464.

Shennan, S. 1989. Introduction: archaeological approaches and cultural identity, in *Archaeological Approaches to Cultural Identity.* One World Archaeology 10. Edited by S. Shennan, pp. 1-32. London: Unwin Hyman.

Steele, J. and S. Shennan eds. 1996. *The Archaeology of human agency: Power, Sex and Tradition.* London: Routledge.

Steguweit, L. 1997. *Steinartefaktefunde aus saalezeitlichen Schottern bei Lübbow (Lkr. Lüchow-Dannenberg) - Sammlung H. Leunig.* Magisterarbeit, University of Tübingen.

Steiner, W. 1979. *Der Travertin von Ehringsdorf und seine Fossilien.* Wittenberg: Ziemsen Verlag.

Thieme, H. 1983. *Der paläolithische Fundplatz Rheindahlen.* Doctoral dissertation, University of Köln.

Thieme, H. and S. Veil 1985. Neue Untersuchungen zum eemzeitlichen Elefanten-Jagdplatz Lehringen, Ldkr. Verden. *Die Kunde* N.F. 36, 11-58.

Thissen, J. 1998. Die aktuellen Ausgrabungen 1995-1997 in der Ziegeleigrube Dreesen, Fundschicht B1, und neuste Ergebnisse zur menschheitsgeschichtlichen Entwicklung. *Rheindahlen Almanach 1998*, 18-39.

Thissen, J. and R. W. Schmitz 1997. *Neue Ausgrabungen in Rheindahlen.* Unpublished lecture February 1997, University of Tübingen.

Tode, A. 1982. *Der altsteinzeitliche Fundplatz Salzgitter-Lebenstedt.* Fundamenta A 11. Köln: Böhlau Verlag.

Tode, A., F. Preul, K. Richter, W. Selle, K. Pfaffenberg, A. Kleinschmidt and E. Guenther 1953. Die Untersuchung der paläolithischen Freilandstation von Salzgitter-Lebenstedt. *Eiszeitalter und Gegenwart* 3, 144-220.

Toepfer, V. 1978. Die altsteinzeitliche Fundstelle Hundisburg nach 75 Jahren. *Jahresschrift des Kreismuseums Haldensleben* 19, 5-18.

Toepfer, V. 1981. Das Acheuléen auf dem Boden der Deutschen Demokratischen Republik. *Anthropologie* 19, 55-77.

Tuffreau, A. 1989. Le gisement paléolithique moyen de Champvoisy (Marne), in *Paléolithique et Mésolithique du Nord de la France: Nouvelles recherches.* Publications du Centre d'Etudes et de Recherches Préhistoriques. Edited by A. Tuffreau, pp. 69-77. Villeneuve d'Ascq: CERP.

Tuffreau, A. and J. Sommé (ed.) 1988. *Le gisement paléolithique moyen de Biache-Saint-Vaast (Pas-de-Calais) Volume 1: Stratigraphie, environnement, études archéologique.* Mémoires de la Société Préhistorique Française 21. Châlons-sur-Marne: Paquez & Fils.

Valoch, K. 1977. Die Mikrolithik im Alt- und Mittelpaläolithikum. *Ethnographisch-Archäologische Zeitschrift* 18, 57-62.

Valoch, K. 1984. Le Taubachien, sa géochronologie, paléoécologie et paléoethnologie. *L'Anthropologie* 88, 193-208.

Valoch, K. 1988. *Die Erforschung der Kůlna-Höhle 1961-1976.* Anthropos 24 (N.S. 16). Brno: Moravské muzeum.

Veil, S. 1983/84. Review of A. Tode's Der altsteinzeitliche Fundplatz Salzgitter-Lebenstedt. *Die Kunde* N.S. 34/35, 277-282.

Veil, S., K. Breest, H.-C. Höfle, H.-H. Meyer, H. Plisson, B. Urban, G.A. Wagner and L. Zöller 1994. Ein mittelpaläolithischer Fundplatz aus der Weichsel-Kaltzeit bei Lichtenberg, Lkr. Lüchow-Dannenberg: Zwischenbericht über die archäologischen und geowissenschaftlichen Untersuchungen. *Germania* 72, 1-66.

Wagner, E. 183. *Das Mittelpaläolithikum der Großen Grotte bei Blaubeuren (Alb-Donau-Kreis).* Forschungen und Berichte zur Vor- und Frühgeschichte in Baden-Württemberg 16. Stuttgart: Theiss Verlag.

Wagner, E. 1996. Eine mittelpaläolithische Blattspitze von Mundelsheim, Lkr. Ludwigsburg. *Fundberichte aus Baden-Württemberg* 21, 7-11.

Weber, T. 1986. Die Steinartefakte des Homo erectus von Bilzingsleben, in: *Bilzingsleben III.* Veröffentlichungen des Landesmuseums für Vorgeschichte in Halle 39. Edited by D. Mania and T. Weber, pp. 65-231. Berlin: Deutscher Verlag der Wissenschaften.

Weber, T. 1996. Merkmalanalytische Untersuchungen an den Artefakten von Zwochau, Tagebau Delitzsch-Südwest, Sachsen. *Arbeits- und Forschungsberichte zur Sächsischen Bodendenkmalpflege* 38, 57-68.

Weissmüller, W. 1995. *Sesselfelsgrotte II. Die Silexartefakte der unteren Schichten der Sesselfelsgrotte.* Quartär-Bibliothek 6. Saarbrücken: Saarbrücker Verlag.

Wiegers, F. 1928. *Diluviale Vorgeschichte des Menschen.* Stuttgart: Enke Verlag.

Wenzel, S. 1996. Die Funde aus dem Travertin von Stuttgart-Untertürkheim und die Archäologie der letzten Warmzeit in Mitteleuropa. Doctoral dissertation, University of Köln.

Wetzel, R. 1958. *Die Bocksteinschmiede.* Stuttgart: Kohlhammer Verlag.

Wetzel, R. and G. Bosinski. 1969. *Die Bocksteinschmiede im Lonetal.* Verlag Müller and Gräff: Stuttgart.

Zotz, L. 1955, *Das Paläolithikum in den Weinberghöhlen bei Mauern.* Quartär-Bibliothek 2. Bonn: Röhrscheid Verlag.

Zotz, L. 1959. *Kösten, ein Werkplatz des Praesolutréen in Oberfranken.* Quartär-Bibliothek 3. Bonn: Röhrscheid Verlag.

DISCUSSION

Avraham Ronen: I would like to draw your attention to some peculiarity of the Levant as compared to Europe. It seems that we have a smaller problem distinguishing between Lower and Middle Palaeolithic, because there are no bifaces in the Middle Palaeolithic. We do not have Middle Palaeolithic with handaxes or with bifacial leaf-shaped points. In Europe things are more complicated.

Gerhard Bosinski: I would like to point out the High Lodge problem. There is a certain agreement that the Middle Palaeolithic starts about 300,000, around the beginning of Stage 8. Personally, I know nothing comparable to High Lodge earlier than Stage 8. All the earlier finds are dominated by small flakes, denticulates and notches, together with handaxes, cleavers, and perhaps some pebble tools. For me, High Lodge is unique. If anybody knows of a comparison to High Lodge in the Lower Palaeolithic, I will be glad to learn about it.

Nicolas Rolland: I sympathize with Dr. Bosinski because I share the same feeling about High Lodge. But I have not seen the industry in detail. But I have seen some material in Italy, like Visoliano, so called Tayacian and Venosa-Loretto, which are given approximately the same age as High Lodge. These are flake industries. We have found a handaxe recently in Visoliano, in the upper phase. The industry is very small, and this is a fact to keep in mind. It has a lot of retouch, but this retouch is not steep retouch. They are trying, without succeeding, to produce step flaking. The Yabrudian has a very early step flaking. In Europe, I think, one of the very earliest occurrences is in the upper culture horizon at Hoxne, where we have a flake industry with a lot of step flaking and very few handaxes. I do not know whether that is Stage 7 or 9. There is also Riparo Paglicci, which has it but it has been given a Rissian date without any specifications. Otherwise there is Lebenstedt, but can anybody date that without fauna? I think that in general I can agree with you about High Lodge, and I would like to see whether the retouch is really step flaking or not.

Avraham Ronen: You are touching here on a problem of beginnings, and beginnings are notoriously delicate to see. At Tabun, it is difficult to know if you are at the end of the Acheulean or already at the beginning of the Yabrudian I would like to learn from my European colleagues the following: we have these curious blade complexes Pre-Aurignacian and Amudian. Could they perhaps be linked with early modern humans, coming out of Africa sometime between 300,000 and 200,000 years ago? What are the oldest blade occurrences known today in Europe?

Gerhard Bosinski: Why are we trying to see similarities between blades and *Homo sapiens*? I can not accept this. There is no proof that blades and human beings of modern type have any connection. Our blades are found in the beginning of Stage 8, of course, but nobody should argue that this is linked with some brain development. There are just blades.

Nicholas Conard: I think the question is very simple. Until we find skeletal remains that suggest that there were modern humans in Europe in that early date, we must reject this idea. Let us restrict ourselves to the Stage 5 blade assemblages. At present we have a number of Neanderthal remains, but no modern human remains. Without the modern fossils, I see no way to make the argument.

Gerhard Bosinski: Avraham, you have stated that in the Levalloiso-Mousterian there is no visible difference between Kebara and Qafzeh. In this case you deny a connection of human types and industries, but on the other hand you are trying to construct something with blades.

Joachim Schäfer: Concerning the beginning of the blade industry. I think that we have to try to divide the blade industries we know from the last interglacial, with lateral retouch, and the Levallois blade industries like Markleberg. In this case the beginning of the blade industries in Europe is in the last interglacial but not in Stage 8. And to Avraham, we have in central Europe Aurignacian assemblages with no blades. The oldest blade industry is in Tadjikistan, about 200,000 years BP.

Nicolas Rolland: About the 'Out of Africa'. What is interesting, as Tony Marks showed at a Congress in Japan a few years ago, is that archaeologically speaking, we have very little indication of movement out of Africa at the crucial time. There are clear differences between the Middle Palaeolithic of North and East Africa and the Levant.

We have a very interesting incidence of an intrusion, possibly from Africa, but that seems to be a discrete event with Gesher Benot Ya'aqov, with cleavers and a specialized technology, which are different from those of Ubeidiya. It does not surprise me in the Levant, because if you just look at the vegetation map, it is a crossroad. Still, there are two points to make. If we look at the circumstances of the environment, glacial, interglacial etc., Africa stands as a huge reservoir of human population. I am not the one who invented that. On the other hand, we find discoveries of bone harpoons in eastern Zaire, perhaps 80,000 years ago, that have never made it beyond the region of origin.

Avraham Ronen: Concerning the Tony Marks paper, there is no archaeological evidence for an out-of-Africa trail in the Negev and Sinai. Recently, however, Eitan Tchernov claimed that the real bridge between Africa and Asia, for both fauna and humans, was not Sinai or the Suez area but, rather, Bab el Mandab. That area is archaeologically still unknown.

Nicolas Rolland: That is a very interesting point, because it is going much further in time. There is some circumstantial evidence for a very narrow cross. It would get beyond the Yemen hill into fairly watered valleys. There is a very early Acheulean site in Djibouti called

Barogali, dated by ESR to 1.4 million years. It could not be more strategically located than there. I am very sympathetic for that view, because when did the Nile start flowing north? Maybe a million years ago, half a million years ago? We do not find Nile Acheulean until about 400 thousand, I think.

Nicholas Conard: Gerhard, how would you deal with the variation we see in the Micoquian and Mousterian in the last glacial cycle? Do you think those terms refer to something coherent and comparatively homogeneous and identifiable as meaningful archeological entities?

Gerhard Bosinski: Of course, things changed very much in 30 years and there is much more detailed stratigraphy, especially on the early parts of the so-called 'Last Glaciation'. There are steppe conditions with black humic soils and there are cold periods during Isotope Stage 5.
I feel that your groups are too general to answer these questions. On the other hand, it is not excluded that given more detailed studies, some assemblages will not differ from the early Upper Palaeolithic. For example, I think that you should not combine all the late bifaces. The Rheindahlen biface which you have mentioned is a classical bout coupé type, and is dated to around the first cold stage. So, it would fit into the Mousterian of Acheulean Tradition (MTA). To me, MTA is something fascinating. In central Europe it is marginal but in south-western France there is evolution and you may even map the different stages. The bout coupé-type biface is linked with the widespread MTA type A.
Another thing. I would not take together all these as a unity. I would refer to the Buhlen III b, Ciemna and Germolles. These finds are very similar. It is not only the sharpening blow, but it is the set of types which corresponds. Recently, in building the Les Eyzies Museum, they discovered an abri which delivered finds corresponding to this group.
The Bockstein material is, in a certain way, unique. I classified this material as 'Micoquian' because I did not dare to invent a new term, being a student then. 'Keilmesser' group is perhaps better, but it is not accepted everywhere. Bockstein, Buhlen III b and Koenigsaue are similar, but Bockstein is so special that I think that even if we have only one site, it should be separated.
Another thing is your Mousterian. If you want to differentiate it, I feel that it is something which François Bordes would call Denticulate Mousterian.

Joachim Schäfer: I think we do not have to speak too much about typology. We should speak more about technology, even about raw materiel. If you have quartz, you will never find a Mousterian of Ferrassie type. And if you have a bad raw material, you very soon will get Denticulated Mousterian.

Gerhard Bosinski: I agree that for the type Ferrassie they will not use quartz.

Avraham Ronen: But a Denticulated Mousterian is not always a matter of raw material.

Nicolas Rolland: Yes, they do not care about that. Denticulates are made on any kind of material. And then there is the industry of Grotta dei Moscerini, which is made out of shells. And you get a beautiful Quina like Mousterian. You have to see it to believe it. Even if you take some things like Chez-Pourre and La Chapelle aux Saints you get some quartz, and you get some typical Quina scrapers.
I am not saying that it does not matter, I will be the first one to maintain that raw material is very important. But there is a range of variation, of tolerance, and we are talking about statistical averages. I have seen in Strasburg the material from a new site in Alsace, Mutzig, in eastern France. They have used basalt because that region is very poor in good raw material. I would say that it is a Typical Mousterian with a certain number of recognizable denticulates. It is not published yet.

Nicholas Conard: In response to Gerhard's question. My grouping of the Late Mousterian Middle Palaeolithic into one category was in no way intended to mean that it is homogeneous. On the contrary, my problem was that it is so heterogeneous, I can not put it into any sort of meaningful subdivisions. So I think we perhaps share the point of view that something has to be done to establish recognizable units within these general kinds of assemblages. But I do not see clear ways to draw the lines between, let's say, different divisions within the Keilmesser group. And concerning the MTA, I agree that it is well-defined in France, but east of the Rhein I can not find any examples of it. I would like, as much as anyone, to see clear spatial-temporal units in the Middle Palaeolithic, I just cannot see them.

Avraham Ronen: When you say heterogeneous, do you mean intra-site or inter-site heterogeneity?

Nicholas Conard: Both.

Gerhard Bosinski: You misunderstood what I wanted to say. I see a unity in the Horizon Buhlen III b, but I am not sure of the connection to Koenigsaue or Bockstein. On the other hand, I think that the Rheindahlen biface belongs to the MTA. The Keilmesser group is much too large to represent a unit. We do not have enough sites, and not enough stratified sites. Avraham, you just mentioned that the Yabrudian is Lower Palaeolithic. This is in contrast to the way we use the term Lower Palaeolithic. In my view, the Yabrudian belongs to the Middle Palaeolithic, which started about 300,000 years ago and the Yabrudian fits well into it. Perhaps this is a good possibility to meet here and establish uniformity.

Avraham Ronen: I would like to clarify this remark of yours. Why is it suddenly Middle Palaeolithic? It was Lower Palaeolithic for Dorothy Garrod, so why has it become Middle Palaeolithic? Simply, because of wrong chronology. When the Yabrudian was erroneously dated to the last interglacial, Jelinek, Bar-Yosef and others claimed that it was a sufficient reason to place it with the Middle, and not with the Lower Palaeolithic. The same entity has been shifted because of chronology, regardless of what it really consisted of. In the Levant we enjoy a clear barrier: there is no Middle Palaeolithic with bifaces. Yabrudian is

Lower Palaeolithic.

Daniel Kaufman: I wonder about this terminology of Lower, Middle and Upper Palaeolithic which has been used ever since the beginning of prehistoric research. I wonder, have these terms really lost their relevance? In this part of the world we are getting more data with good strong evidence. If we qualify the Yabrudian as Middle Palaeolithic, it may partly overlap with something we refer to as Acheulean, which we place in the Lower Palaeolithic. How can we have Lower Palaeolithic overlapping in time with Middle Palaeolithic? I really wonder if what have been heuristic devices for helping us to sort and classify archeological assemblages, have gone beyond their time.

Nicolas Rolland: You mentioned the point that you cannot have two periods overlapping in time. What is the purpose of that division? It is a periodization. Gordon Childe mentioned the concept of homotaxial, that things really happened simultaneously everywhere. Take Scandinavia, for example. During their Late Neolithic, they were imitating weapons of the central European Bronze Age, so you really have a time lag. But with formal terminology and too narrow definitions, we are going to create problems. We have to move with the time. We might now give working definitions, since we have much more material to accommodate to see whether there is a certain coherence. You have the same problem in East Asia. At some sites handaxes appear at the beginning of the Upper Pleistocene, totally out of the blue, and there are large blades and step flaking.

Nicholas Conard: I just want to avoid having numerous groups that are characterized only by one locality. I agree with you. The Buhlen Keilmesser complex is very distinctive, partly due to raw materials and partly due to other factors. For instance, Wallertheim D, I will recognize that anywhere if I ever see it again. I am, however, very hesitant to have dozens of Middle Palaeolithic groups. And for simplicity's sake, rather than splitting it into too many categories, I prefer to keep the categories broad.

Gerhard Bosinski: This is another subject. We may compare it to certain Upper Palaeolithic phenomena, like Kostenki 1, the top-most level, which corresponds to Avdeevo in many details and there were very few other sites. Concerning the length of Palaeolithic time, we do not know many sites which have really existed at the same time.
As far as I know, African and Euro-Asian Lower Palaeolithic is always like Evron and Ubeidiya, while the Middle Palaeolithic is like Yabrud and Tabun. The Middle Palaeolithic is characterized by a marked variation of types including bifaces and flake types. When I learned that you call the Yabrudian Lower Palaeolithic, you should consider that this is so only in this part of the world. To a certain degree, it is in the old François Bordes tradition. We are lucky to meet here, and we should try to deal with this question.

Mina Weinstein-Evron: If the Yabrudian is 400,000 years old, would it then be Lower Palaeolithic?

Gerhard Bosinski: The Yabrud material is Middle Palaeolithic. The variability of scrapers and the presence of points make it a Middle Palaeolithic. How old - I do not know!

Jan Burdukiewicz: Concerning the definition of Middle Palaeolithic, I would like to add another definition, used by W. Chmielewski who thinks that the Middle Palaeolithic started at 600,000 years. He supposed that in fact, no Lower Palaeolithic existed in Europe.
Going back to the discussion on the chronology question, we mix two terms - chronology and typology. Maybe it is the same as the Neolithic in the Levant and in Europe. In Europe the Neolithic starts much later than here.

Joachim Schäfer: I think, Gerhard, that you have to specify the technological and typological features which make the Yabrudian Middle Palaeolithic for you. Now there is a time scale, and with this time scale you return to the Bordian system. We are archaeologists, and we have to work with typology and technology, not with a time scale. What are your features for Middle Palaeolithic?

Gerhard Bosinski: It is the variability of retouched flake tools.

Jan Burdukiewicz: Maybe we can say that the archaeological time scale is local, not global like the geological time scale.

Nicolas Rolland: I think that the criteria we select should be archaeological, not chronological. Once I did a little exercise about what I would consider to be Middle Palaeolithic techniques over time, at assumed duration. I compared that to the Lower Palaeolithic and what I got was a trajectory that was geometric, non-linear in terms of increased frequency. Something is happening in this continuum somewhere between 300 and 400 thousand years, and there is acceleration of innovations which in terms of hominid behaviour are very important. Why is it happening at that time? These are issues which we have to investigate.

TOWARD MODERN HUMANS: YABRUDIAN AND MICOQUIAN, 400 – 50 kyears ago.
Edited by A. Ronen and M. Weinstein-Evron. BAR S850, 2000.

STRATIGRAPHY AND PALAEOLITHIC OF THE MIDDLE AND UPPER PLEISTOCENE IN THE SAALE-ELBE REGION

Dietrich Mania

As a result of my field work conducted in the region between the Thuringian Forest (Thüringer Wald) and the Elbe river, the climate record of the last 500,000 years was established. Four sediment sequences, of great significance for the Pleistocene stratigraphy of the European continent, were discovered (Fig. 1):

- The terrace-travertine sequence of Bilzingsleben, located in northern Thuringia (Mania, 1994).
- The cyclic sequence deriving from the southeast basin of the Helmstedt 'Sattel' (anticline) in the Northern Harzvorland (Schöningen) (Thieme and Mania, 1994).
- One sequence of limnic, glacial and periglacial sediments, deriving from the basin of the Geisel valley in the Eastern Harzvorland.
- One cyclic sediment sequence deriving from the basin of the Aschersleben lake, located in the northeastern Harzvorland. This sequence (Mania, 1999) is completed by some other special sequences (travertine from Burgtonna in Thuringia, limnic-telmatic series in the Geisel valley), representing the climatic record from the Eem to the Holocene (Mania, 1999a).

All the four sequences are connected with Palaeolithic horizons. Additional finds complete and confirm the above mentioned series, e.g., some travertines from Ehringsdorf, the Middle Pleistocene of Neumark-Süd, in the Geisel valley (Mania and Mai, 1969), a series at Lengefeld-Bad Kösen (Mania, 1994), a limnic Eem - early Weichsel series from Gröbern (Litt, 1994).

The stratigraphical sequences of Bilzingsleben and of Schöningen consist of 6 cycles, each following upon the Elster glaciation (Figs. 2, 6, 7 and 22). They correspond to 6 large climate cycles. Each cycle covers about 80,000 to 120,000 years. Three cycles containing 3 interglacials were discovered for the period between the Elster and the Saale glaciation (so-called Holstein complex). One interglacial separates the Saalian (Drenthian, Saalian *sensu stricto*) from the Warthian. All the three represent the Saalian complex. The Eem interglacial, the Weichsel glacial and the Holocene interglacial are attributed to the 5th and the 6th cycle.

The site of Bilzingsleben

The terrace-travertine sequence of Bilzingsleben was subjected to a special geological mechanism which was regulated by cyclic climatic changes. It is characteristic of the river valleys of the Mittelgebirge forelands and it can be subdivided into four phases:

1. *Early glacial phase*. Relatively humid cold climate. Fluviatile erosion, transport and accumulation of gravels are synchronous processes in the course of which the valley was eroded with a gravel cover developed on its bottom.
2. *Pleniglacial phase*. Arid and cold climate. The processes described above have come to a standstill. Permafrost prevails. Solifluction and aeolic events occur. Loess is formed.
3. *Late glacial phase*. The climate becomes more and more humid and warm. The river is refilled with water and cuts into the gravel cover. To a lesser degree, sands and gravels are again transported and accumulated. During a period of rewarming and reforestation, however, these processes are stopped again.
4. *Interglacial phase*. Warm temperate climate. Thick travertine covers are formed. The interglacial weathering processes favour the production of forest soils (Parabraunerden).

The terrace, situated 45 m above the recent bottom of the Wipper valley near Bilzingsleben, carries the glacial series of the Elster glaciation (so-called first glaciation terrace). The subsequent terrace-travertine series lies at 32 - 35 m above the recent bottom of the valley (Bilzingsleben I). The terraces situated at 26 - 27 m, 20 - 22 m, 16 - 18 m and 8 - 10 m contain the cycles of Bilzingsleben II-V. The base of cycle VI lies below the present bottom of the valley (Fig. 2).

The terrace situated at 16 - 18 m is identified with the 'second glaciation terrace' and, in the Saale region, it carries the groundmoraine of the Saale glaciation. Thus, the travertine complexes of Bilzingsleben cycles I-III demonstrate three completely developed interglacials of the period between the Elster and the Saale glaciation. They are separated from each other by glacials with permafrost and loess formation. Their climate is influenced by Mediterranean and subcontinental conditions. The oldest interglacial of Bilzingsleben I corresponds to the Holstein. The interglacial of Bilzingsleben II contains a rich Palaeolithic find horizon. The interglacial of Bilzingsleben III can be correlated with the Dömnitzien (Erd, 1978; Cepek, 1986).

Besides species typical of a mixed oak wood, Mediterranean elements such as *Celtis australis* and *Syringa josikaea* are characteristic elements of the vegetation of the travertines of the interglacials. Most plant remains were found in Bilzingsleben II. They belong to a Buxo-Quercetum, Buxo-Syringetum and a Berberidion. Other exotic species are *Buxus sempervirens*, *Pyracantha coccinea*, *Juniperus sabina*, *Vitis vinifera* and the subcontinental species *Potentilla fruticosa* (Mai, 1983).

Thermophile molluscan faunas contain south and southeast European elements and the guiding species *Helicigona banatica* prevails (Mania, 1983). A *Palaeoloxodon antiquus*-fauna with *Trogontherium cuvieri* and *Arvicola cantianus* (Fischer et al., 1991) was determined from Bilzingsleben II.

With the help of $^{234}U/^{230}Th$ and ESR dating, this middle interglacial of the Holstein complex is dated to be 320,000 - 412,000 years old. Consequently, it can be correlated with Isotope Stage 11.

The predominant plant species found in the Intra-Saalian interglacial (Bilzingsleben IV) were determined as *Quercus*, *Corylus*, *Celtis* and *Syringa josikaea*. The molluscan fauna contains, however, no *Helicigona banatica*.

The cultural horizon in the travertine of Bilzingsleben II (Mania, 1990) originates from a camp site located on the border terrace of a small lake (Fig. 3). 1,400 m^2 of the living floor and of an alluvial fan were excavated as far, the latter containing the refuse of the settlement. About 40 - 50% of the cultural objects originating from the living floor were found in their primary position. They are indicative of structures and find associations deriving from different activity areas.

The oval to circular outlines of three dwellings with a diameter of 3 to 4 m were brought to light. Hearths were discovered in front of them. Within some working zones, a great number of workshops with anvils of pebbles, slabs of stone or large bones in their centre appeared. A pavement with a diameter of approximately 9 x 9 m was found southeast of the dwelling structures. The pavement consists mainly of pebbles which were alien to the site, and had to be transported to the place. It seems to have served a special cultural function.

The cultural remains found at the site consist of:

- large pebble tools (choppers, chopping tools, hammerstones);
- small special tools made of flint (knives, scrapers, bifacial points, Tayac points, Quinson points, borers, denticulates and notches) (Fig. 4);
- large special tools made of bone: scrapers with retouched edges 10 - 80 cm long, chopper-like and chisel-like tools, handaxe-like tools, cudgels, dagger-like tools, work-supports (U. Mania, 1995);
- large, split rods made of elephant tusks with their pointed ends polished;
- cudgel- and mattock-like tools made of red deer antlers; some bone tools display deliberate, regular, directed lines engraved on their surface (Mania and Mania, 1988);
- remains of wooden tools: spatula-like tools, remains of two spears 2.3 - 2.5 m long, some tools with hooks and holes at their butts.

The human remains recovered include 23 fragments of two skulls of a typical *Homo erectus* (Vlcek, 1980; Mania et al., 1994) (Fig. 5, 1).

The Site of Schöningen

In the course of 5 years, between 1992-1996, the sediments discovered in the brown-coal open quarry of Schöningen were studied geologically and stratigraphically. These investigations revealed the same 6 sediment sequences as in Bilzingsleben, recognized in channels which were in superposition from west to east (Figs. 6 and 7). The oldest channel at Schöningen I was cut into the Elster glacial series. Schöningen I-III was covered by the groundmoraine of the Saale glaciation *sensu stricto* (Drenthian). Schöningen IV, containing an Intra-Saalian interglacial, is younger than the Drenthian moraine. Schöningen V is mainly composed of Eemian and early Weichselian sediments, followed by Schöningen VI with late Weichselian and Holocene limnic series.

The geological succession had a strikingly different course from that found in Bilzingsleben:

1. *Late glacial phase*. In the course of the dissolution of the permafrost, the flowing waters eroded a channel into which sands and gravels were deposited. When rewarming and reforestation, as well as suberosion had set in, the channel formed a basin.
2. *Interglacial phase*. Thick limnic-telmatic series developed in the lake basin.
3. *Early glacial phase*. Sands and silts were washed in with the return of cold climate. Organogeneous deposits from the interstades were left behind.
4. *Pleniglacial phase*. With a prevailing cold-arid climate, permafrost developed in the soil and aeolian loess was deposited. This cycle ends with a surface of denudation.

According to palynological diagrams (Urban, 1995), the interglacial of Schöningen I can probably be assigned to the Holsteinian. Schöningen II consists of 5 special cycles made up of limnic-telmatic sediments. The pollen succession of this sequence cannot be correlated with other well-known interglacials. It is designated 'Reinsdorf Interglacial' (Thieme and Maier, 1995). The limnic series of Schöningen III is identical with the pollen composition of the Dömnitzian (Urban et al., 1991).

A thick soil complex was recognized in Schöningen IV. In Schöningen V, a pedocomplex of the Eemian and the early Weichselian appeared. About 1000 m away, Schöningen V was limnic, exposing gyttjas, peats and travertines. The Holocene in Schöningen VI had the same features.

The interglacial of Schöningen II is identical with Bilzingsleben II. It displayed two find horizons. The older one was discovered in the special sequence 1 of the interglacial optimum containing the cultural remains. These show a striking similarity to the culture of the Bilzingsleben *Homo erectus*.

The youngest horizon was discovered in the special sequence 4 originating from a late interglacial continental forest steppe phase. At present it is being excavated by H. Thieme. He discovered a butchering place within the peat of a shore zone there. More than 90% of the hunted animals found are wild horses. Artefacts of flint with traces of cutting function, as well as tools made of spruce (Thieme and Maier, 1995) were uncovered: there are 8 spears 2 - 2.3 m long and one throwing stick 0.75 m in length. They are similar to the fighting clubs known from the Australian aborigines.

The Site of Markkleeberg

The site of Markkleeberg was discovered in the gravel pits and browncoal open casts located south of Leipzig. Thousands of artefacts made of flint and characteristic of an evolved Acheulean industry were found (Grahmann, 1955; Baumann and Mania, 1983) (Fig. 8). This site has also become important because the artefact complex could exactly and unmistakably be attributed to the early glacial phase of the *sensu stricto* Saalian (ca. 280,000 years ago).

The artefacts were found at the base of the gravels originating from the early Saalian terrace of the Pleiße and Elster. They derived from workshops found on the bottom of the valley with plenty of flint pebbles constituting the raw material needed for the manufacture of artefacts. They are the denudation residue from the Elster groundmoraine, which later, at places, formed the slope of the valley.

The groundmoraine of the Saale glaciation (*sensu stricto*) covered the gravels. In the surrounding area, smaller find spots came to light, displaying the same stratigraphical position. The last site which has been discovered is Eythra (Eissmann et al., 1991).

The Site of Neumark-Nord (Geisel valley)

In the course of mining works conducted in the brown coal open quarry of Neumark-Nord, an interglacial lake of 600 x 400 m was completely and thoroughly explored from 1986 to 1996. It was studied geologically, palaeontologically and archaeologically. The limnic-telmatic sediment sequence has proved the existence of an Intra-Saalian interglacial.

The sequence was discovered in a basin formed by the groundmoraine of the Saale glaciation *sensu stricto* (Figs. 9 and 10). Two periglacial series, each 6 m thick, were superposed. A soil complex of two strongly developed soils, formed as humus gleys, separated both sequences. It is assigned to the early Weichselian glacial. The lower soil can perhaps be attributed to the Eemian. The upper periglacial series represents the characteristic loess sequence of the Weichsel glacial. The lower periglacial series equally consists of loess, and contains 4 generations of frost wedges up to 8 m deep, forming a structured ground which was identified over the entire lake basin. These features indicate an intense, arctic climate and cannot be assigned to the first early Weichselian stage. They indicate instead a fully developed glacial. Thus, it is only possible to assign this horizon to the younger glacial of the Saalian complex, the Warthian.

TL-dates of this horizon (Karelin, 1997) yielded 204,000 years for the interglacial optimum, with the lower loess sequence ending at about 130,000 years. The basal part of the upper loess is dated to 70,000 BP, so the interglacial can be correlated with Isotope Stage 7. It corresponds to the interglacials identified in the cycles of Bilzingsleben IV and Schöningen IV. The pollen succession resembles the Eem type. It constitutes a special feature of this interglacial, which may have led to mistaken interpretations as regards the Eem interglacial occurrences. The interglacial of Neumark-Nord, different from all the interglacials so far known from the Elbe-Saale region, is to a large degree influenced by subcontinental conditions.

In the neighbouring area of the lake, during the phase of a mixed forest of oak, *Taxus* and hazel, an *Acer tataricum* - oak steppe forest, associated with 45% of west Asian-Pontic - southeast European steppe species, was determined (Mai, 1990). The molluscan fauna also proved that the arboreal species were associated with pontic steppe species. *Helicigona banatica* is missing.

The complete interglacial ecosystem found at the lake with fauna and cultural remains can be reconstructed from a great number of skeletal finds of *Palaeoloxodon antiquus, Bos primigenius,* fallow deer, red deer, as well as some skeletons of *Dicerorhinus kirchbergensis* and *Panthera (Leo) spelaeai,* from plenty of fish bones, small vertebrates, and plant remains such as seeds, fruits, leaves, trunks and stubs of trees.

Traces of Middle Palaeolithic humans included some butchering sites containing skeletons of *Bos primigenius, Palaeoloxodon antiquus* and *Dicerorhinus kirchbergensis,* which were excavated (Fig. 12, 1). Some knapping areas with discoidal cores and flakes made of flint were found on the shore. The artefacts (Fig. 11) associated with the dissected animals exhibited flaked knives, used for butchering purposes, and denticulates and notches, which were used for wood working. These finds represent a special economic activity (Mania et al., 1990; Mania, 1992).

The Site of Ehringsdorf

The travertines of Ehringsdorf (Fig. 13) are also assigned to the Intra-Saalian interglacial (Mania, 1993). The flora shows a Mediterranean-subcontinental influence, with the characteristic *Syringa josikaea* also occurring in the travertine of Bilzingsleben IV. The artefacts (Fig. 14) derive from different living-floor horizons, exhibiting partly Acheulean and partly Mousterian (Mousteroid) characteristics. The human remains are attributed to an archaic *Homo sapiens* (Vlcek, 1993) (Fig. 5, 2).

The Site of Aschersleben

The sequence deriving from the basin of the Aschersleben lake, supplemented by the Eemian travertine sequence of Burgtonna (Thuringia) and a late glacial profile coming from the Geisel valley (Mania et al., 1993), shows that the last main climatic cycle, covering the last 125,000 years, is composed of at least 15 subordinate oscillations (Figs. 15 and 22). This is one of the most complete profiles of the

Upper Quaternary studied on the continent (Mania 1967; 1975; 1991; 1999a). The sedimentation was subject to a cyclic mechanism:

1. *Cold phase*. Cold-arid climate. Denudation, predominantly deflation, loess transport and loess accumulation. Formation of a denudation area.
2. *Transitional phase*. Cool, relatively humid. Gravels and sands are washed into the basin. Thawing of the permafrost.
3. *Warm phase*. Thick vegetation cover leading to the formation of forests. Suberosion. Lake basin with limnic-telmatic organogenic sediments.
4. *Transitional phase*. Cool, relatively humid. Formation of a permafrost. Solifluction. Frost structures. Net of ice-wedges.
5. *Cold phase*. Cold-arid - see 1 above.

A total of 15 sediment cycles including the Eemian and the Holocene were identified, which correspond to as many minor climatic cycles. They are subject to the large cycle climate development. The course of the small cycles was investigated geologically, geomorphologically and palaeontologically with the help of the molluscan fauna, ostracods, vertebrate and plant remains, as well as with palynological studies. According to these results, the cycle contained the following phases:

1. *Interglacial, Eemian*. Warm-temperate. Close to the end, two cool oscillations. Isotope stage 5e.
2. *Early glacial phase I*. Between approximately 100,000 and 55,000 years BP, 4 cycles containing 5 stadial phases and 4 interstades. During the first two interstades forest steppes, chernozem steppes and a boreal to cool-temperate continental climate developed. The two younger interstades demonstrate the expansion of the chernozem steppes with a smaller wood cover. Boreal, relatively dry climate. The phase ends with an intensely cold, relatively humid stage with a distinct ice-wedge horizon. Large ice advance in the region of the Baltic Ocean. Approximately 65 - 55,000 years BP. Corresponds to Isotope Stage 4.
3. *Early glacial phase II*. Four cycles with 4 interstades. They have an increasingly continental, boreal-subarctic climate. Chernozem steppe, some small woods scattered about. Ends around 22,000 years BP.
4. *Pleniglacial phase*. Extremely cold-arid. Probably at least 3 cycles of loess steppes with intervals of subarctic climate. Main phase of loess formation. Maximum expansion of the inland ice sheet (Brandenburg stage). The retreat of the ice begins with several oscillations (Frankfurt and Pommern stage). Ends around 15,000 years BP. Corresponds to Isotope Stage 2.
5. *Late glacial phase*. Four cycles with 2 intervals and 2 interstades (last interstade: Alleröd, with tephra deriving from the Laach lake, Eifel found in the Aschersleben lake and in the Geisel valley). Increasingly warmer and more humid. Thawing of permafrost occurring in several phases. Prevalent grass steppes during the intervals. Immigration of birch and pine woods during the interstades. First subarctic and later boreal climate during the warm phases. End of the Younger Dryas, 8,000 BC (dendrochronological age = 9,700 years BC).
6. *Interglacial Holocene*. In the second interstade (Aschersleben lake Ib), discovered within a peat on a border zone, three Middle Palaeolithic archaeological horizons in superposition were excavated at the site of Königsaue (Mania and Toepfer, 1973; Mania, 1975). These finds are important because, on the one hand, they represent three distinct assemblage groups and, on the other hand, these groups are contemporaneous, geologically speaking. The three lithic groups are: a Micoquo-Prondnikian in the lowest horizon, Königsaue A ('Keilmessergruppen') (Figs. 16 and 17); a typical Mousterian in the middle horizon Königsaue B (Fig. 18); and a special Micoquo-Prondnikian industry in the uppermost, Königsaue C horizon (Fig. 19).

The Mousterian is mainly distributed over western Europe, and belongs to the culture of the Neanderthal man. The 'Keilmessergruppen' cover the Central and East European territory (Fig. 12, 2). Possibly, its cultural carrier is modern *Homo sapiens* from the Levant, having come across the Neanderthals in Central Europe.

Some other Micoquo-Prondnikian sites were discovered in the middle Elbe-Saale region (Fig. 1): Gera is assigned to the first interstade, Petersberg near Halle/Saale is assigned to the third or fourth interstade (Mania and Toepfer, 1973; Toepfer and Mania, 1973). The 'Keilmesser' complex of Lichtenberg in Lower Saxony (Veil, 1994) belongs to this phase rather than to a younger one. The same is probably valid for the 'Keilmesser' assemblage of Gamsenberg near Oppurg, in East Thuringia (Schäfer et al., 1991). The final stage of the Middle Palaeolithic in the Saale-Elbe region is represented by the 'Blattspitzen' group (Fig. 20) of Ranis, East Thuringia (5th interstade about 50,000 years BP) (Mania, 1975; Hülle, 1977). The rich Aurignacian living floor horizon of Breitenbach near Zeitz, in East Thuringia (Mania, 1975), as well as the Aurignacian horizon of Ranis (Fig. 21), is assigned to the 6th or 7th interstade (approximately 32,000 years BP).

A rich Perigordian/Gravettian assemblage with Font Robert points, collected in an open-air site near Bilzingsleben, is attributed to the 8th interstade, at about 27,000 - 25,000 years BP. A great number of Magdalenian find spots derive from the late glacial sections preceding the Alleröd interstade (e.g., the site of Nebra, Mania, 1999b). Finally, the so-called 'Federmesser' groups increase in the Alleröd, for example in Weinberg near Schadeleben, at the Aschersleben lake. They are found in the Alleröd soil horizon, which is an initial black soil containing tephra from lake Laach.

Translated by Ursula Mania.

Author's address: Forschungsstelle Bilzingsleben, Ibrahimstr. 29, FSU, Jena, Germany.

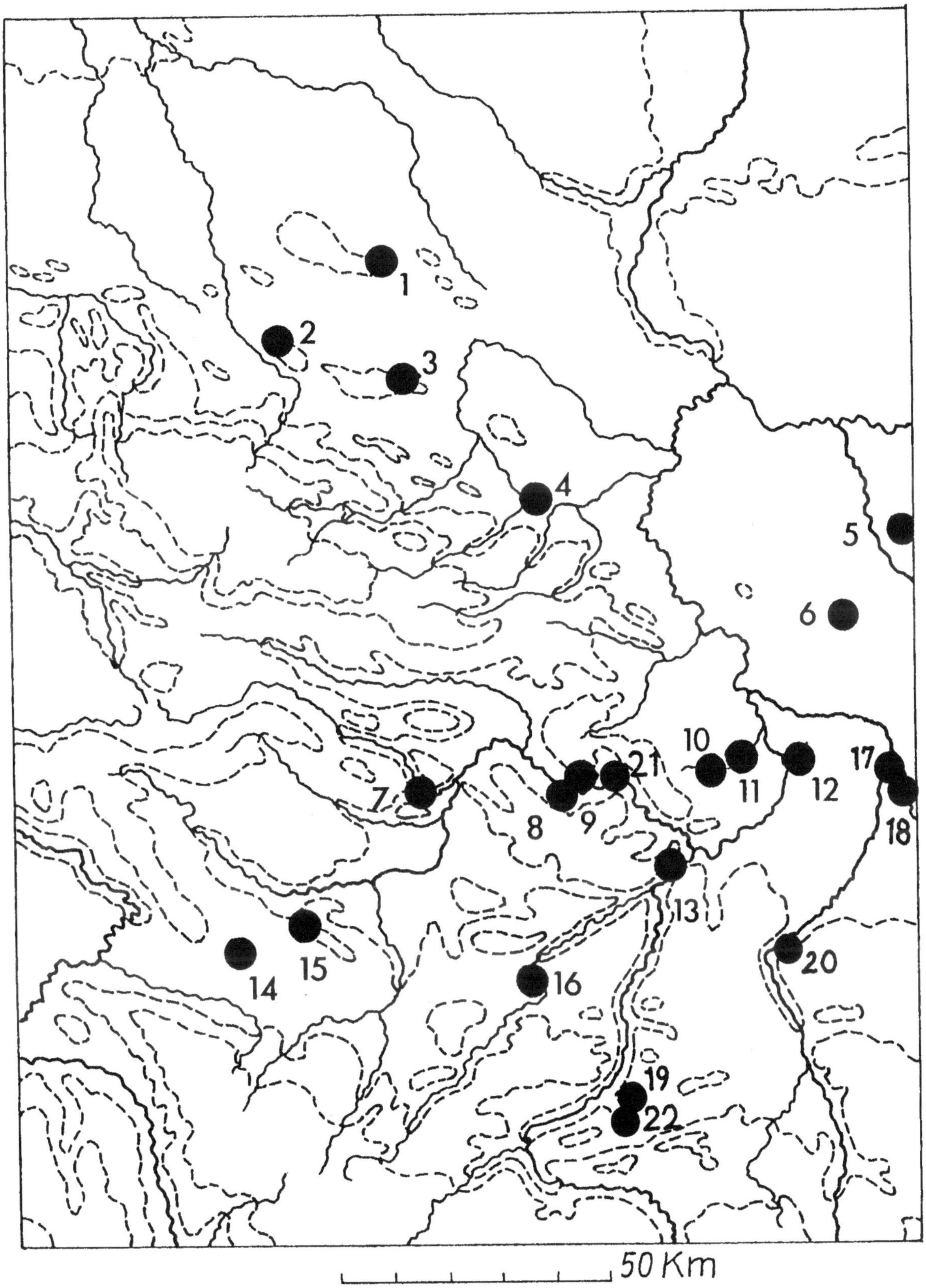

Fig. 1. The middle Elbe-Saale region with Palaeolithic and Pleistocene sites. 1. Schöningen; 2. Osterode/Fellstein (travertine); 3. Schwanebeck/Huy (travertine); 4. Aschersleben lake (Königsaue); 5. Gröbern; 6. Grabschütz; 7. Bilzingsleben; 8. Memleben; 9. Wangen; 10. Neumark-Süd (Geiseltal); 11. Neumark-Nord (Geiseltal); 12. Wallendorf; 13. Lengefeld-Bad Kösen; 14. Brüheim (travertine); 15. Burgtonna; 16. Weimar, Taubach, Ehringsdorf; 17. Markkleeberg; 18. Eythra; 19. Oppurg; 20. Breitenbach; 21. Nebra; 22. Ranis.

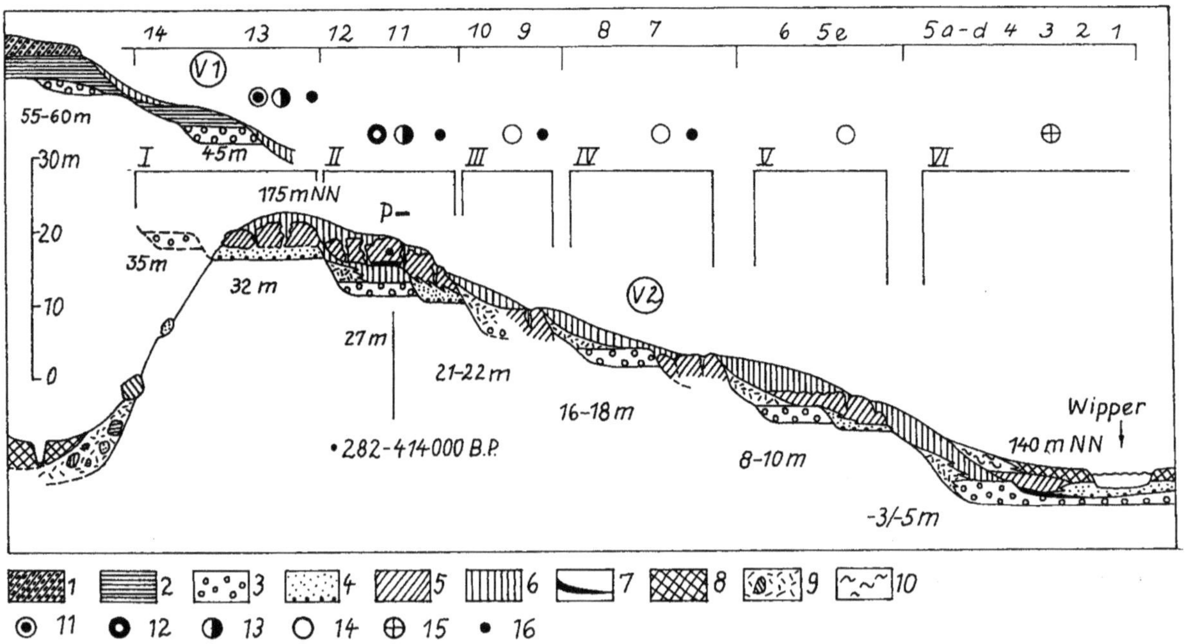

Fig. 2. Bilzingsleben. Terrace-travertine sequence. 1. groundmoraine (Elsterian); 2. glaci-limnic Warvit (Elsterian); 3. gravels (glacial); 4. sandy gravels (interglacial); 5. travertine; 6. loess; 7. muds and peats; 8. silt (Auelehm); 9. sediments of solifluction; 10. humic slope debris; 11. *Theodoxus serratiliniformis* fauna; 12. *Corbicula fluminalis* fauna; 13. *Helicigone banatice* fauna; 14. other warm climatic fauna of forests; 15. Holocene forest fauna; 16. *Celtis australis*. (P = Palaeolithic finds horizon. V1, V2: the first and the second glaciation terraces. I-IV cycles. 1-14 OIS stages);

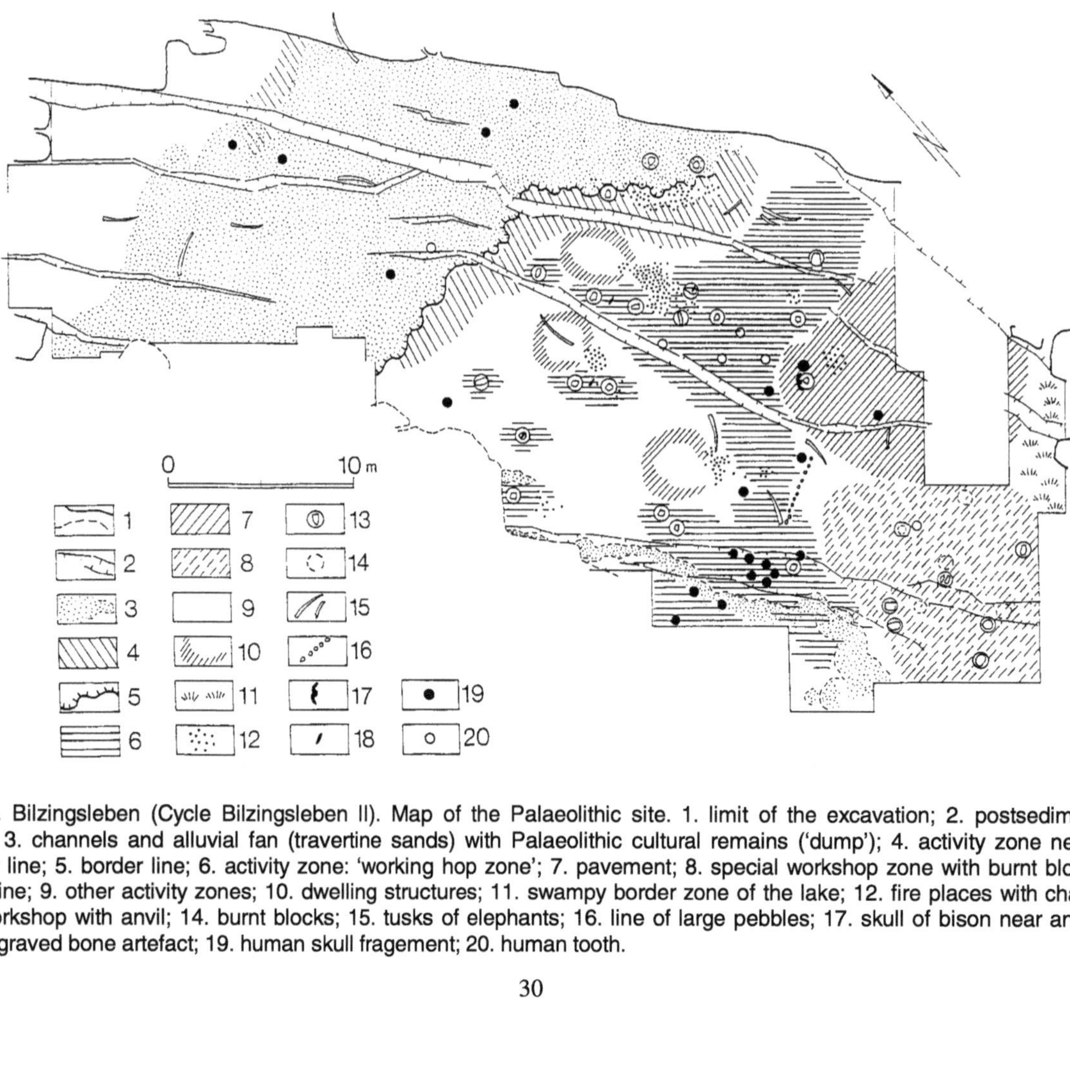

Fig. 3. Bilzingsleben (Cycle Bilzingsleben II). Map of the Palaeolithic site. 1. limit of the excavation; 2. postsedimentary faults; 3. channels and alluvial fan (travertine sands) with Palaeolithic cultural remains ('dump'); 4. activity zone near the border line; 5. border line; 6. activity zone: 'working hop zone'; 7. pavement; 8. special workshop zone with burnt blocks of travertine; 9. other activity zones; 10. dwelling structures; 11. swampy border zone of the lake; 12. fire places with charcoal; 13. workshop with anvil; 14. burnt blocks; 15. tusks of elephants; 16. line of large pebbles; 17. skull of bison near an anvil; 18. engraved bone artefact; 19. human skull fragement; 20. human tooth.

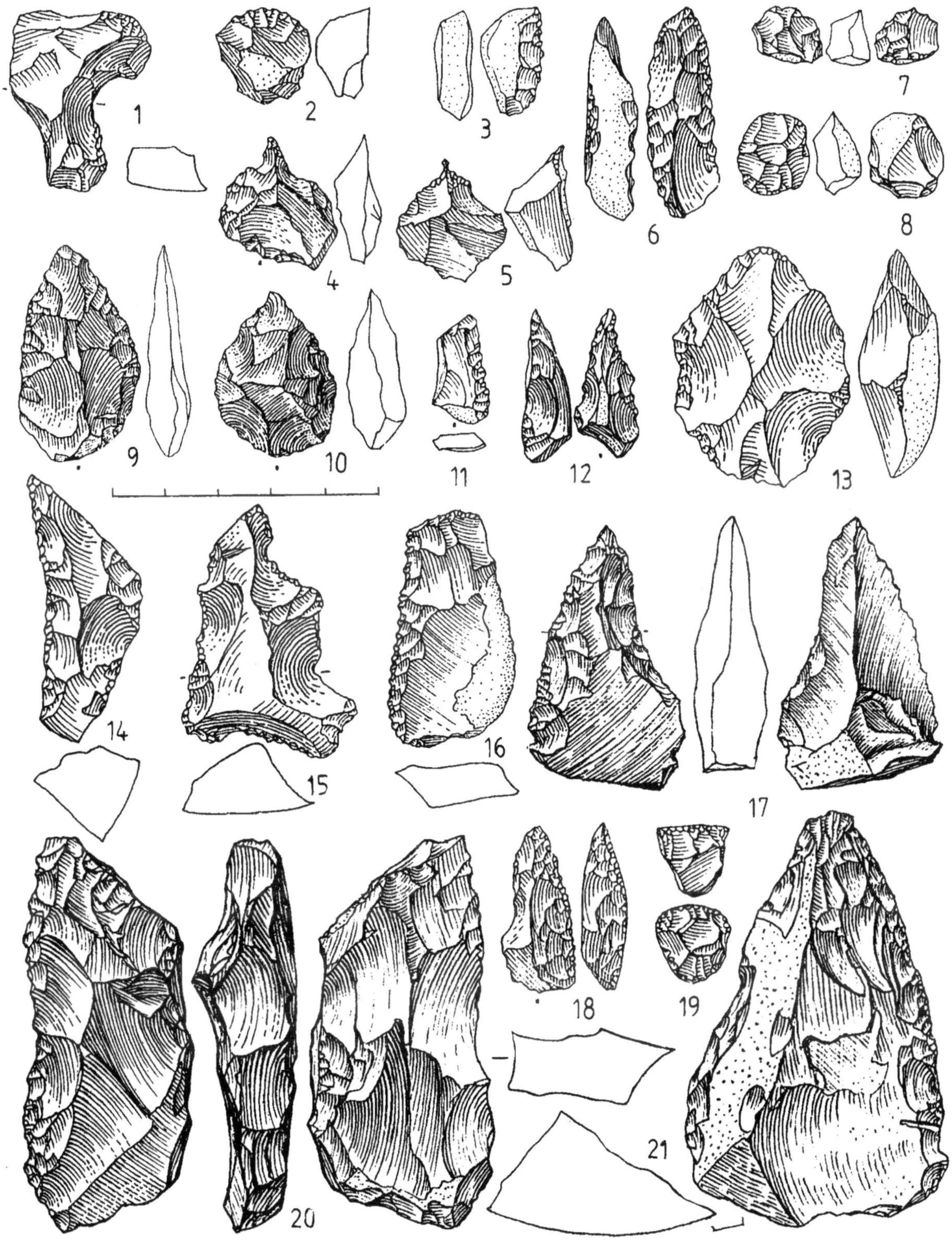

Fig. 4. Bilzingsleben. Special tools of flint stone.

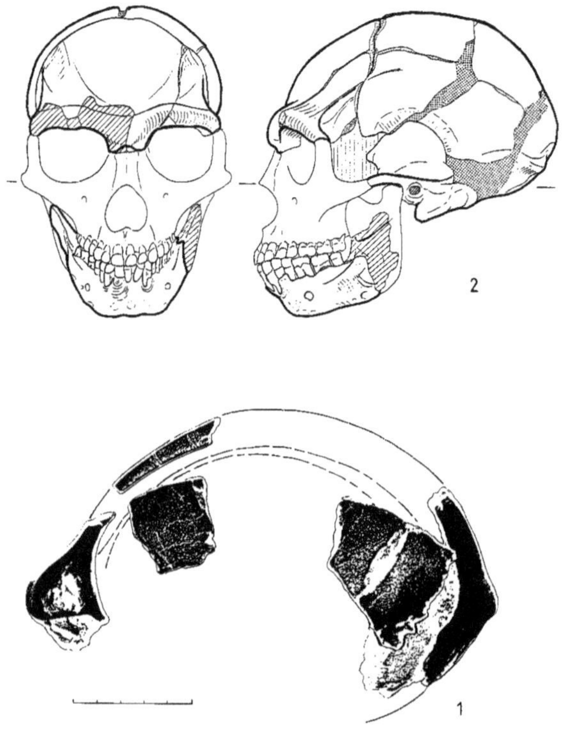

Fig. 5. 1. Bilzingsleben. Human skull (after Vlcek, 1993). The profile shows the characteristic features of *Homo erectus*; 2. Ehringsdorf. Reconstruction of the skull (archaic *Homo sapiens*) (after Vlcek, 1993).

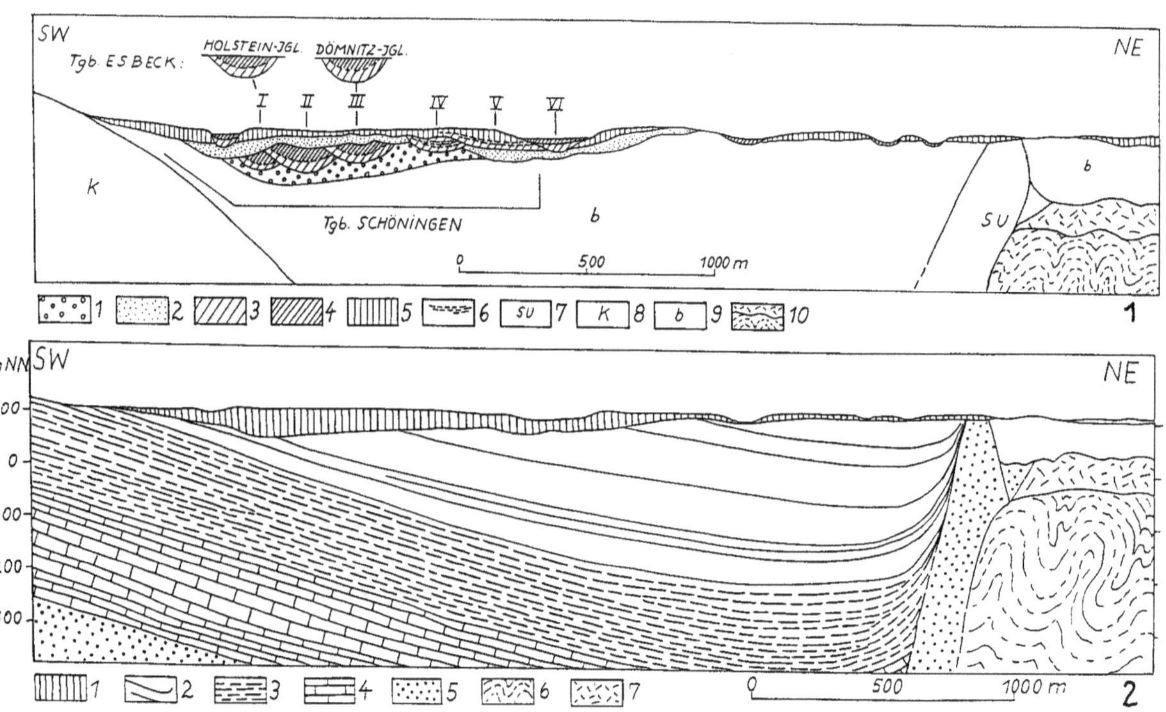

Fig. 6. Schöningen, northern Harzvorland. 1. The Quaternary sequence with the cycles I - VI. 1. the glacial series of the Elsterian; 2. the glacial series of the Saalian (Drenthian); 3. basin silt; 4. organic sediments (muds and peats); 5. loess; 6. pedocomplexes; 7. Buntsandstein; 8. Keuper; 9. Tertiary; 10. diapir of Permian salt.
2. Profile of the southwestern basin of the salt diapir (Helmstedter Sattel). 1. Quaternary sediments; 2. Tertiary; 3. Keuper; 4. Muschelkalk; 5. Buntsandstein; 6. salt; 7 gyps.

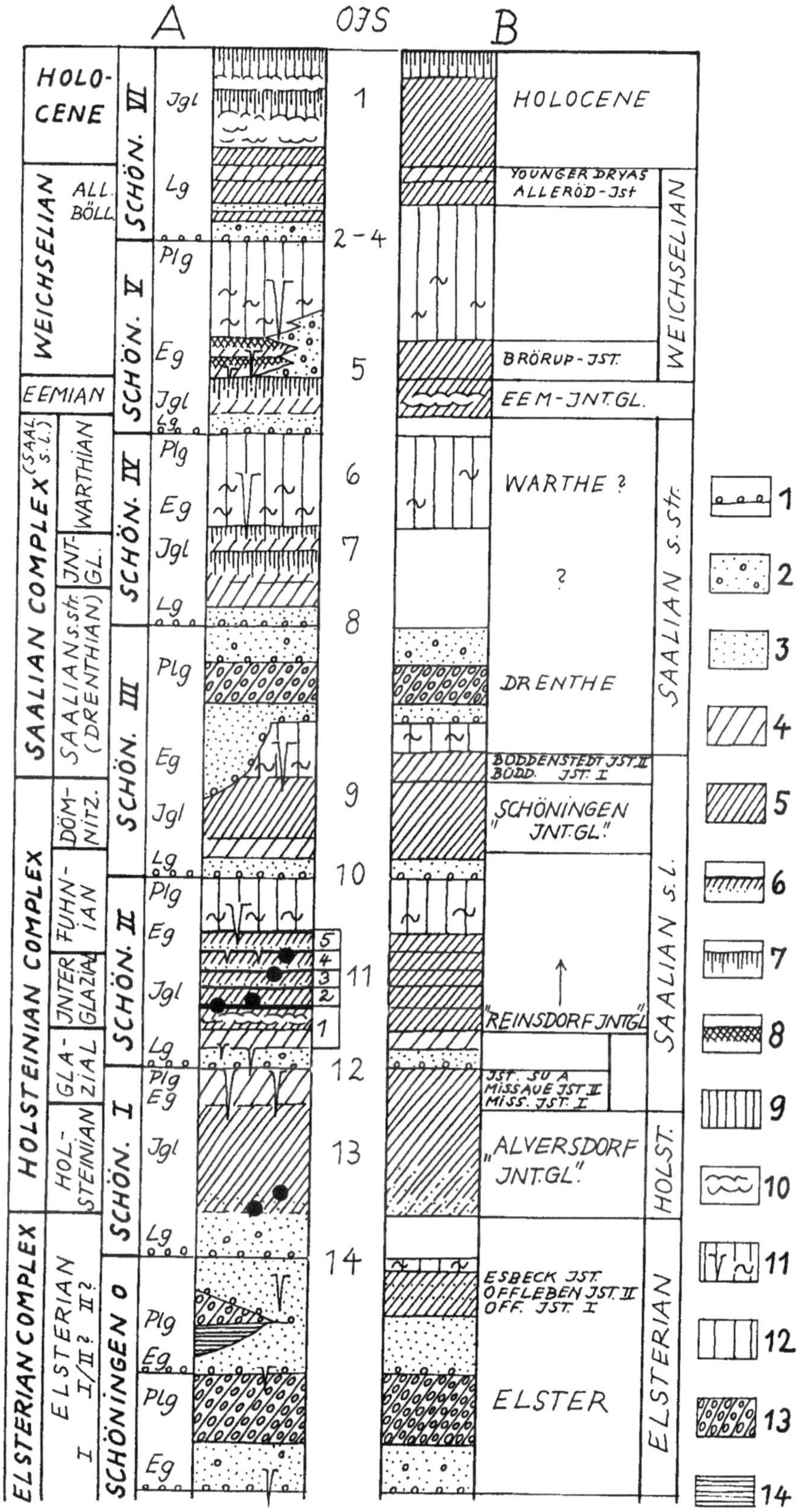

Fig. 7. Schöningen. The Quaternary sequence. A. the geological stratigraphy (D. Mania) ; B. the palynological investigations (B. Urban). 1. surface of denundation ; 2. sandy gravels or sands ; 3. sands; 4. limnic silts; 5. organic sediments (muds and peats); 6. special limnic sequences of cycle Schöningen II; 7. soils: pseodogley, lessive; 8. humic soils; 9. Holocene 'Auelehm'; 10. travertine; 11. solifluction loess; 12. loess; 13. groundmoraine; 14. glacilimnic silt. OIS - Oxygen Isotope Stages 1-14.

Lg, Igl, Eg, Plg - late glacial, interglacial, early glacial, pleniglacial phase.

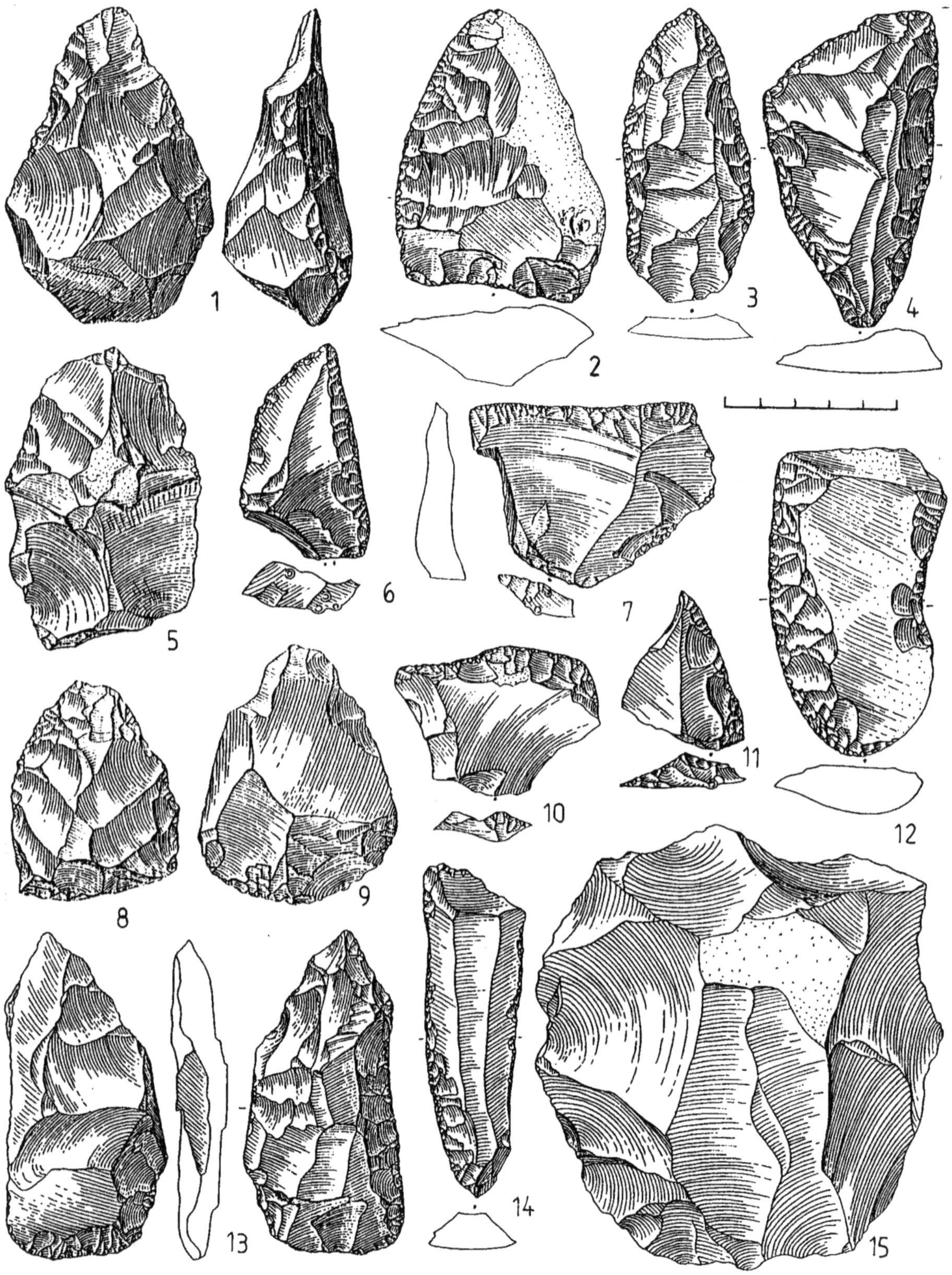

Fig. 8. Markkleeberg. Acheulean flint artefacts.

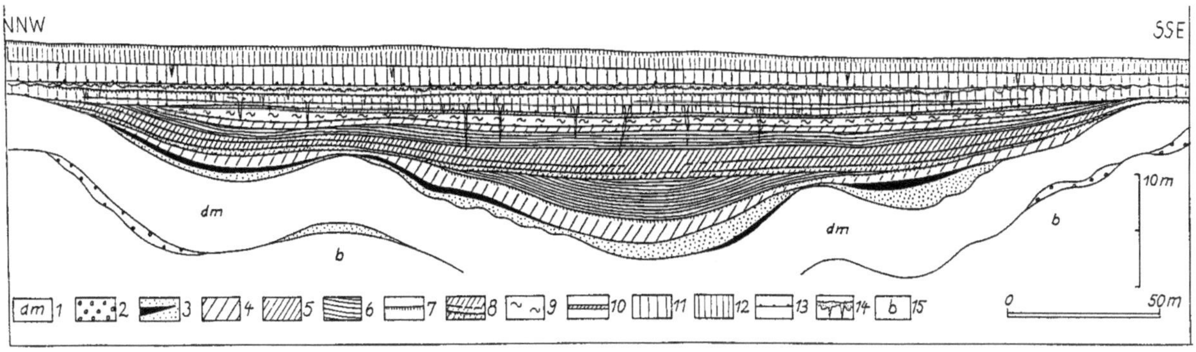

Fig. 9. Neumark-Nord (Geisel valley). The sequence of the lake basin. 1. groundmoraine of the Saalian (Drenthian); 2. gravels; 3. glacilimnic and glacifluviatile sediments; 4. basin silt; 5. silty muds; 6. varved muds; 7. swampy soils, humic soils and recent soil; 8. litoral zones; 9. redeposited muds; 10. special limnic horizon of a late interglacial oscillation; 11. solifluction loess; 12. loess; 13. denundation surface; 14. frost structures, frost wedges; 15. Tertiary sediments.

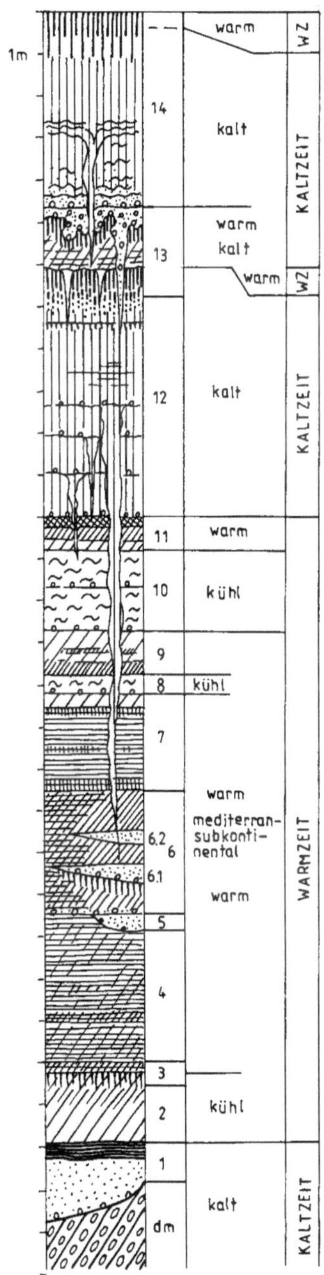

Fig. 10. Neumark-Nord (Geisel valley). The profile of the basin sediments. dm groundmoraine of the Saalian. 1. glacifluviatile and glaci-limnic sediments; 2. limnic basin silt; 3. swampy soil, peaty muds; 4. varved muds; 5. channels with sands; 6. the middle silty muds with litoral horizons (6.1., 6.2); 7. upper varved muds; 8. redeposited sediments; 9. silty muds; 10 and 11. mud and peat; 12. lower periglacial sequence with loessic sediments; 13-14. upper periglacial sequence with humic soils (13), sandy gravels and loessic sediments.

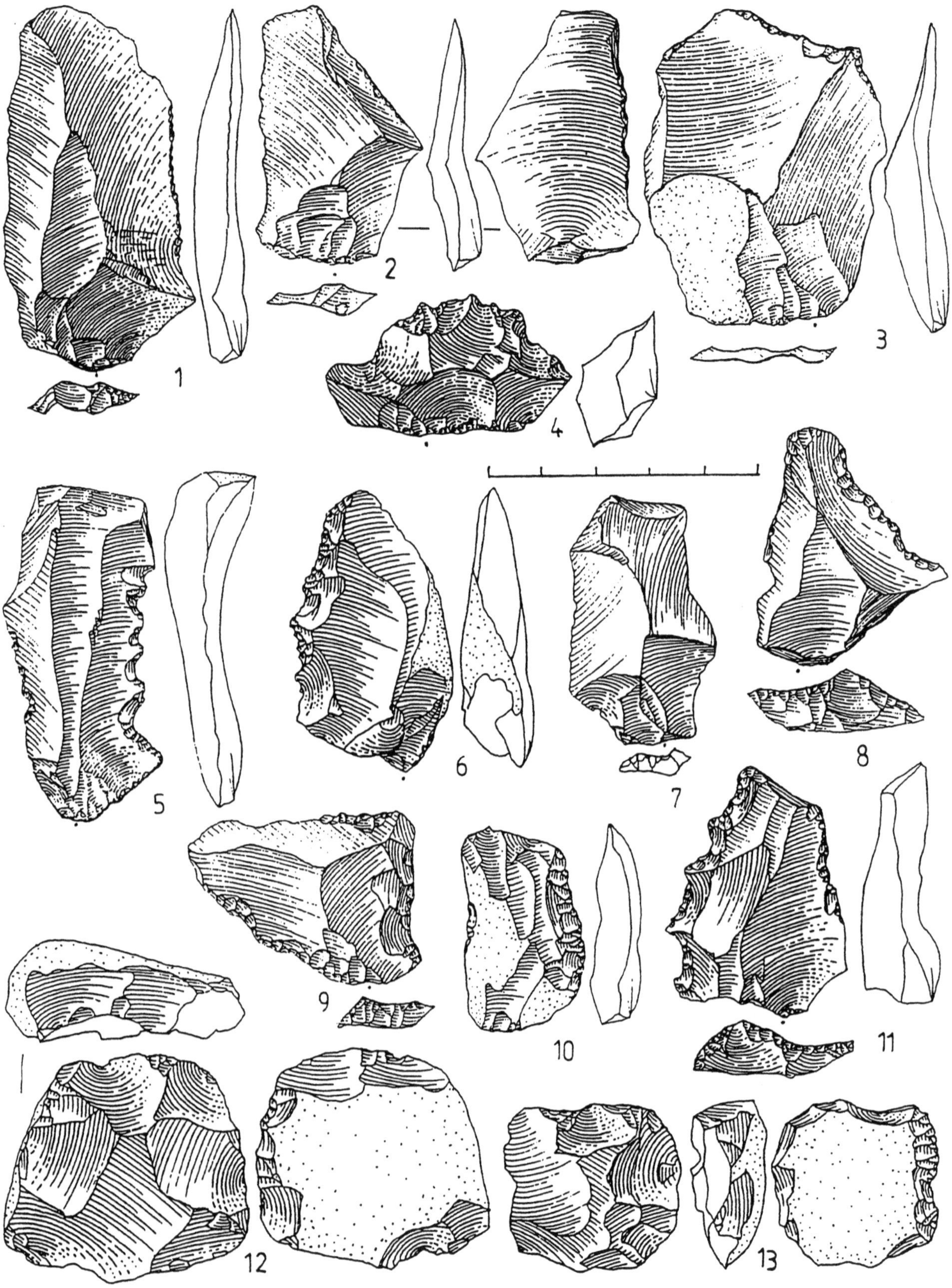

Fig. 11. Neumark-Nord. Flint artefacts. 1, 2. butchering sites with *Bos primigenius* and *Dicerorhinus kirchbergensis;* 3-13 workshops near the border zones.

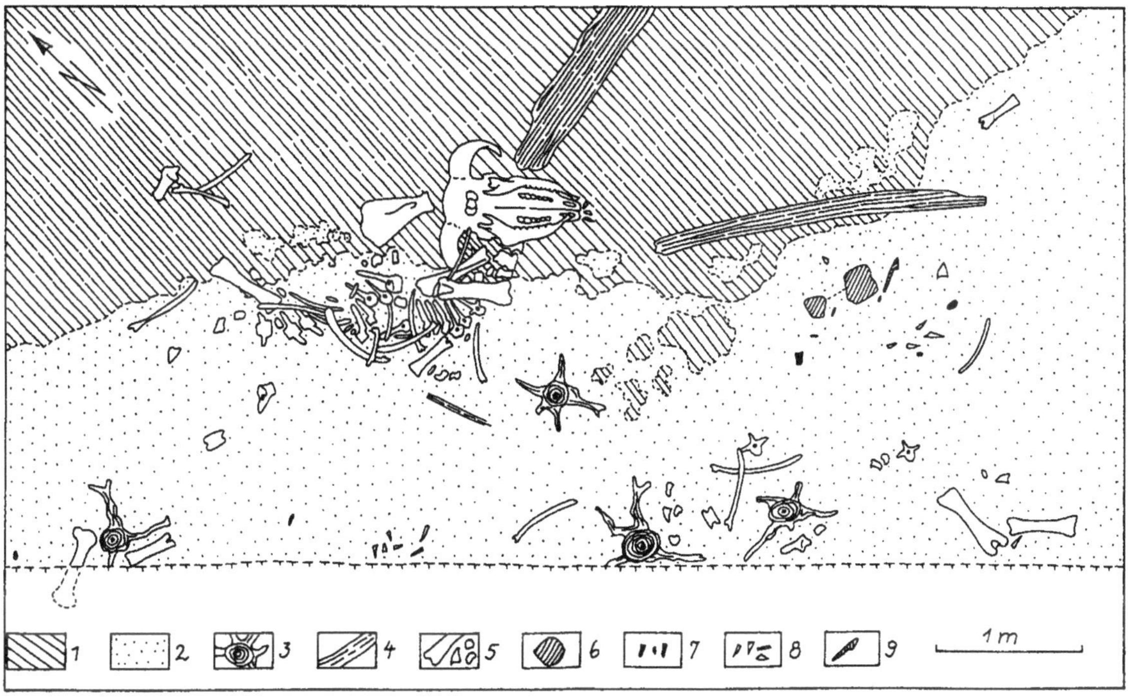

1

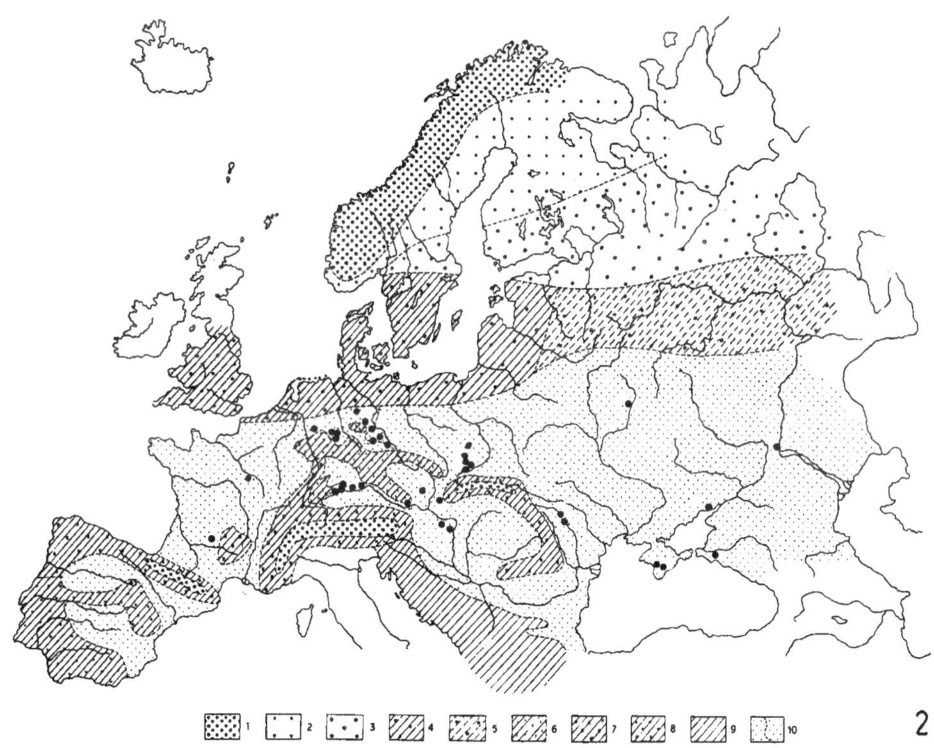

2

Fig. 12. 1- Neumark-Nord. Butchering site with the skeletal remains of *Bos primigenius*. 1. muds; 2. border sands; 3. trees; 4. remains of trees; 5. bones; 6. anvil with hammerstone, core and flakes; 7. flint flakes and flake tools; 8. bone splinters; 9. tool from a splinter of task.

2- Distribution of the Micoquo-Prondnikian in Europe. Conditions of an early Weichselian interstade.

1. Tundra; 2. tundra with shrubs and subarctic woods of birch; 3. tundra with spruce and birch woods; 4. woods of spruce, birch, pine and larch; 5. wood steppes with pine; 6. forests of spruce, pine, birch and fir; 7. forests of spruce and fir with species of the oak mixed forest; 8. oak mixed forest with pine; 9. forests of oak and hornbeam with beech and spruce; 10. steppes (after Mania and Toepfer, 1973).

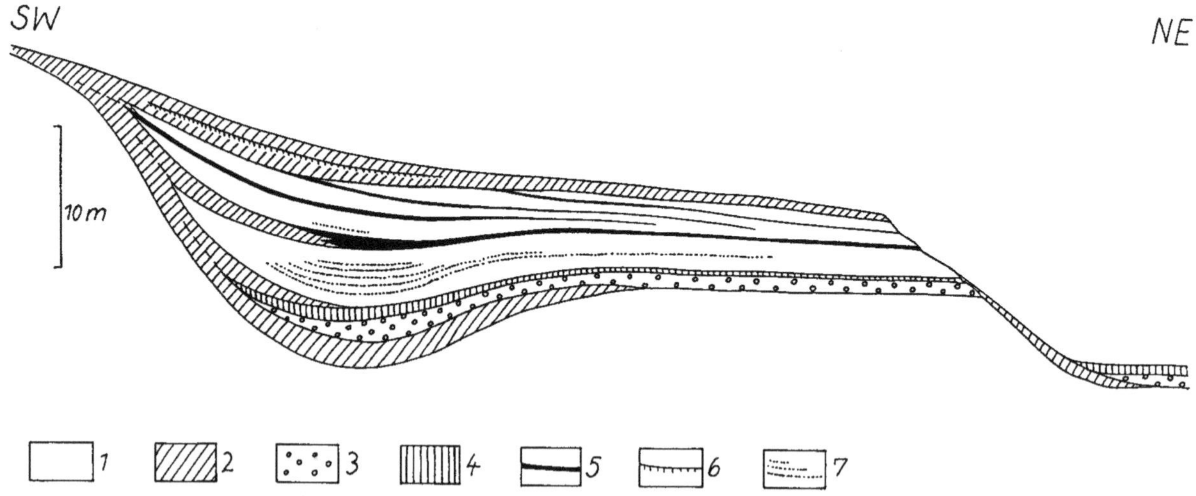

Fig. 13. Ehringsdorf. Profile of the travertine complex. 1. travertine; 2. slope debris, silt; 3. gravels of the Ilm river; 4. 'Auelehm'; 5. silt horizons with humic soils; 6. upper Pleistocene soil; 7. cultural layers (after W. Steiner).

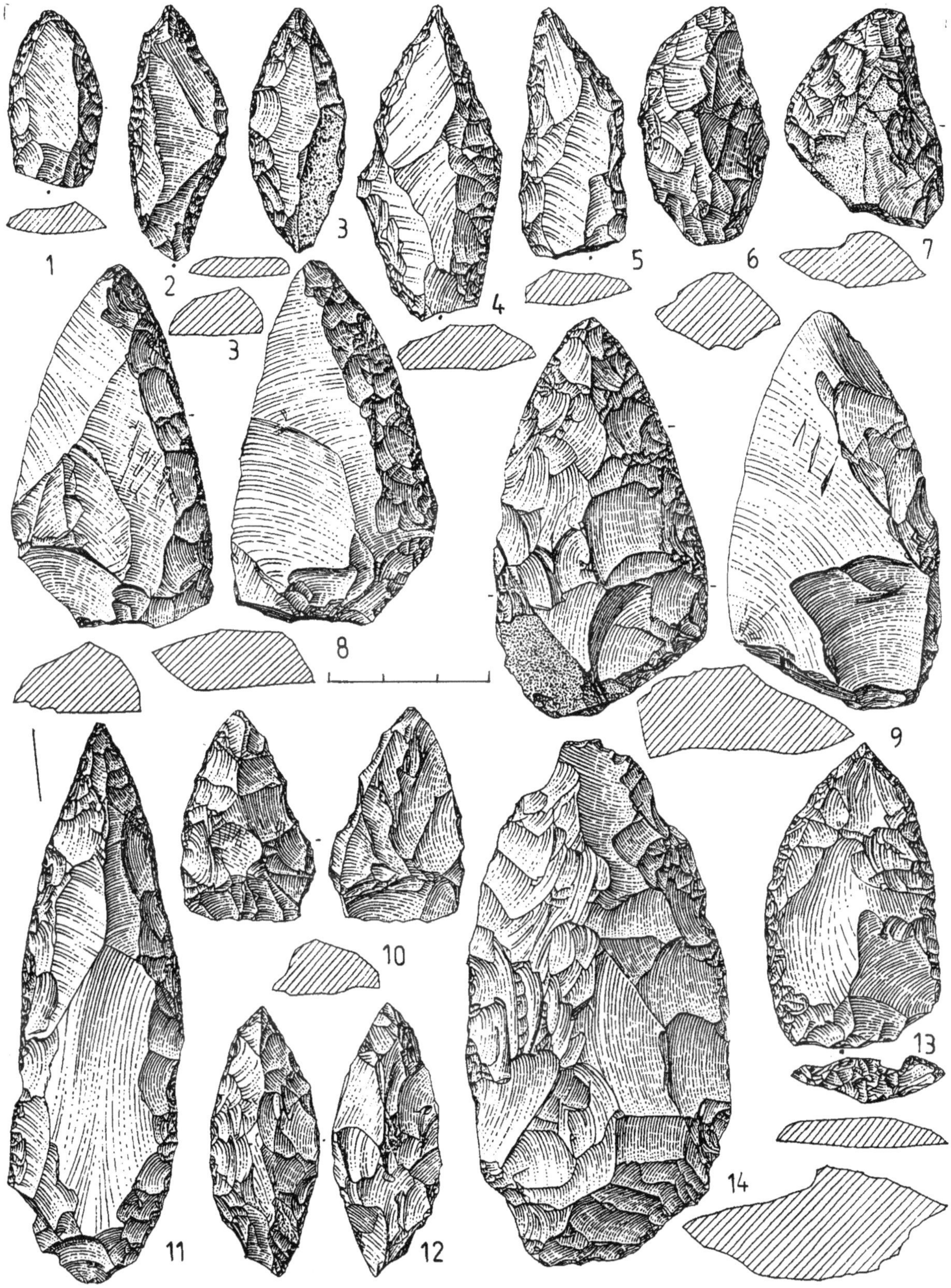

Fig. 14. Ehringsdorf. Artefacts from flint and other material (after R. Feustel).

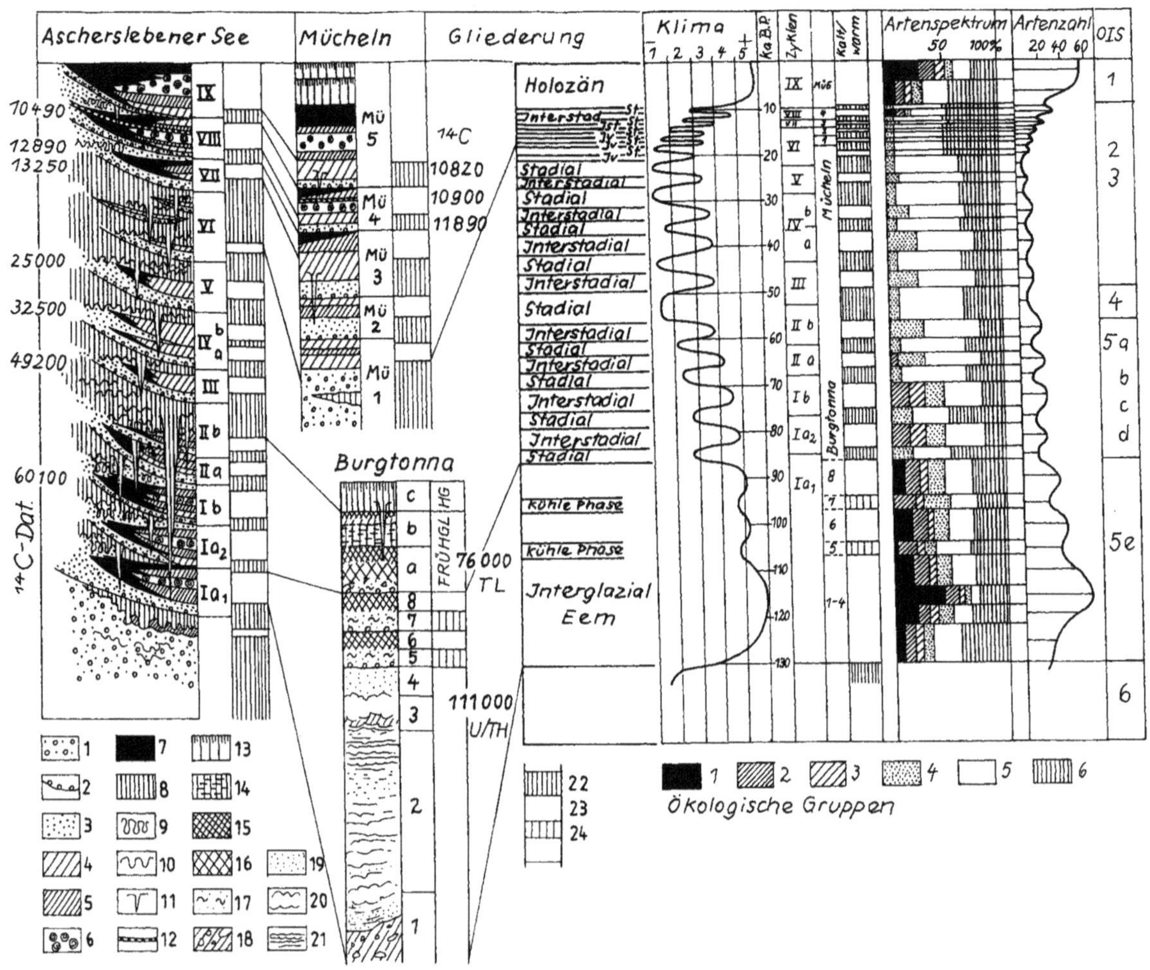

Fig. 15. Aschersleben lake. The climate record of the last 125,000 years. The profile of the Aschersleben lake is completed by the series of Burgtonna and Mücheln (Geisel valley). 1. sand and sandy gravel ; 2. denundation surface ; 3. sands ; 4. silty muds; 5. clayey muds; 6. calcareous muds; 7. peats; 8. solifluction sediments; 9. structures of gravitation (Tropfenboden); 10. cryoturbation; 11. icewedges; 12. tephra from the Laacher Sea ; 13. Auelehm with soil (Mücheln) and loess with soil (Burgtonna); 14. solifluction loess; 15. humic soil; 16. redeposited humic material; 17. redeposited travertine sediments; 18. slope debris; 19. travertine sands; 20. travertine; 21. structured travertine with leaves. On the right side: succession of the molluscan fauna ecological groups: 1. forest fauna; 2. fauna of the wood steppes; 3. fauna of the gallery forests; 4. fauna of the steppes and the common fauna of the open landscape; 5. mesophile fauna; 6. acquatic fauna.

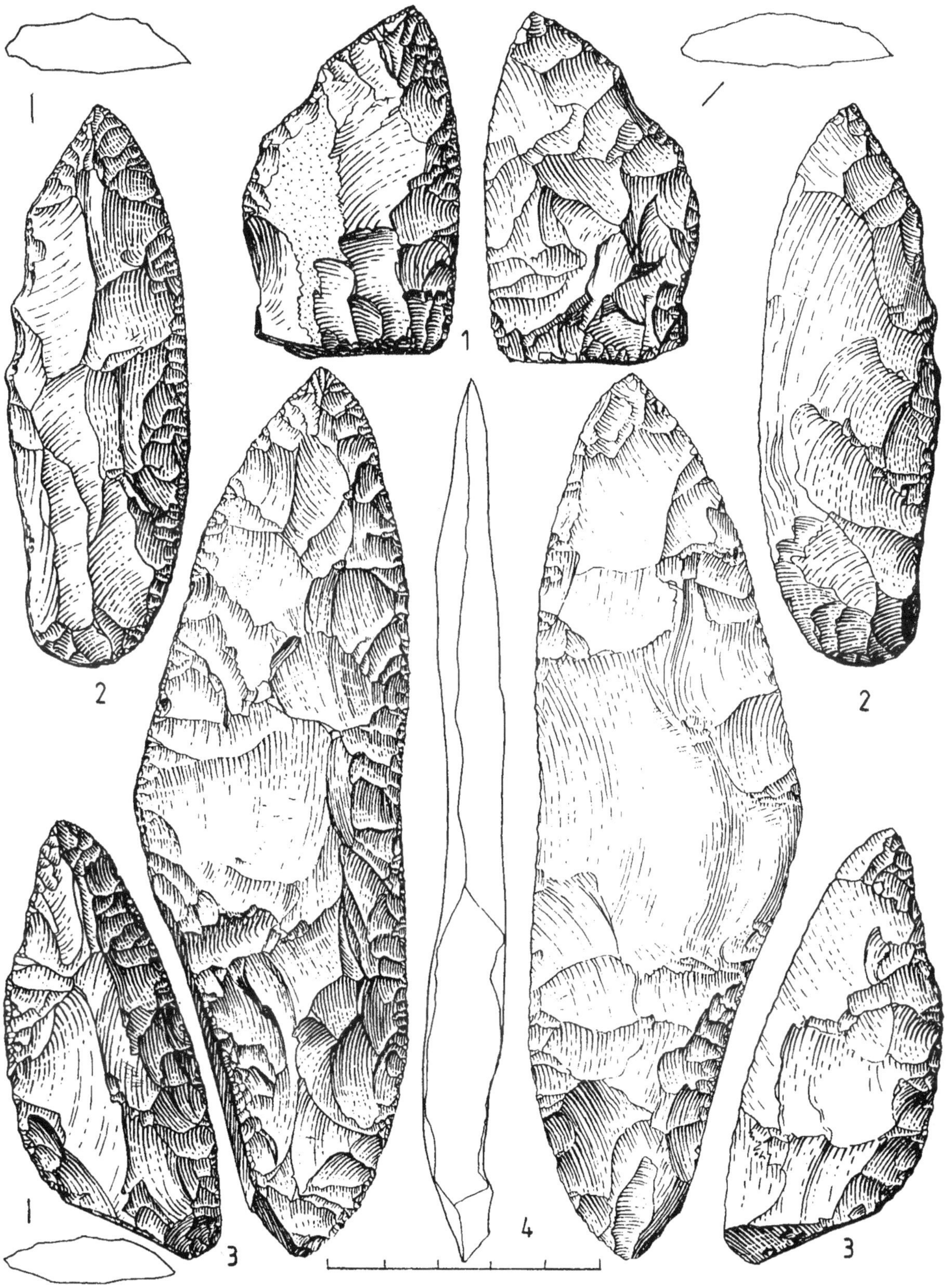

Fig. 16. Aschersleben lake. Königsaue. Artefacts of the Micoquo-Prondnikian (Königsaue A).

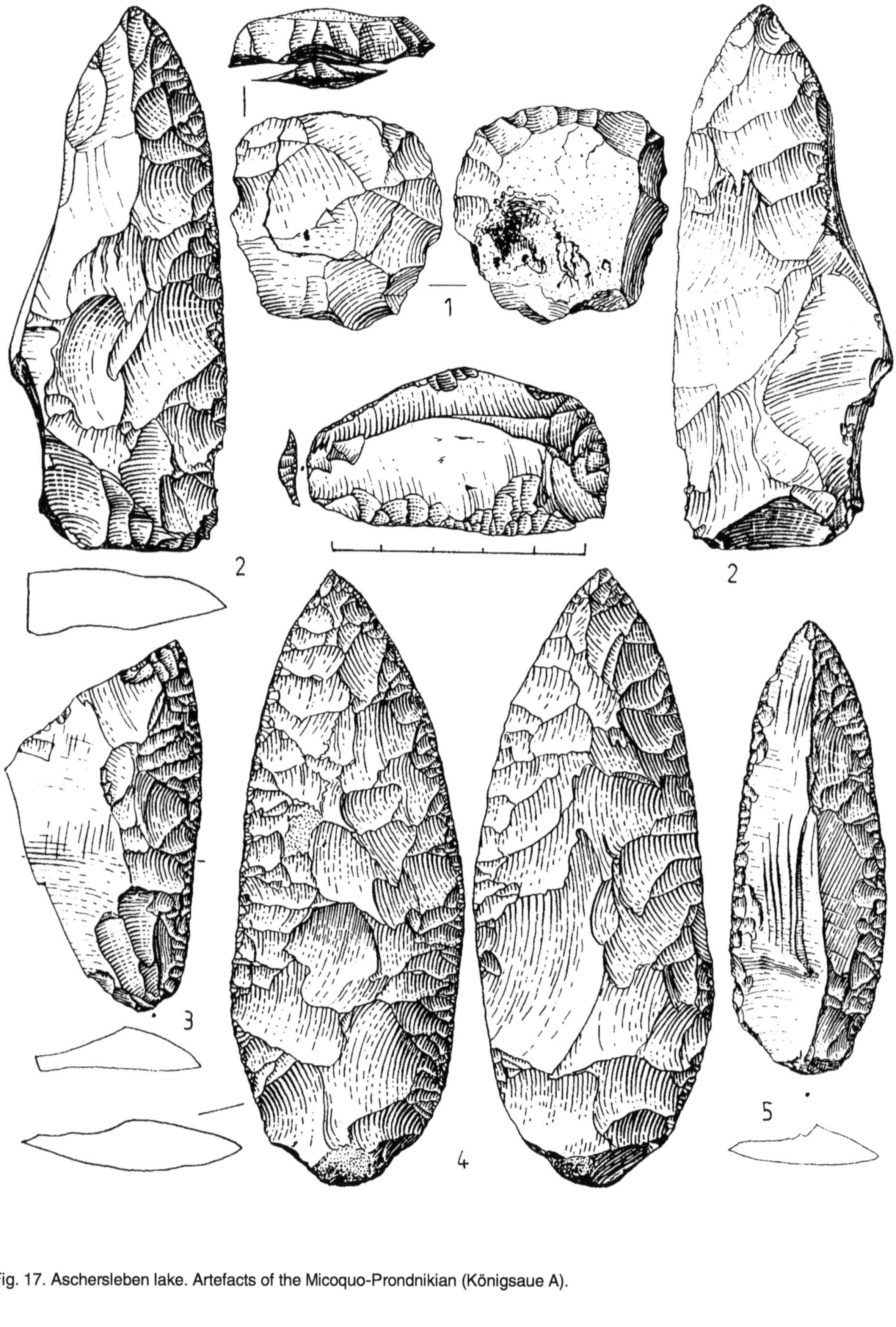

Fig. 17. Aschersleben lake. Artefacts of the Micoquo-Prondnikian (Königsaue A).

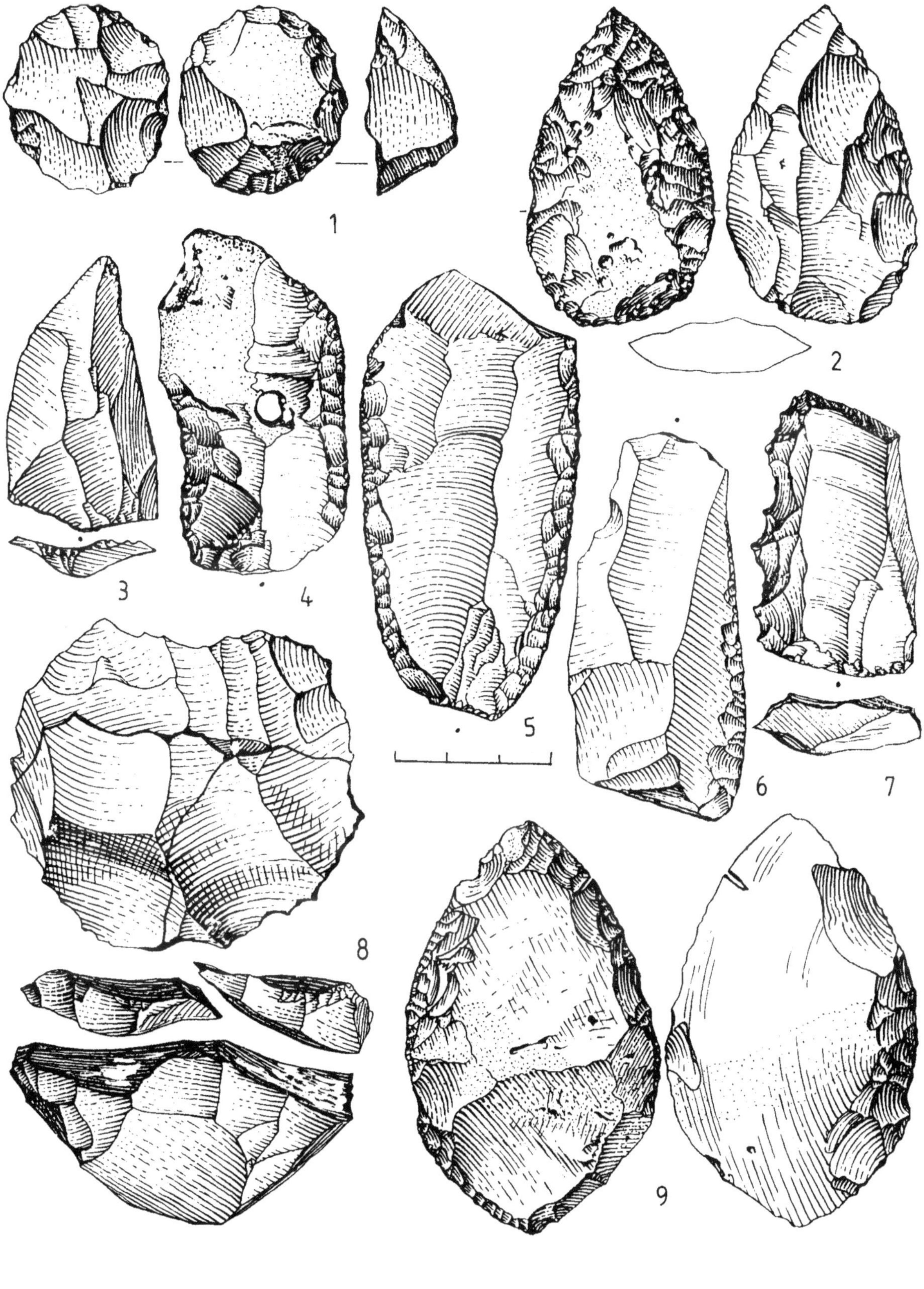

Fig. 18. Aschersleben lake. Königsaue. Artefacts of the Mousterian (Königsaue B).

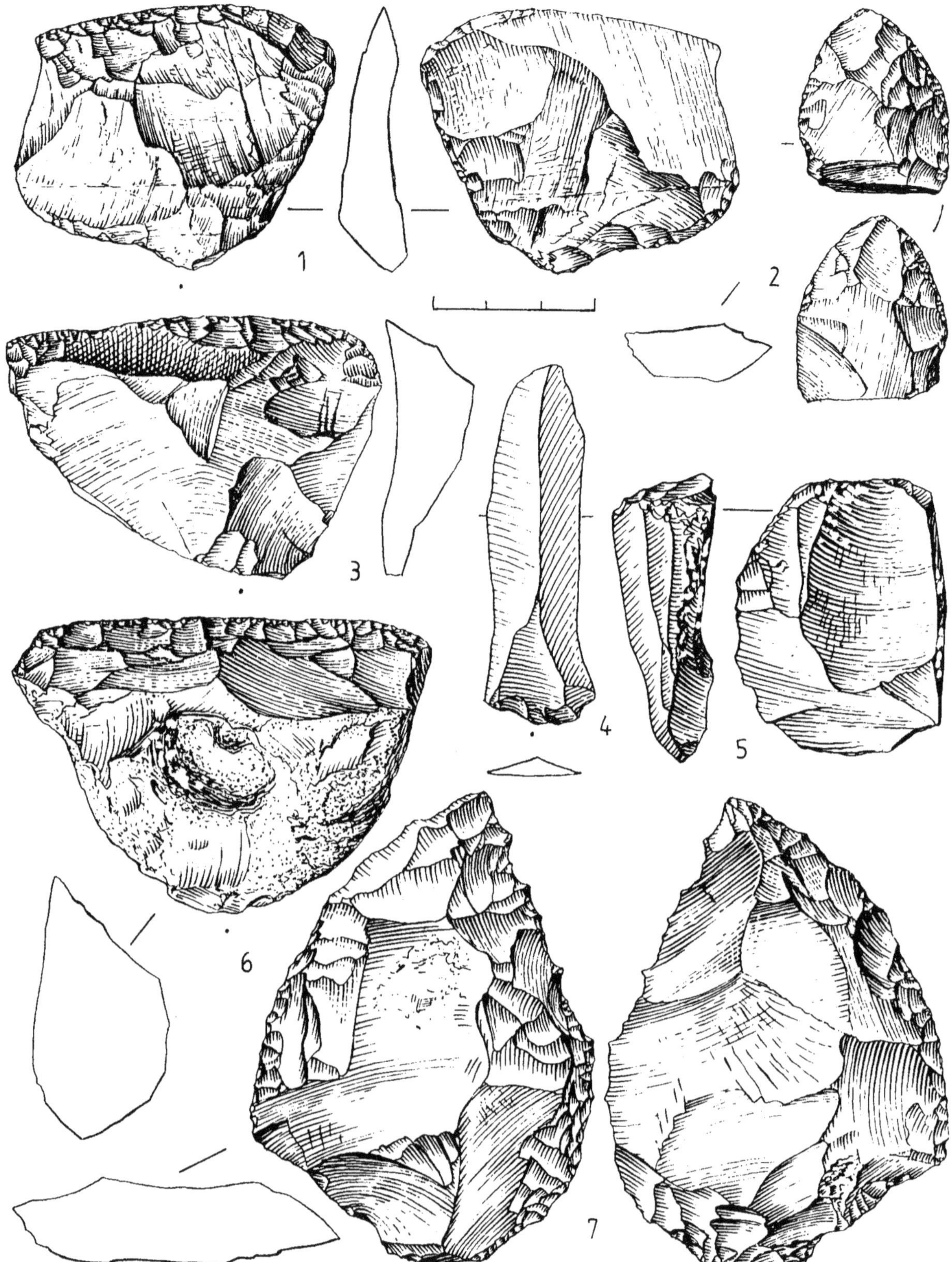

Fig. 19. Aschersleben lake. Artefacts of the Micoquo-Prondnikian (Königsaue C).

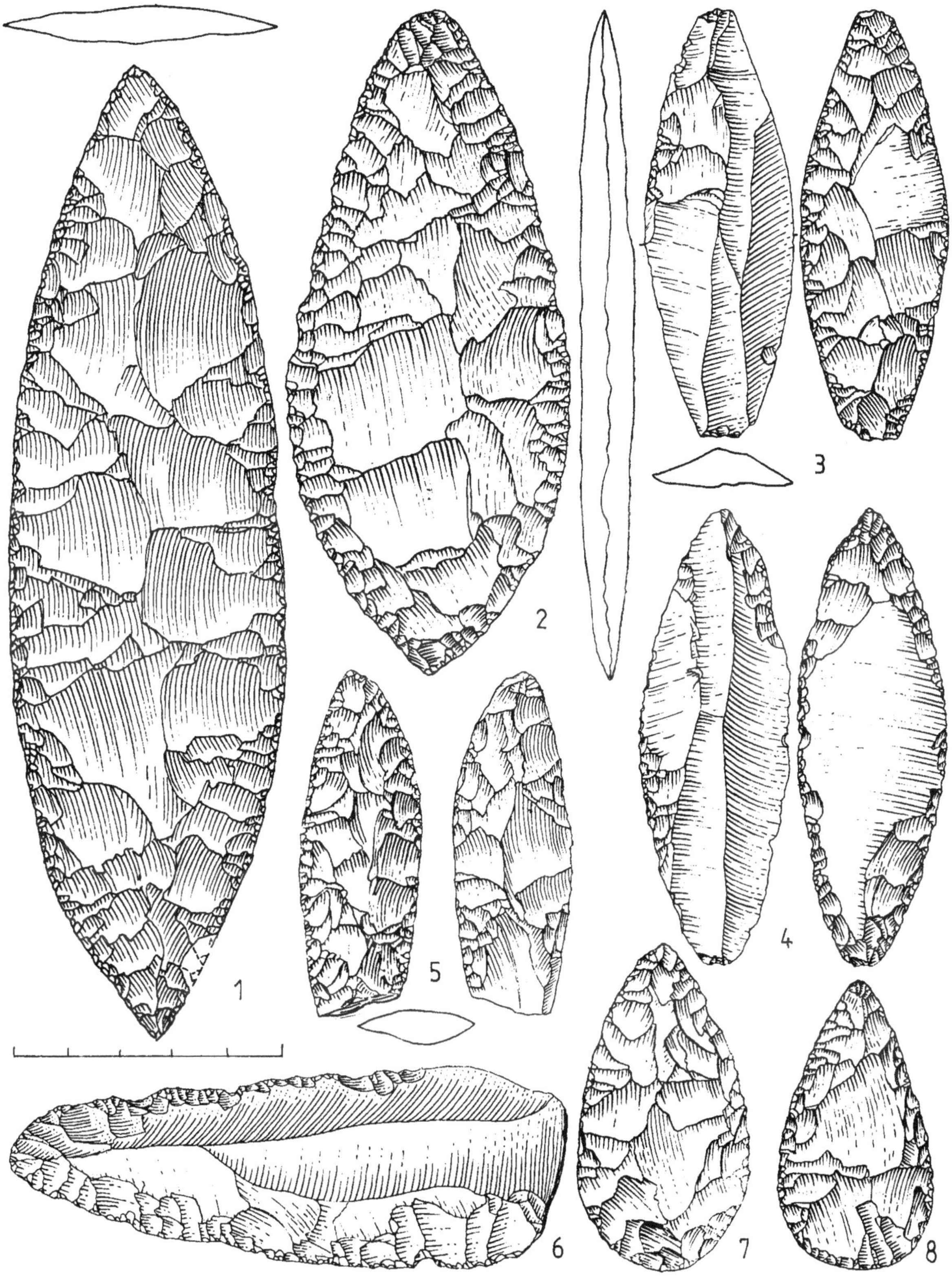

Fig. 20. Ranis, Ilsenhöhle. Artefacts of the 'Blattspitzengruppe' (lower horizon: Ranis 2).

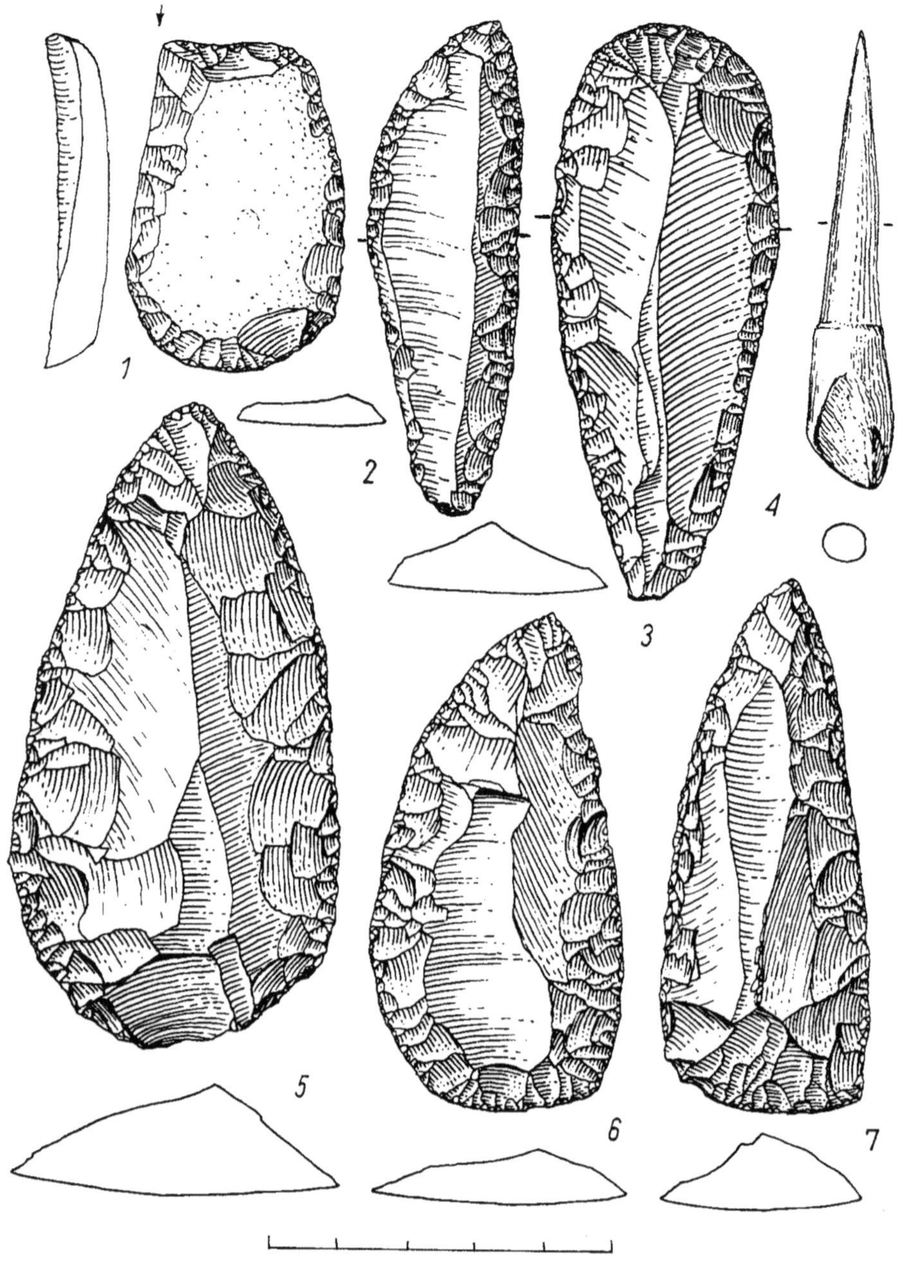

Fig. 21. Ranis, Ilsenhöhle. Artefacts of the Aurignacian (upper horizon: Ranis 3).

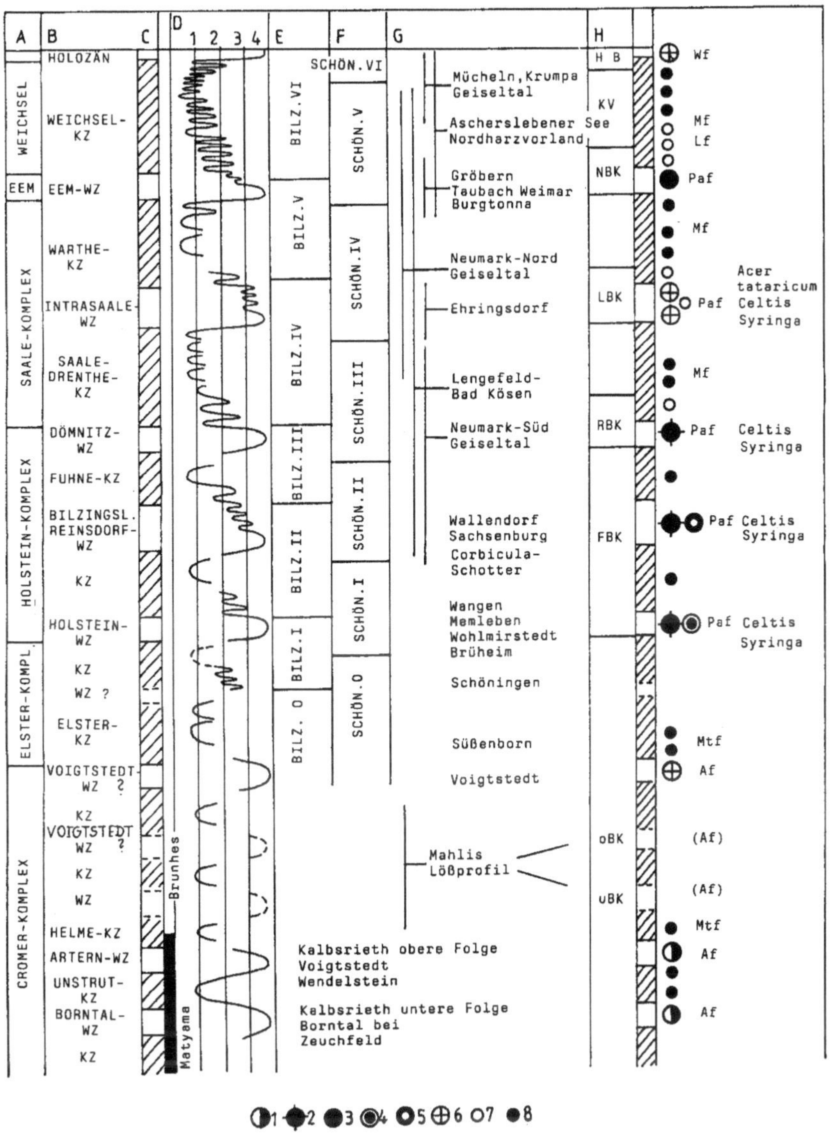

Fig. 22. Elbe-Salle region: the climate record. 1-8 molluscan fauna. 1. *Fagotia acicularia* fauna, 2 the middle Pleistocene *Helicigona benatica* fauna; 3. the upper Pleistocene *Helicigona benatica* fauna; 4. *Thadoxux serretiliniformis* fauna; 5. *Corbicula fluminalis* fauna; 6. commen middle European forest fauna; 7. steppic fauna; 8. tundra fauna. vertebrate fauna: Af *Archidiskodon* fauna, Mtf *Perelaphes trogontherii* fauna, Paf *Palaeoloxodon antiquus* fauna, Mf *Mammuthus* fauna, Lf *Lagurus* fauna, Wf Holocene forest fauna.
Pedocomplexes: uBK, oBK lower and upper P of Mehlis, FBK P of Freyburg, LBK P of Langanbogen, NBK P of Naumburg, KV soil of Kösen, HB Holocene soil.

Bibliography

Baumann, W. and D. Mania, 1983. *Die paläolithischen Neufunde von Markkleeberg bei Leipzig*. Veröff. Landesmus. Vorgesch. Dresden 16, 280 p., Berlin.

Cepek, G. A. 1986. Quaternary stratigraphy of the German Democratic Republic. *Quat. Sci. Rev.* 5, 359-364.

Eissmann, L., W. Bernhardt and A. Rudolph, 1991. Eine mittelpaläolithische Fundstelle von Eythra bei Leipzig. *Veröff. Naturkundemuseum* 9, 33-54.

Erd, K. 1978. Pollenstratigraphie im Gebiet der skandinavischen Vereisungen. *Schriften Geol. Wiss.* 9, 99-119.

Fischer, K., E. W. Guenther, W. D. Heinrich, D. Mania, R. Musil and T. Nötz, 1991. *Bilzingsleben IV.* Veröff. Landesmus. Vorgesch. Halle 44, 248p., Berlin.

Grahmann. R. 1955. The Lower Palaeolithic site of Markkleeberg. Transact. *Americ. Philos. Soc. Philadelphia N.S.* 45/46, 501-687.

Hülle, W. M. 1977. *Die Ilsenhöhle unter Burg Ranis/Thüringen.* 203 p., Stuttgart – New York .

Karelin, P. 1997. *Untersuchungen zur Datierungsanwendung des 325°C-TL-Peaks an Quarzen aus dem nordischen Verreisungsgebiet Deutschlands.*

Unveröff. Diss. Univ. Heidelberg.

Litt, T. 1994. *Paläoökologie, Paläobotanik und Stratigraphie des Jungquartärs im nord-mitteleuropäischen Tiefland.* Diss. Botanica 227, 185 p., Stuttgart.

Mai, D. H. 1983. Die fossile Pflanzenwelt des interglazialen Traver von Bilzingsleben (Thüringen). *Veröff. Landesmus.Vorgesch. Halle* 38, 45-129.

Mai, D. H. 1990. Zur Flora des Interglazials von Neumark-Nord, Kr. Merseburg. *Veröff. Landesmus. Vorgesch. Halle* 43, 159-160.

Mania, D. 1967. Das Jungquartär aus dem ehemaligen Ascherslebener See im Nordharzvorland. *Petermanns Geograph. Mitteilungen* 111, 257-273.

Mania, D. 1975. Stratigraphie, Ökologie und Paläolithikum des Weichselfrühglazials im mittleren Elbe-Saalegebiet. *Swiatowit* 3, 81-138.

Mania, D. 1983. Die Molluskenfauna des mittel-pleistozänen Travertin-komplexes bei Bilzingsleben und ihre ökologisch-stratigraphische Aussage. *Veröff. Landesmus. Vorgesch. Halle* 36, 131-155.

Mania, D. 1990. *Auf den Spuren des Urmenschen. Die Funde von Bilzingleben.* 283 p. Berlin.

Mania, D. 1991. Eiszeitarchäologische Forschungsarbeiten in den Tagebauen des Saale-Elbe-Gebietes. *Veröff. Mus. Ur-u. Frühgesch.* 25, 78-100.

Mania, D. 1992. Neumerk-Nord - ein fossilreiches Interglazial im Geiseltal. *Cranium* 9, 53-76.

Mania, D. 1993. Zur Paläontologie der Travertine von Weimar-Ehringsdorf. *Weimarer Monographien zur Ur- und Frühgeschichte* 30, 26-42.

Mania, D. 1994. Die Terrassen-Travertin-Sequenz von Bilzingsleben. *Ethnograph.-Archäol. Z.* 34 (1993), 554-575.

Mania, D. 1997. Das Quartär des Saalegebietes und Harzvorlandes unter besonderer Berücksichtigung der Travertine von Bilzingsleben. *Bilzingsleben* V, 23-104.

Mania, D. 1999a. 125,000 Jahre Klimaentwicklung im mittleren Elbe-Saalegebiet. *Hercynia* (in press).

Mania, D. 1999b. Nebra - eine jungpaläolithische Frei-landstation im Saalegebiet. *Veröff. Landesmus. Vorgesch. Halle* 1999 (in press).

Mania, D. and D. H. Mai, 1969. Warmzeitliche Mollusken und Pflanzenreste aus dem Mittelpleistozän des Geiseltales (südlich von Halle). *Geologie* 18, 674-690.

Mania, D. and U. Mania, 1988. Deliberate engravings on bone artefacts. *Rock Art Research*, 5, 91-107.

Mania, D., Mania, U. and Vlcek, E. 1994. Latest finds of skull remains of *Homo erectus* from Bilzingsleben. *Naturwissenschaften* 81, 123-127.

Mania, D., M. Seifert and M. Thomae, 1993. Spät- und Postglazial im Geiseltal (mittleres Elbe-Saalegebiet). *Eiszeitalter u. Gegenwart* 43, 1-22.

Mania, D., M. Thomae, M. Altermann, M. Seifert, D. H. Mai, R. Fuhrmann, E. Pietrzeniuk, W. D., Heinrich, H. J. Döhle, T. Litt, J. Goersdorf, J. Erfurt, K. U. Heussner and Th. Webber, 1990. *Neumark-Gröbern. Beiträge zur Jagd des mittel-paläolithischen Menschen.* Veröff. Landesmus. Vorgesch. Halle 43, 319p., Berlin.

Mania, D. and V. Toepfer, 1973. *Königsaue - Gliederung, Ökologie und mittelpaläolithische Funde der letzten Eiszeit.* Veröff. Landesmus. Vorgesch. Halle 23, 164 p., 73pl., Berlin.

Mania, U. 1995. The utilisation of large mammal bones in Bilzingsleben - a special variant of Middle Pleistocene Man's relationship to his environment. In: Ullrich, H. (ed.), *Man and Environment in the Palaeolithic.* ERAUL 62, 239-246, Liège.

Schäfer, D., K. D. Jäger and M. Altermann, 1991. Zur Stratigraphie periglazialer Decken im thüring-ischen Bergland (Oppurg, Landkreis Pössneck) *Archäol. Korrespondenzblatt* 21, 323-334.

Thieme, H., and R. Maier, (eds.), 1995. *Archäologische Ausgrabungen im Braunkohlentagebau Schöningen, Landkreis Helmstedt.* 191 p., Hannover.

Thieme, H. and D. Mania, 1994. Schöningen 12 - ein mittelpleistozänes Interglazialvorkommen im Nordharzvorland mit paläolithischen Funden. *Ethnograph. -Archäol. Z.* 34 (1993), 610-619.

Toepfer, V. and D. Mania, 1973. Paläolithische Funde vom Petersberg nördlich von Halle (Saale). *Jahresschrift mitteldeutsche Vorgeschichte* 57, 49-69.

Urban, B. 1995. Vegetations- und Klimaentwicklung des Quartärs im Tagebau Schöningen. In: Thieme, H. and Maier, R., (eds.), *Archäologische Aus-grabungen im Braunkohlentagebau Schöningen, Landkreis Helmstedt*, pp. 49-56, Hannover.

Urban, B., R. Lenhard, D. Mania and B. Albrecht, 1991. Mittelpleistozän im Tagebau Schöningen, Ldkr. Helmstedt. *Zeitschr. Deutsche Geol. Ges.* 142, 351-372.

Veil, S. 1994. Ein mittelpaläolithischer Fundplatz aus der Weichselkaltzeit bei Lichtenberg, Ldkr. Lüchow-Dannenberg. *Germania* 72, 1-66.

Vlcek, E. 1980. Die mittlepleistozänen Hominidenfunde von der Steinrinne bei Bilzingsleben. *Veröff. Landesmus. Vorgesch. Halle* 32, 91-130.

Vlcek, E. 1993. Fossile Menschenfunde von Weimer-Ehringsdorf. *Weimarer Monographien zur Ur- und Frühgeschichte* 30, 220p., Stuttgart.

DISCUSSION

Nicholas Conard: Are there any hearths in Schöningen?

Dietrich Mania: There are very nice fireplaces at this site, up to 150 centimeters in diameter. The ground beneath is burnt, and there is charcoal.

Nicholas Conard: In the chronostratigraphy of Ascherslebener See you show relatively little sedimentation during the interglacial complex. In all the phases of Isotope Stage 5, you have only one cycle in your schema. Why is there so little sedimentation in that period?

Dietrich Mania: Sedimentation is the result of some environmental phenomena and climatic influence. In the interstages we have no permafrost, and the formation of a small flat basin with limnic sedimentation. In the stages we do have permafrost and deflation, and there is a denudation phase in each sequence.

Nicholas Conard: In the reconstruction you showed, you have the opposite phenomenon that we have in other parts in Germany. The interglacial complex or the early Weichsel has relatively little sedimentation. It seems very different from phenomena that we see in sediment traps in other regions.

Dietrich Mania: The formation of the basin begins in the Eemian and in the cores we have only 1-2 m sediments. We have pollen analysis for the Eemian. We also have pollen diagrams for the other phases. These pollen associations are mixed with Tertiary pollen from the shores. It is not possible to separate the Tertiary and Quaternary pollen in this association.

Nicholas Conard: It is interesting that from 130,000 to 80,000 on your scale, there is relatively little sedimentation. In sediment traps in the Rhineland, that is the period when a great amount of sedimentation took place. The difference in your reconstruction strikes me as unusual.

Joachim Schäfer: In your stratigraphy you have one warm period between Warthe and Drenthe, and three interglacials between Saale and Elster. Does this mean that the Elster becomes stage 14?

Dietrich Mania: No. I do not think so. It is very difficult to compare the oceanic and the continental zones. Do not forget, the sequences are not complete.

Nicolas Rolland: Just a brief comment, discussing the season of occupation. This seems to resemble the Palaeoindian Folsom case. This situation continues until the last century in the Northern Plains and southern Canada. And the reason for well-selected hunting in winter, is because the herds aggregate in the deep river valley, where they are protected against the cold. That is why you get larger aggregates of hunters in winter. In the summer they break into smaller groups. This went on for thousands of years, and should be considered, as well as the quality of the game.

MIDDLE PALAEOLITHIC PHENOMENA IN HUNGARY

Viola T. Dobosi

Introduction

The Carpathian Basin has been, in almost all periods, the meeting point of widely different cultural units. This is partly the result of its central position, as well as its favourable orographical endowments which offer varied ecological conditions. While generally true, this statement is not very applicable to the Lower Palaeolithic period. There is only one authentic site known from this era, and this is identified with the Pebble Industries. Furthermore, we have no evidence of the Lower Palaeolithic handaxe tradition as yet.

It is most unfortunate that no traces of the ancestors of the local Middle Palaeolithic industries have been discovered from the long period of the Riss Glaciation. However, we can reasonably postulate a latent continuity of the Vértesszőlős-type local industry, at the same spot or in its wider surroundings. We would not be much surprised to find a genetic link between the Vértesszőlős chopper industries and the pebble-working Taubachian-type industry of Tata.

After a long hiatus following Vértesszőlős, which is not documented by authentic finds, we can observe a demographic boom and accelerated development in Hungary by the end of the Riss/Würm (Eemian) interglacial, which corresponds to general tendencies of European development during this period. The variety of this period was pointed out by M. Gábori in 1976.

The following Palaeolithic cultures are known to have existed in the territory of Hungary between the end of the Riss/Würm interglacial and the cold maximum of the Würm I period:

- Classical (western European) Mousterian
- Bifacial industries rooted in handaxe tradition
    - Jankovichian
    - Bábonyian
    - Szeletian (lower)
- Middle Palaeolithic Pebble Industries
    - Taubachian
    - Southeast European Charentian

Classical (Western European) Mousterian

The Hungarian material pertaining to this complex was originally classified in research carried out between the two world wars. Classification was based completely on the 'French system'. Its type locality, Subalyuk, was excavated in 1932. A revised classification of the material obtained at this site was recently undertaken by Mester (1994). This summary is partly based on his results.

Sites (Fig. 1)

Subalyuk cave, Kecskésgalya cave, Farkaskő rock shelter, Mexikó cave and Büdöspest cave. The main site is the lower cultural layer of Subalyuk, while the classification of the other materials is based on less adequate evidence (a small number of pieces or mixed material).

Topography

Sites are located in the southern, eastern and north-eastern margins of the Bükk Mountains. This is a central area of the north-Hungarian mid-mountain range which is comprised primarily of limestone. The area is characterized by caves opening into valleys, which themselves lead towards the plains. The maximal height of the mountains does not exceed 1,000 m. The altitude at which caves are located varies. Caves are typically situated near the valley base (e.g., Subalyuk, Büdöspest), with a few higher-lying sites such as Kecskésgalya, at 380 m a.s.l.

The altitude at which cave sites were selected during the period under discussion was more or less incidental: prehistoric people were looking for suitable caves irrespective of the caves' positions. Other orographical demands of the population are not known. We have no information regarding open-air sites which were selected according to the specific needs of the population.

Chronology

The most complete sequence of layers is found in the Subalyuk cave. The different sedimentological layers were formed under different climatic conditions. On the basis of correlations with open-air sequences, the cave sediments were assigned to the interval occurring between Oxygen Isotope Stages 5d and 4 of the Emiliani division of the Würm period (Ringer, 1993, Fig. 31).

Ecology

Botanical data. In the lower cultural layer of Subalyuk, the dominance of deciduous trees with *Abies* denotes a moderate, mixed forest indicating a climate similar to the present one. In the upper layer, only charred remains of conifers were found, mainly *Pinus cembra* (20 specimens).

Faunal data. (Fig. 2). Both of the archaeologically significant layers contained species characteristic of mid-mountain and plain environments. The latter species were apparently transported to the cave as booty. The former, particularly the cave bears (of which abundant remains were found), were probably the natural fauna of the caves.

Prehistoric humans in this period apparently specialized in the hunting of *Capra ibex*, the mountain goat. Mottl made the following important observation regarding this specialization: 'From the lower reddish layer of the Subalyuk cave, artefacts of the classical Mousterian culture and several hundred remains of one species of the mountain goat were found. It is typical of these finds that all skeletal elements belonging to the same anatomical position were chopped in the same manner. It is evident from this observation that the main booty of the prehistoric humans of the classical Mousterian in the Bükk mountains was *Capra ibex*. This characteristic way of chopping the bones was, in a way, harmful for scientific elaboration of the material, as we have very few intact bones among the material' (Mottl, 1938, 288).

Culture

Only 447 tools were found in the 14 layers of the Subalyuk cave. There were 126 items in the lower complex, which were typical of the 'abundant scraper' Mousterian period. The finish of the artefacts matches those found in other classical Mousterian sites: The ratio of faceted bases is more than 50 %, the form of flakes is close to the standard of blades, and the ratio of scrapers is very high (Mester, 1994, 63). In the most recent elaboration of the classical site, Mester (1994, 69) classified the lithic assemblage of the upper layer as belonging to the Charentian industries. In our opinion, the evidence is not convincing enough.

In Subalyuk, local and regional silex varieties of rock dominate the raw material utilized for toolmaking. Kadic (1938, 15) concluded that the majority of the raw material originated from the Triassic layers of the Bükk Mountains. On the basis of field surveys, the information can be completed accordingly: 'The considerable amount of broken pebbles originated presumably from the Pleistocene terraces near Cserépfalu. Hornstone nodules can be observed in the Middle Triassic Ladinian occurrences in the region of Bükkzsérc and Cserépfalu. Therefore, the majority of the raw material types could have been collected in the immediate vicinity of the site: In the load of the Hór brook and in the Triassic layers nearby. The source of chalcedony quartz is farther: It can be found as surface material on the Tóhegy hill near Egerbakta. The presence of Szeleta raw material relates to the Bükk mountains where its traditional source is located. Chalcedony can be found in the rhyolite hills near Cserépfalu, although it occurs mainly in the region of Kács. The obsidian originated in the Tokaj area' (Dobosi, 1986, 250). Obsidian is the only evidence for long-distance raw material procurement, which can be established with sufficient certainty here, as its source lies 80 - 90 km from the site.

A dotted flint scraper made of Swieciechów flint was found in the Sólyomkút rock shelter. Although Swieciechów (Poland) lies at a distance of 400 kms from the site, it cannot be concluded that the scraper is evidence of long-distance raw material trade in the Middle Palaeolithic, as it was discovered in uncertain circumstances. In the other sites, the total number of artefacts does not reach 70 pieces.

Connections

To our current knowledge, the population characterized by a classical Mousterian industry arrived at the sites in present-day Hungary without known antecedents. We also know nothing about its descendant cultures. So far we are not able to follow its route of arrival, nor do we have evidence of its departure from this territory.

Bifacial Industries Rooted in Handaxe Tradition

The identification of this cultural entity has been achieved only during the past few years. The Jankovichian industry is the longest known. Ever since 1913, when artefacts were discovered in the Jankovich cave excavations by Jenő Hillebrand, the cultural assignment of this type of material has been the subject of debate. The oldest name used for this complex was 'Transdanubian Szeletian'. Veronika Gábori-Csánk introduced the name 'Jankovichian culture' into the Hungarian Palaeolithic literature (Gábori-Csánk, 1993).

The other bifacial industry, the Bábonyian culture, was identified as the result of the investigations of Ringer (1983). This new culture was delineated mainly on the basis of his own field surveys and excavations. He also reclassified some sites formerly held to be 'Grobgerätige Mesolithikum' as belonging to the Bábonyian complex.

In spite of the divergence in development, a common origin can be postulated for the leafpoint industries of the Middle Palaeolithic. In the opinion of some scholars, a strong genetic link can be postulated between the three phyla, i.e., Jankovichian, Bábonyian and Szeletian, especially the latter two. Ringer hypothesized that the Bábonyian evolved towards Szeletian (Ringer, 1989).

Svoboda and Simán raised the possibility of a different interpretation:
'... we may conclude that what Ringer calls 'Bábonyian' is actually a series of Szeletian workshops, which may include different phases with stronger or weaker Middle Palaeolithic traditions and which are related to the more or less permanent hunters' camps in the nearby caves.' (Svoboda and Simán, 1989, 307). In addition to the distance from the source, the extent of similarity depends on the type of raw material as well. It is highly probable that differences on a facies level can be simply explained by the character of the raw materials used and the resulting differences in technique and workmanship.

Jankovichian

Sites

Jankovich cave, Kiskevély cave, Pilisszántó II rock shelter, Szelim, Bivak, Budapest-Remete caves, and the Csákvár rock shelter.

Topography

The north-south line of the Hungarian reach of the Danube river separates the country into two parts which are different in climate, as well as surface morphology. While classical Mousterian sites were found in the northern parts of the eastern area, the Jankovichian sites are centered in the northern mid-mountain zone of the western areas (i.e., Transdanubian). According to our present knowledge, the Jankovichian culture was also 'cave-dweller', i.e., open-air sites are not known from this complex.

The famous Lovas mine of colourant materials has been assigned to several related cultural complexes during its scientific 'career'. On the basis of a single leafpoint the tool kit (more than 70 pieces of worked bone and antler) was classified as Transdanubian Szeletian and later as Jankovichian. However, a new radiocarbon dating, so far unpublished, and the decorated bone seem not to support this classification. The assemblage should be re-evaluated.

Chronology

On the basis of carefully analysed data, Gábori-Csánk dated the Jankovichian culture to the early phase of the Würm Glaciation, which ranged from the long introductory phase (including the Brörup interstadial) until the cold maximum of the Würm I (Gábori-Csánk, 1993, 88).

Ecology

Botanical data. From the charred wood remains found in the caves, a vegetation history was reconstructed which is valid for the Carpathian Basin in general and the Bükk Mountains in particular. Botanical information gathered from the distribution area of the Jankovichian culture supports the common view (established on the basis of other evidence) that the changes in the vegetation of the period followed shortly after changes in climate.

Faunal data. The same generalities can be related to the fauna. At most of the old excavations, the exact stratigraphical position and the actual number of finds cannot be reconstructed. Thus species of fauna which would otherwise not be found in the caves (for instance, the horse, mammoth, and woolly rhinoceros, among others) can definitely be considered booty animals. Some of these were large herbivores from a forest environment, while others were habitants of the grooved steppe. The general fauna list contains no individual numbers of animal remains and therefore is not suitable for detecting hunting specialization.

Culture

In the six sites of the Jankovichian culture, not more than 150 archaeological artefacts were found. Thus the scientific arguments regarding the definition of this culture are of equal importance to typological arguments. With respect to several sites, they are even more important. On the basis of some 70 typical bifacial and approximately 80 other tools, the industry was defined by Gábori-Csánk in the following way:

'The basic features of the culture are rooted, partly in a Levallois-technique Mousterian and bifacial leaf-point tradition, and, beside these two Middle Palaeolithic components, the Upper Palaeolithioc features are almost completely missing. The Levallois character is reflected indirectly by the index IF = 27.5, while the Acheulean-Micoquian tradition is demonstrated directly by Iat = 35.6 and IB = 32.8 respectively' (Gábori-Csánk, 1993, 83).

The excellent workmanship of the stone tools can be attributed first and foremost to the quality of the raw material. Apart from local (Gerecse) radiolarite, there are other silicas of undefined sources. Two raw materials, present at the Jankovich cave, indicate long-distance procurement: Obsidian and Szeletian felsitic porphyry from north-eastern Hungary.

Connections

The Jankovichian culture appears to be related to younger industries preserving Acheulean traditions. In this it is similar to the Bábonyian, and apparently has a similar origin. According to Gábori-Csánk, during the Middle Palaeolithic the western part of the Carpathian Basin was a 'precipitation area' for groups arriving from south-western and north-western directions. This huge semi-circular area includes the north-western part of Yugoslavia, the south-eastern territories of the Alps, Transdanubia and western Slovakia (Gábori-Csánk, 1993, 118). Concerning the Jankovichian connections, Simán notes that 'Its dating is rather uncertain, based on revised cave sequences and the critical separation of mixed faunal remains. It is sure, however, that it is not older than the Bohunician (i.e., Moravian bifacial industry), but rather contemporary or perhaps even younger' (Simán, 1990, 16).

Bábonyian

Sites

These include several localities in the Sajóbábony and Miskolc environs, as well as Mályi-Öreghegy. The site Korlát-Ravaszlyuktető has been known for a long time by archaeologists. Almost all cultures from Chellean till Neolithic have been assigned to this site. Likewise, it may also be classified, in part, as Bábonyian. According to Simán, '...the indirect evidence seems to suggest that the workshop was used in the Upper Palaeolithic and had northern contacts. Its significance, however, was not essential in raw material catering' (Simán, 1995, 57).

Though the occurrence of bifaces can be presumed until the Bronze Age, at Korlát the occurrence of *bona fide* Middle Palaeolithic stone tools can be also postulated (Simán, 1995, 41). This Middle Palaeolithic element may be seen as consistent with the Bábonyian culture.

Topography

According to our present knowledge, the Bábonyian culture was located in a 20 x 10 km patch in the western margin of the Sajó river. This culture is found on the terraces bordering the Bükk Mountains, in open-air sites at 180 - 190 m a.s.l.

Chronology

In the quarry of the brickyards at Mályi-Öreghegy, Á. Ringer found a soil complex which can be interpreted as a stratotype of the last interglacial period. Oxygen isotope studies more specifically set the date around the optimum of the Last Interglacial, in Oxygen Isotope Stage 5e. The oldest implements of the Bábonyian culture were found in this soil layer. This culture existed in the above mentioned area during the early Würm (Ringer, 1993, Fig. 31).

Ecology

There are no floristic or faunal data. On the basis of the stratigraphy of Mályi-Öreghegy, the ecological conditions were hypothesized by Ringer as follows: 'On the basis of the exposure studied in the Mályi brickyards, it seems certain that the formation of this forest soil took place as a sub-Mediterranean reddish brown forest soil. Later this soil was infiltrated by clay. Finally, signs of degradation appeared and the soil complex terminated as a podzolic brown forest soil with pseudogley. This process reflects the changes in climate from a mild Mediterranean - sub-Mediterranean influence to mild temperate and finally, cold temperate.' (Ringer, 1993, 83).

Culture

The number of tools gathered from the 5 main sites and 10 other surface-collecting points is around 300 (Ringer, 1983). Apart from very archaic bifacial forms, there are flat knives, scrapers and points present, as well as some bulky Upper Palaeolithic types. According to Ringer, this means both a chronological sequence and an interior development. One of the archaic features of the industry is the rough, handaxe- or chopper-type workmanship. This type of finish is mainly related to the quality of the most widespread raw material used, i.e., Szeletian felsitic porphyry which has specific textural-structural characteristics.

Connections

Ringer associated this industry with the Middle Palaeolithic complexes built on Acheulean tradition, particularly the Bockstein and Altmühl type industries which have identical or similar typological spectra (Ringer, 1983, 120). The ramifications of the Bábonyian in terms of its relation to contemporary and later cultures (i.e., the 'Szeletian problem') is treated by Ringer (this volume).

Middle Palaeolithic Pebble Industries - the Taubachian

The term 'Taubachian' was introduced by D. Collins in 1969. The interpretation presented here is based on Valoch's summary of layer 11 of the Kůlna cave (Valoch, 1984). According to Valoch, 'Taubachian' is understood as representing a microlithic industry which existed in the second half of the Eemian interglacial and used pebbles as raw material. In Hungary, the representative site of this culture is Tata.

Sites

Tata is located at Tata-Porhanyóbánya. It is one of the first authentic Palaeolithic sites discovered in Hungary. It was first excavated by Tivadar Kormos in 1906, then in 1958-1959 by L. Vértes. During the past two years, work to conserve the site produced an additional number of significant finds. It must be mentioned here that some of the pre-Würm sediments in the Bükk caves yielded several atypical quartzite flakes which could fit the Taubachian culture on the basis of chronology, raw material and base form. However, the level of information on these finds is not yet adequate enough to determine their cultural affiliation.

Topography

The only site belonging to this culture is located in the northeastern part of Transdanubia, in the Tata depression separating two areas of the Transdanubian mid-mountain range. At the foothill region of the Gerecse Mountains, spring activity was considerable during both the Pleistocene and the Holocene, and was accompanied by sedimentation of calcareous tuff. This process was not only important for surface morphology and hydrology of the period, but also had significant consequences for prehistoric habitations. Sites of two important cultures were found in the calcareous tuff basins, one of which was Tata. This is an important example of the manner in which habitation sites were selected by the Taubachian population. The site was settled in a protected calcareous tuff basin beside temperate spring waters, close to raw material sources.

Chronology

We have a number of contradictory 'absolute' chronological data from the site, including a misleading, too young, ^{14}C date (Vértes, 1964). A more recent Th/U date places the occupation around 100 ky BP, which seems realistic in light of the age of similar industries (Schwarcz and Skoflek, 1982).

Ecology

Although a detailed interpretation cannot be presented here, abundant evidence from several branches of science was found on the site. Of general relevance are palaeobotanical and palaeontological finds. These can be traced from the Riss/Würm interglacial through the Würm cold introductory phase, and changes in climate can be followed accordingly. The overall view is coloured by the tempering effect of the warm springs, which affected mainly the 'micro' range of species (malacology, lower plants) and essentially modified the faunal assemblage. The existence of cave bear and hyena in the faunal assemblage may indicate that the Taubachians hunted in the vicinity of the Gerecse caves. Remains of juvenile mammoths (6 individuals) indicate, according to the original study on the fauna, hunting specialization in this species (Kretzoi, 1964, 117).

Culture

Vértes referred to the Middle Palaeolithic pebble industries as 'epichopper-cultures'. There is a lengthy hiatus between the hypothesized Lower Palaeolithic origins and the Middle Palaeolithic occurrence. The pebble industries during the Mindel/Riss and the Riss periods must have developed latently and reached versatile forms. However, the selection of raw material and the survival of certain basic forms appear evident (Vértes, 1970).

About 60% of the Tata lithic industry was based on silex pebbles. The average size of tools is 30(!) mm, the width of the implements being 65% of the length. Hence, tools were both small and bulky. In the type list, scrapers made from pebble segments dominate together with the characteristic Tata scraper, which is a development of the chopping tool base form. An outstanding item in the inventory is the polished mammoth tooth lamella, interpreted by Vértes as a 'chouringa' (cultic object of Australian aborigines, Vértes 1964, 139).

The raw material originates in the Tertiary gravel beds around the site. A study of the original composition of the gravel beds indicates that the small flint pebbles used for the Tata tools were arbitrarily selected.

Connections

The technology displays a distant relation to all pebble-using Middle Palaeolithic assemblages. Vértes suggested closer links to the north-Italian Micropontiniano industries (Grotta Guattari).

Southeast European Charentian

The La Quina type Mousterian industries (i.e., Charentian) within the Mousterian complex were defined by Bordes (1953; 1961). These are characterized by a specific type list and corresponding technological indices. By the excavation and study of Érd site, Gábori-Csánk, suggested a specific local variant of this technology under the name of 'Southeast European Charentian' (Gábori-Csánk, 1968). The Charentian culture has been shown to exist during the Hungarian Palaeolithic from the earliest archaeological finds. While the pebble tools of the Kiskevély cave (where excavation began in 1912) were assigned to this complex in a recent revision of the material (Gábori-Csánk,1993), determination of the Charentian traits was made possible only by the abundant and authentic material found at Érd.

Sites

Evidence of the Charentian culture was found at Érd, the Szelim cave, and Kiskevély cave.

Topography

This cultural unit is restricted to the northeastern corner of Transdanubia. The two caves have already been mentioned above in connection with other complexes. Érd is an open-air site. Its special setting and the fact that it has been selected for habitation, provide valuable insights regarding the settlement strategies of the Middle Palaeolithic population. The Palaeolithic hunters set up camp on the southern margin of the Érd plateau, at the point at which two valleys intersect. Habitation lasted for a long time, during several geochronological periods (Gábori-Csánk, 1968, 10).

Chronology

The age of the upper cultural layer is 38,100 years ago. The lower cultural complex was divided into four levels, the oldest of which is dated to 44,300 and >50,000 years. Based on biochronological grounds, the Érd hunters lived during the 'Altwürm' period as defined by the presence of *Asinus hydruntinus* (Gábori-Csánk, 1968, 107-110).

The Kiskevély cave sequence can be fairly well correlated with Érd. The archaeological material is rather poor, but the fauna can be dated to the same faunal phase (i.e., the Subalyuk fauna phase of the early Würm in which *Asinus* and *Ursus spelaeus* dominated 'together with other herbivores: *Equus*, *Alces* and *Su'*. (Dobosi-Vörös, 1994, 35).

Ecology

Botanical data. The flora illustrates a gradual climatic deterioration of during the onset of Würm. *Larix* and *Picea* were replaced by *P. cembra*, and the number of other arboreal plants decreased.

Faunal data. The fauna of Érd was exceptionally rich. In addition to the 15,000(!) determined bones, there are further 30,000 unidentified different bone fragments. Apart from general species typical of the Early Würm, *Asinus hydruntinus* is an important chronological marker. The abundance of *Ursus spelaeus* (exceeding 90% in some levels) is surprising at an open-air site, and indicates a marked hunting specialization (Gábori-Csánk and Kretzoi, 1968).

Culture

Altogether, 780 tools were found throughout the entire settlement at Érd. Together with the implements obtained at the other two sites (Szelim and Kiskevély), the total number of tools known from this cultural complex is ca. 900. The dominating tools are pebble segments and scrapers made from these segments. The raw material is quartzite of inferior quality derived from local gravel beds. The size range of the implements was apparently determined by the size of the pebbles, which are from 8 - 10 cm long. The size seems to be the most marked difference from the Tata-type industry. Other differences exist in the workmanship and tool-type distribution, apparently due to differences in the quality of the raw material.

Connections

The closest connections to the site were found in the

southeastern Alps and the Croatian-Slavonian sites. The settlement of Érd is the northernmost representative of the south-east European Charentian complex (Gábori-Csánk, 1968).

Conclusions

According to our present knowledge, the Middle Palaeolithic sites are located in the northern parts of Hungary, mainly the northeast, in a band 200 kms long. This band is divided into two uneven parts by the Danube river as it turns towards the south. Sites are spread from the northeastern parts of Transdanubia to the Hernád valley. There are both cave sites and open-air sites. The cave sites are located on the mid-mountain ecological niche. The open-air sites are found in calcareous tuff basins, at valley heads along the margin of the plateau and on hill-tops along the margin of the hills.

Chronologically, the Middle Palaeolithic period lasted from the end of the Riss/Würm interglacial until the cold peak of the Würm I period. This time interval is usually characterized better on the basis of the rich and well-elaborated evidence from natural sciences rather than contradictory 'absolute' chronological dates. It is most remarkable that during a limited time span and within narrow geographical boundaries, many varieties of Middle Palaeolithic industries were present which differed in both character and origin.

These industries can be associated with three large tool-working traditions: (1) Classical Mousterian; (2) Bifacial-handaxe industries; and (3) Epichopper industries. The classical Mousterian is typical in spite of the great geographical distance from its central territory. The bifacial and epichopper industries produced several unique features which varied as a function of geographical and chronological distance from the 'base industry'. Surviving types, technology and other specializations clearly distinguish the origin, delineate the distribution, as well as restrict the circle of related, analogous sites. Local and regional deposits played the main role in determining which raw materials were used. Two long-distance raw materials, at least (obsidian and Szeletian felsitic porphyry), reached the northeast Transdanubian sites, probably via the route leading along the southern margin of the north-Hungarian mid-mountain range.

Middle Palaeolithic human remains are known from two localities in Hungary: The remains of two individuals, which unfortunately have not received enough attention, were found in Subalyuk cave, and several teeth came from the Remete-Upper cave.

Important observations concerning the way of life during this period pertain to hunting specialization for: (1) the mountain goat at Subalyuk cave; (2) mammoth calves at Tata; and (3) cave bears at the open-air site of Érd. The probable social role of the above mentioned 'chourunga' of polished mammoth tooth plate from Tata-Porhanyóbánya (Vértes, 1964, 139) should be stressed.

In the light of historical processes, the Hungarian Middle Palaeolithic seems a development in its own right. We can observe in this age the duality of discontinuity and continuity. The main indications for discontinuity are:

- None of the immediate local antecedents of the Middle Palaeolithic industries in Hungary - Classical Mousterian, Jankovichian and Taubachian - are known.
- Likewise, there is no evidence regarding the survival and local development of the Middle Palaeolithic industries during the Upper Palaeolithic.
- The relations between the Bábonyian and the Szeletian, argued on typological grounds by Ringer (1989), still await confirmation by excavating undisturbed Bábonyian sites.

Continuity can be observed in some fortunate cases, such as the Érd site, where the same culture existed for a long period of time. There is evidence for a prolonged local development of raw-material management and technology.

The connection of the Middle Palaeolithic industries with the subsequent Upper Palaeolithic industries in Hungary remains unclear. The oldest Upper Palaeolithic culture in Hungary is the Aurignacian, present at two cave sites in a typical, already developed form. The arrival and subsequent route of the Aurignacian, as well as its temporal dimensions, are known in broad lines, and it is probable that this culture represented only a short interlude in the Hungarian Palaeolithic.

There are two additional sites worth mentioning here. At Sajószentpéter-Margitkapu, a small amount of material was obtained which allows only the superposition of certain Middle and Upper Palaeolithic industries to be established (Ringer, 1993, 76-78). The other site, Eger-Kőporostető, is located in the centre of the area inhabited by the Bükk Palaeolithic cultures at the southern margin of the mountains. So far, it has not been possible to accurately relate this industry to any Middle or Upper Palaeolithic facies, or to any Middle/Upper Palaeolithic transitional phase (Dobosi, 1995).

The Hungarian Middle Palaeolithic is characterized by an unexpected variety of finds and a high quality manufacture. It offers a good example of accelerated development. We are confident that more detailed descriptions of this period will become available in the near future.

Author's address: National Hungarian Museum, H-1370 Budapest, Hungary. tdv@hnm.hu

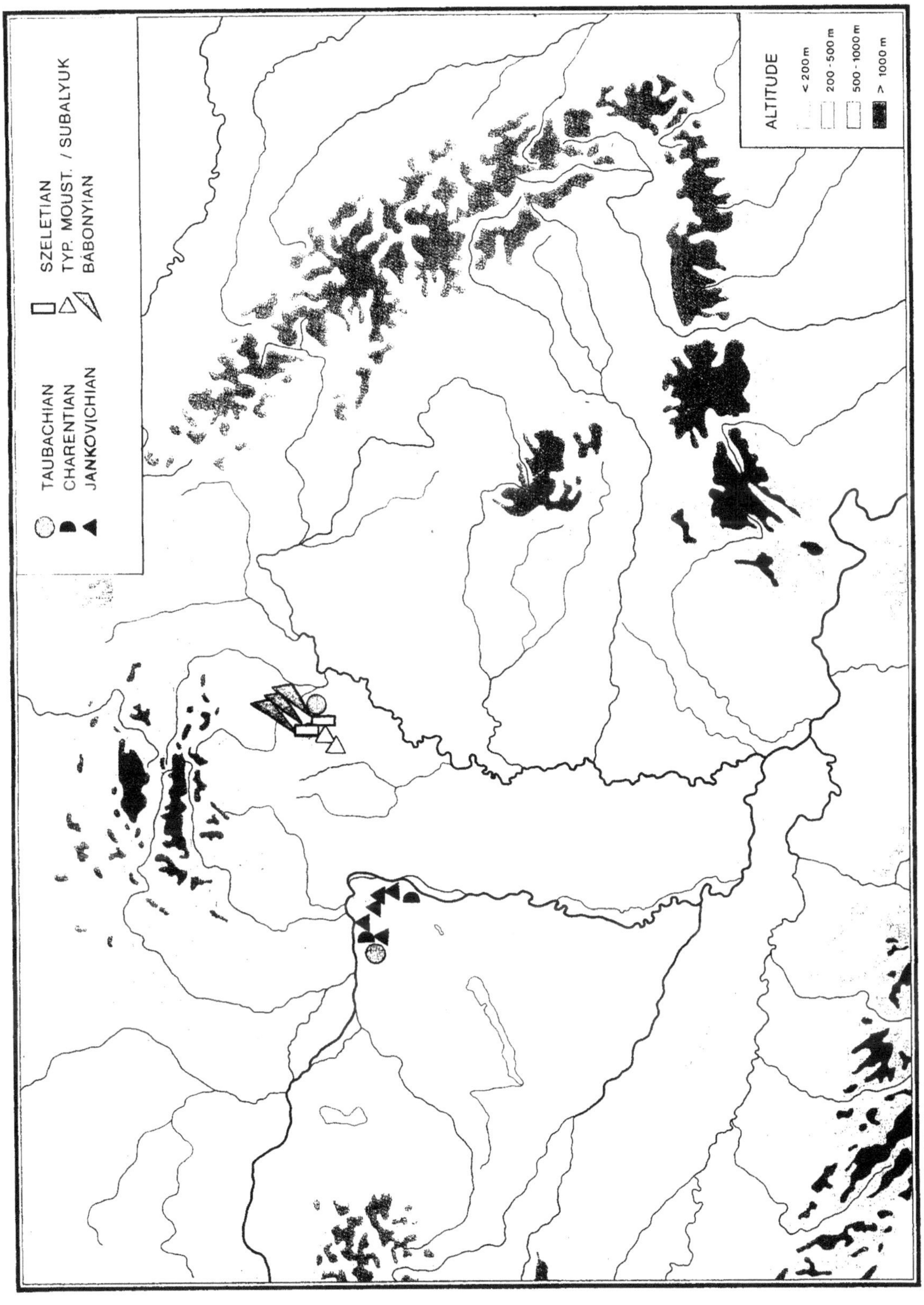

Fig. 1. Location map.

CULTURE	SITES	Canis	Crocotta	Lynx	Leo	URSUS	Mammut	Coelodonta	Equus	Asinus	Alces	Rangifer	Megaloceros	Cervus	Capreolus	Rupicapra	Capra	Bison	Sus	BIRDS	Vulpes	Alopex	Martes	Meles	Putorius	Lepus	Castor	Hystrix
		FAUNA																										
		BIG GAME																			SMALL GAME							
		CARNIVOROUS					HERBIVOROUS														CARNIVOROUS					RODENTS		
JANKOVICHIAN	JANKOVICH	○	+	○	+	○		○				○	○			○		○						+				
	KISKEVÉLY	○		○	+	●		○		○	○	○		●	○	○		○			+		+			+		+
	REMETE - FELSŐ	○			+	●	○	○	○			○	○	○			○	○		○	+			+		+		
TAUBACHIAN	TATA	○	+			○	○	○	●	○			○	○				○	○	○				+		○	○	
CHARENTIAN	ÉRD	○	+		+	●	○	○	○	○		○	○	○		○	○	○	○		+		+	+	+	+		
CLASSICAL MOUSTERIAN	(3) SUBALYUK lower	●	+	○		○		○	○					○		○	○	○			+					+		
	(II) SUBALYUK upper	○	+		+	○			●			○	○	○		○	○	○			+	+		+		+		
SZELETIAN	SZELETA lower	○			+	○							○									○						
BÁBONYIAN		~	~	~	~	~	~	~	~	~	~	~	~	~	~	~	~	~	~	~	~	~	~	~	~	~	~	~

○ presumable hunted　　+ non - hunted
● frequent among hunted　　~ no data

Fig. 2. Faunal assemblages of the different cultural entities in Hungary.

Bibliography

Bordes, F. 1953. Essai de classification des industries 'moustériennes'. *Bulletin de la Societé Préhistorique Française* 50, 457-466.

Bordes, F. 1961. Mousterian cultures in France. *Science* 134, 803-810.

Collins, D. 1969. Cultural traditions and environment of early man. *Current Anthropology* 10, 267-316.

Dobosi, V. T. 1986. Raw material investigations on the finds of some palaeolithic sites. In: K. T. Biró (ed.), *International Conference on Prehistoric Flint Mining and Lithic Raw Material Identification in the Carpathian Basin.* Budapest-Sümeg. pp.249-260

Dobosi, V. T. 1995. Eger-Kőporostető. Revision d'une industrie a outils foliacés. In: Les industrie a pointes foliacés d'Europe centrale. *Paleo. Suppl.* I. pp.45-56.

Dobosi, V. T.- Vörös, I. 1994. Material and chronological revision of the Kiskevély cave. *Folia Archaeologica* 43, 9-50.

Gábori, M. 1976. *Les civilisation du paléolithique moyen entre les Alpes et l'Oural.* Budapest.

Gábori-Csánk, V. 1968. *La station du paléolithique moyen d'Érd - Hongrie.* Monumenta Historica Budapestiensis, III. Budapest.

Gábori-Csánk, V. 1993. *Le Jankovichien. Une civilisation paléolithique en Hongrie.* ERAUL 53. Liège.

Gábori-Csánk, V. and Kretzoi, M. 1968. Zoologie archéologique. In: Gábori-Csánk 1968. pp. 223-243.

Kadic, O. 1938. Ősrégészeti eredmények. In: A cserpéfalui Mussolini barlang. (Subalyuk) *Geologica Hungarica* ser. pal. 14, 107-154.

Kretzoi, M. 1964. Die Wirbeltierfauna des Travertinkomplexes von Tata. In: L. Vértes (ed.), *Tata, eine mittelpaläolithische Travertinsiedlung in Ungarn.* Budapest, pp.105-124.

Mester, Zs. 1994. *A bükki moustérieren revizió ja.* PhD Thesis. Budapest.

Mottl, M. 1938. A lerakódások állatvilága. In: A cserépfalui Mussolini barlang. *Geologica Hungarica* ser. pal.14, 205-308.

Ringer, Á. 1983. *Bábonyien: Eine mittelpalaolithische Blattwerkzeugindustrie in Nordostungarn.* Diss. Arch. ser. II. No.11. Budapest.

Ringer, Á. 1989. L'origine du Szélétien de Bükk en Hongrie et son évolution vers le paléolithique suprieur. *L'Anthropologie* 27, 223-229.

Ringer, Á. 1993. *Északkalet-magyarországi geomorfológiai szintek és régészeti adataik.* PhD. Thesis, Miskolc.

Simán, K. 1990. Population fluctuation in the Carpathian Basin from 50 to 15 thousand years BP. *Acta ArchHung* 42, 13-19.

Simán, K. 1995. The Korlát-Ravaszlyuktető workshop site in north-eastern Hungary. *Archaeologia Polona,* 33, 41-58.

Svoboda, J. and Simán, K. 1989. The Middle-Upper Palaeolithic transition in southeastern central Eurpe (Czechoslovakia and Hungary). *Journal of World Prehistory* 3, 283-322

Schwarcz, H. P. and Skoflek, I. 1982. New dates for the Tata, Hungary, archaeological site. *Nature* 295, 590-591.

Valoch, K. 1984. Le taubachian, sa geochronologie paléoécologie et paléoethnologie. *'Anthropologie* 88, 193-208.

Vértes, L. 1964. Die Ausgrabung und die archäologischen Funde. In: L. Vértes (ed.), *Tata, eine mittel-paläolithische Travertinsiedlung in Ungarn.* Budapest, pp.133-250.

Vértes, L. 1970. 'Zitrus' (epi-chopper) Industrien in Ungarn. In: H. Schwabedissen (ed.), *Frühe Menschheit und Umwelt.* Fundamenta, Reihe A, Band 2. Köln, pp.28-33.

TOWARD MODERN HUMANS: YABRUDIAN AND MICOQUIAN, 400 – 50 kyears ago.
Edited by A. Ronen and M. Weinstein-Evron. BAR S850, 2000.

LE PALÉOLITHIQUE MOYEN EN MORAVIE: LES INDUSTRIES LITHIQUES ET LEURS MATIÈRES PREMIÈRES

Martin Oliva

L'histoire de la découverte du Paléolithique moyen en Moravie est assez compliquée. Bien que, déjà à la fin du 19e siècle, Maška et Knies aient signalé dans les grottes de Šipka et de Kůlna des outils datant indubitablement du Paléolithique moyen, cette datation a été contestée, entre les deux guerres, par une grande autorité, le Professeur K. Absolon. Après la guerre, des chercheurs pragois, Prošek et Žebera, ressuscitent le problème de la présence du Paléolithique moyen dans nos pays. Les résultats les plus importants publiés à ce jour, concernent la grotte de Kůlna, près de Sloup dans le Karst morave, fouillée d'abord par M. Kříž et J. Knies, puis, l'aide de méthodes plus modernes, par K. Valoch (1961-1976). La stratigraphie de cette grotte est assez compliquée. A l'entrée sud, environ 12 couches lithologiques différentes ont été mentionnées, le complexe inférieur (14-7) appartenant au Paléolithique moyen. Dans la couche 6 apparaît par endroits du Paléolithique moyen final mais aussi du Gravettien récent et du Magdalénien, et dans les couches supérieures 5-4 du Magdalénien et du Paléolithique final (Valoch, 1988).

Le peuplement le plus ancien de Kůlna est lié à la couche 14. Il date selon toute évidence des phases finales du Rissien. L'ensemble d'une trentaine d'artefacts est caractérisé avant tout par l'emploi du concept Levallois et par ses dimensions plus larges que dans les couches postérieures. La grande faune de ce niveau n'a pas encore été traitée. La microfaune représente une communauté steppique froide.

Sur le dessus reposait presqu'un mètre de sédiments sableux-argileux gris avec éboulis et cailloux, sans doute de l'époque éemienne. Dans les niveaux 11a et b, apparaissent des espèces steppiques comme les bovidés et beacoup d'Ursus spelaeus, alors que dans les niveaux 11c et d plus anciens, prédomine la faune forestière avec le rhinocéros, le mammouth, l'Equus taubachensis et beaucoup de cervidés, renne excepté. Les os sont très fragmentés et portent fréquemment des traces d'activité humaine. Les industries associées sont de type taubachien et contiennent au total douze mille pièces lithiques dont plus de 700 (6%) ayant été retouchées. Au premier abord, un mélange de matières premières diverses attire l'attention. Le hornstein jaunâtre crétacé et le quartz blanc constituent chacun un tiers de ces matières, 15% étant composés de quarzite brun et le reste de différentes roches.

Quant à la provenance des matières premières, seul le psammite (1%) provient des environs immédiats de la grotte, Kůlna étant pratiquement située à la frontière des calcaires dévoniens et des grès culmiens. Les calcaires ne contenant aucune roche siliceuse pouvant servir à la taille d'artefacts, la majorité des roches utilisées a été ramassée dans un rayon de 10-15 km d'où furent extraits des restes d'autres formations géologiques, tout d'abord la spongilite (hornstein créracé) apportée de la vallée de Svitala à l'ouest de la grotte, Fig.2, 9).

Pourtant, même les roches provenant de gisements plus éloignés ne présentent qu'un nombre restreint de pièces. Ainsi la porcelanite (0, 82%) n'est connue de nos jours que dans la Moravie de l'Est, à une soixantaine de kilomètres de Kůlna. Il s'agit pour la plupart d'artefacts tout faits, avec un nombre particulièrement élevé de pièces à retouche plate (Fig.2, 1-5) et un seul petit nucléus (Fig.2, 6). Les artefacts en porcelanite constituent les seuls petits bifaces ('Fäustel') de cette couche, et peut-être sont-ils témoin d'un emprunt à une autre tradition culturelle, comme celle du Taubachien bifacial du gisement hongrois de Tata (Vértes, 1964). Le silex gris (min. 2 pièces, Fig.2, 10-11) provient de dépôts glacifluviaux datant de l'époque rissienne, à 80 km au Nord au minimum.

Les dimensions des artefacts sont aussi étonnantes que la variété des matières premières. Les 1713 pièces mesurées sont longues et larges de 3 cm en moyenne. 32% des 140 nucléus découverts, discoïdes pour la plupart, sont de taille inférieure à 3 cm. Cependant, Boëda a distingué quelques nucléus levallois récurrents de dimensions minuscules qui rappellent le faciès micro-levallois de Pech de l'Azé en Dordogne, l'Asinipodien de F. Bordes.

Se basant sur la liste-type de F. Bordes, K. Valoch (1988) a classifié 514 artefacts dont la majeure partie se compose de racloirs avec un petit pourcentage de type charentoïde. La deuxième place appartient aux denticulés et encoches dont 7% sont clactoniennes. Les artéfacts datant du Paléolithique supérieur constituent environ 6% et les perçoirs atypiques 2, 8%.

L'inventaire peu nombreux de la couche sus-jacente 9a est une première preuve de la présence du Micoquien à Kůlna comme le démontre la découverte d'une pointe foliacée à base arrondie et d'un 'Faustkeilblatt' (biface-feuille) en quartz. Du point de vue morphologique, on constate l'existence de formes minuscules qui rappellent celles de la couche 11. Cette couche (9a) se distingue par les plus récents éléments interglaciaires (*Dicerorhinus kirchenbergensis*). Les couches micoquiennes sont riches en défenses et molaires appartenant surtout à des rennes et des mammouths. Trois défenses ont été appuyées sur les parois rocheuses de la grotte. Là aussi, les os sont très fragmentés et souvent striés.

L'époque Micoquienne se caractérise par l'utilisation de roches variées mais le silex corné crétacé (spongolite) domine nettement (presque 80%), suivi par le quartz (6%). Même ici, on trouve presque 40 pièces de cristal de roche

(Fig.3, 2 des couches remaniées), dont un biface fruste en quartz enfumé. Vues les dimensions parfois considérables des artefacts, cette matière première a été vraisemblablement importée des hauteurs tchéco-moraves, pas encore peuplées pendant cette période. L'examen minéralogique entrepris récemment avec l'aide de spécialistes (A. Přichystal, S. Houzar, M. Novak) a révelé l'existence inattendue de calcédoines, rappellant les roches siliceuses limniques importées peut-être de la Hongrie du Nord. Les pièces en question sont effectivement identiques aux échantillons de la région de Cserhát-Miskolc-Zemplén, à 250-370km de la grotte de Kůlna (Biró et Dobosi, comm. pers.). Cette provenance a également été confirmée par des spécialistes hongrois et slovaques. Cette matière première n'etait jusqu'à présent pas connue en Moravie et en Slovaquie occidentale et centrale. Cependant, les prospections tout-à-fait récentes de Neruda et Válek ont démontré la presence très limitée de matières premières apparemment identiques à proximité de la grotte de Kůlna, aux alentours du village de Němčice, laissant ainsi ouverte la question d'origine. Il s'agit de quelques dizaines de pièces, dont la composition est assez curieuse surtout si l'on accepte leur origine lointaine: on observe une dominance d'éclats non-retouchés et de quelques gros nucléus peu travaillés. Il est à remarquer la présence d'ébauches d'outils bifaciaux, qui sont restés inachevés comme le prouve la qualité manifestement insuffisante de la pièce initialisée (Fig. 5, 2-3). Les outils parachevés restent relativement rares (Fig. 4, 1; 5, 1, 4-6). D'autres roches exotiques ou inhabituelles furent également découvertes, comme un biface en andésite (Fig. 6, 4), un biface en quartzite brune très fine (Fig.6, 3), inconnu pour le moment de la Moravie, de la quartzite limonitique (Fig.6, 2) et des silex nordiques (Fig.6, 1).

La taille des artefacts est supérieure à celle du Taubachien sous-jacent: (longueur 4.1 cm, largeur 3.5 cm, épaisseur 1.3 cm). Selon Boëda (1995), l'inventaire de Kůlna est le meilleur exemple du concept discoïde (non-levallois).

Quant à la typologie, je renvoie à la communication de Valoch dans ce volume. L'émergence de types très prononcés du Paléolithique supérieur (grattoirs) et de nucléus laminaires non-levallois bien préparés par les enlèvements bifaciaux (Valoch, 1988) rend la collection du niveau 6a digne d'attention (Figs. 18-20; 25, 3). Le silex corné crétacé domine avec 76%, suivi par le quartz local (6%), le silex corné jurassique (4%) et quelques autres roches, telles que le cristal de roche (11 pièces, 2, 75%, Fig. 3, 1-2), la radiolarite (3 pièces, Fig. 3, 3), le silex erratique (Fig. 3, 4) et de la calcédoine jaune. La sédimentologie et la microfaune sont typiques des steppes froides, existant probablement avant l'amélioration climatique du Wurm moyen.

La couche 7a contenait 5 restes de Néandertaliens: fragments du maxillaire droit, de l'os pariétal et 3 dents de lait.

Une machoire néandertalienne a été trouvée dans la petite grotte de Švédův stůl près de Ochoz dans le Karst morave. Les fouilles dirigées ici par B. Klima dans les années 50 ont révélé deux niveaux du Paléolithique moyen, une couche éemienne contenant quelques outils atypiques et une couche datant vraisemblablement de l'interstade Hengelo, ayant donné une industrie sporadique proche du Micoquien (Klima et al. 1961). La mandibule trouvée par un amateur en 1905 appartient probablement à la couche supérieure.

Une autre grotte morave riche en Paléolithique moyen, surnommée Šipka ('la Fléchette'), est située dans la Moravie du Nord. Elle avait été fouillée au 19e siècle par K.J. Maška, qui avait trouvé dans un foyer, un fragment de la mandibule d'un enfant agé d'environ 10 ans. La couche inférieure a livré une riche industrie fortement cryoturbée en hornstein local classifiée comme le Moustérien à denticulé avec de nombreux types de Paléolithique supérieur (14%, notamment des burins). La méthode Levallois et la retouche plate sont pratiquement absentes (Valoch 1965). La faune est dominée par des jeune ours des cavernes. De la grotte voisine Čertova dira ('Trou du Diable') provient une pointe moustérienne en radiolarite importée (Fig. 8, 3).

A peu près 10 km à l'ouest de la grotte de Kůlna s'étend la vallée de la rivière Svitava où les Néandertaliens de Kůlna ramassaient le hornstein crétacé. Cette région a produit une centaine de gisements où l'on a découvert des outils de caractère Micoquien évolué pour la plupart et une quantité énorme de nucléus et de débitage brut (Oliva, Štrof 1985). Presque tous les sites, même les plus riches (Bořitov, Ráječko ou Doubravice), sont situés sur des élévations et offrent une large vue, parfois marqués comme ils le sont sur les cartes touristiques. Par contre, les gites de hornstein ne sont généralement pas situés aux mêmes endroits. Il ne s'agit donc pas des simples ateliers de taille mais de sites à fonction multiple. Parmi les quelques centaines d'outils apparaissent surtout des racloirs (Fig. 7, 5, 10), parfois avec la retouche Quina (Fig. 7, 7), des encoches et denticulés, des petits bifaces, des bifaces-couteaux (Fig.7, 11), des pointes foliacées etc. On remarque un nombre assez élevé de types du Paléolithique supérieur d'apparence archaque (Fig. 7, 1-2, 4, 6, 8). Les nucléus discoïdes et irréguliers mais aussi prismatiques montrent une forte préparation bifaciale sans qu'on puisse parler de la méthode Levallois (Oliva, 1987; 1991). La même situation se répète dans tous les sites plus riches de la région et aussi dans l'ensemble stratifié de la grotte de Kůlna (couche 6a). Parmi les matières premières constituées pour la plupart de silex corné local (spongolite) on trouve également de la radiolarite (Fig. 7, 1-4), des silex nordiques (Fig. 7:5-6), du hornstein jurassique (Fig. 7, 7, 10), de l'opale importée des hauteurs Tchèco-Moraves (Fig. 7, 9), etc.

Dans la Moravie du Sud, se trouve un complexe industriel insuffisamment précisé, appelé Krumlovien (Valoch, 1971). Les outils tels que les éclats retouchés, les racloirs grossiers, les encoches, les denticulés et les galets aménagés sont dominants, alors que les bifaces, les polyèdres et les grattoirs sont rares. Les nucléus sont pour la plupart irréguliers mais aussi prismatiques non-levallois. A côté du silex corné local apparaït le spongolite et un biface à dos (Fig. 8, 1) confectionné à partir d'andésite, probablement de la Slovaquie centrale. Les stations principales se trouvent autour de Vedrovice, Maršovice et

Dolni Kounice dans la région de Krumlovský les. Malheureusement, seul un petit ensemble d'une briquetterie près de Moravský Krumlov a été daté de l'Éemien grâce à la trouvaille d'un éclat *in situ*. Les autres documents proviennent de la surface où ils sont souvent, mais pas toujours, accompagnés d'outils du Paléolithique supérieur. Il est parfois difficile de distinguer les pièces du soi-disant Krumlovien des ébauches des nucléus et des outils provenant des ateliers de taille d'âge plus récent. Le terme Krumlovien n'est pas utilisé ni même utilisable dans d'autres régions. Il désigne vraisemblablement les outils plutôt grossiers de différentes phases et peut-être même de plusieurs faciès du Paléolithique moyen et supérieur influencés par l'occurrence locale de matériaux lithiques (Oliva, 1996).

Dans cette région, on a découvert deux autres industries de surface de caractère différent, fabriquées du même hornstein jurassique: la première provient du Kupařovice II et montre la connaissance du concept trifacial (Fig.9), la seconde de Jamolice, à 10 km environ à l'ouest des gites de cette matière première, utilise nettement la méthode levallois récurrente (Fig.10). Les lames et les outils retouchés sont très rares. A Kupařovice II apparaissent quelques petits bifaçoïdes de très mauvaise facture. Un troisième ensemble a été découvert, depuis peu, dans la couche inférieure à Vedrovice Ia (Fig. 8, 4-9). C'est la première industrie stratifiée du Paléolithique moyen de la Moravie du Sud et de la partie adjacente de la Basse Autriche. Elle est située dans le loess sous-jacent au sol du Würm moyen et de l'Aurignacien. Cette collection malheureusement trop pauvre, comprend des racloirs et des encoches mais aussi un éclat et une pointe levallois, tous de dimensions réduites (Oliva, 1993).

Un ensemble de petite taille, confectionné de matières premières diverses, dont la radiolarite et le quartz, a été trouvé par K. Žebera à Předmosti II dans la Moravie de l'est. L'industrie comporte des petits racloirs, quelques uns, rares, avec retouche plate, des denticulés, des encoches et des éclats retouchés. Les matières premières sont représentées comme suit: quartz (couche inférieure 49%, couche supérieure 60.5%), quartzite (11.6% et 14.5%), silex (8.7% et 5.7%), radiolarite (3.8% et 2.7%) et roches diverses (26.7% et 26.6%) (Svoboda et al., 1996, 159). J. Svoboda, qui a récemment poursuivi les fouilles, situe cette industrie, apparamment de type Taubachien, dans le dernier interglaciaire.

Un peu plus au Sud, près de Karolin, j'ai trouvé deux bifaces de type limande assez plats (Fig.11), en radiolarite (Oliva, 1981). Leur forme rappelle celle d'un autre biface plat de taille plus réduite (Fig. 8, 2), confectionné en silex erratique, et trouvé à Lubná dans la même région. Il montre une altération beaucoup plus prononcée de la surface que les outils aurignaciens des gisements voisins. Un nucléus levallois ovalaire et deux bifaces subtriangulaires en silex ont été ramassés sur des graviers datant probablement de l'époque rissienne de la rivière Odra dans la Moravie du Nord (Svoboda et al., 1991).

Pour récapituler la variabilité culturelle ou faciale du Paléolithique moyen en Moravie, il est clair que le Micoquien est l'époque la mieux documentée et sujet de nombreuses publications, bien que des analyses plus poussées de l'économie lithique et des chaînes opératoires aient à peine commencé. Le Micoquien est un faciès essentiellement non-levallois avec des outils bifaciaux et un taux assez important de formes du Paléolithique supérieur dans sa phase récente. Il date du Wurm ancien (W I et W II dans le sens français).

Nettement plus ancien est le Taubachien, lui aussi très peu levallois, mais de petite taille et également fabriqué à partir de matières premières très diverses, indépendamment de sa situation géographique. Des habitats situés dans le voisinage de sources minérales, c'est-à-dire aujourd'hui dans les travertins, spécialement en Slovaquie, sont considérés comme une manifestation très typique de ce faciès. Les outils à taille bifaciale apparaissent occasionnellement. Généralement, l'utilisation du concept Levallois sur notre territoire était sporadique. Quelques bifaces témoignent de l'existence d'une sorte d'Acheuléen (Určice: Valoch, 1980) ou du MTA (Karolin).

Il est intéressant de constater que l'occupation intense du Paléolithique moyen apparaît seulement dans les régions dotées de sources lithiques. Les sites localisés en dehors de ces régions emploient toujours des matériaux de sources les plus proches. Il semble que dans le Paléolithique moyen, les matériaux lithiques aient été assez immobiles ce qui est d'ailleurs une des causes de l'impression de 'non curational technology' propagée par Lewis Binford. Le problème de la répartition des gisements reflète une densité variable de l'occupation, ou peut-être seulement une visibilité différente des vestiges causée par une quantité variable de matières premières, utilisées dans les régions des sources de roches plus ou moins propices à la taille. Dans ce dernier cas, il faudrait supposer l'existence de sites alithiques peu connus. L'image du Paléolithique moyen est plus floue que celle du Paléolithique supérieur où le transport des matériaux rend plus identifiables les sites éloignés des sources lithiques. Mais ce sujet est à traiter avec prudence étant donnée les nombreuses spéculations possibles.

Les échantillons des matières premières importées de distances plus importantes (plus de 20 km) trouvent leur origine surtout dans les régions où le Paléolithique moyen est quasiment inconnu pour le moment. Il s'agit par example des quartz hyalins du Plateau Tchèco-morave ou des porcelanites provenant de la Moravie du Sud-ouest, trouvées dans la grotte de Kůlna. Ceci ne témoigne pas en faveur de contacts entre les groupes mais plutôt prouve qu'il s'agit de pièces récoltées lors des déplacements des chasseurs. Il semble d'ailleurs que les Néandertaliens s'intéressent vivement aux objets curieux (Schafer, 1996), comme le montre la fabrication de bifaces en porcelanite, en cristal de roche et en calcédoine colorée, mais aussi la fabrication de minuscules éclats levallois (d'environ 1 cm), débités, à Kůlna 11 (Taubachien), à partir de petits nucléus, selon la méthode levallois.

Les artefacts en calcédoine identique á celle d'Hongrie septentrionale constituent une question à part. En fait, on aurait affaire à des importations plus lointaines de témoins lithiques dans le Paléolithique moyen (comme par example 2 pièces en silex de Swieciechow à Solyomkut en Hongrie, distant de 300-350 km: Vértes, 1965, 122). Il est tentant d'envisager un transfert successif des roches curieuses, originaires dans la région d'une dense occupation du Bábonyien bifacial (Ringer, 1983). Cependant, vu la taille fruste des pièces en question, nous préférons rester prudent quant à leur origine hongroise et attendre les résultats des prospections géologiques locales. Du point de vue technique il est intéressant de voir que les habitants de la grotte Šipka ont presque négligé les sources de bon silex dans les moraines voisines et ont préféré le hornstein à gros grain. Dans ce cas, peut-être la force de la tradition l'emportait sur les avantages techniques du silex. L'absence du transport plus considérable des matières premières lithiques est - à mon avis - une des causes de l'inexistence des vestiges du Paléolithique moyen dans les enclaves très importantes du Paléolithique supérieur mais éloignées des sources des matières premières, telles que la région de la rivière de Morava, du Danube en Basse Autriche ou du Don en Russie. Les ensembles pauvres en témoins lithiques peuvent être trouvés avec plus de chance dans les grottes dont l'espace limité détermine à la fois l'occupation préhistorique et la prospection archéologique. C'est pourquoi les grottes sont un peu surreprésentées parmi les gisements du Paléolithique moyen en Europe centrale.

Adress de l'auter: Moravske Muzeum, 65937 Brno, Czech Republic

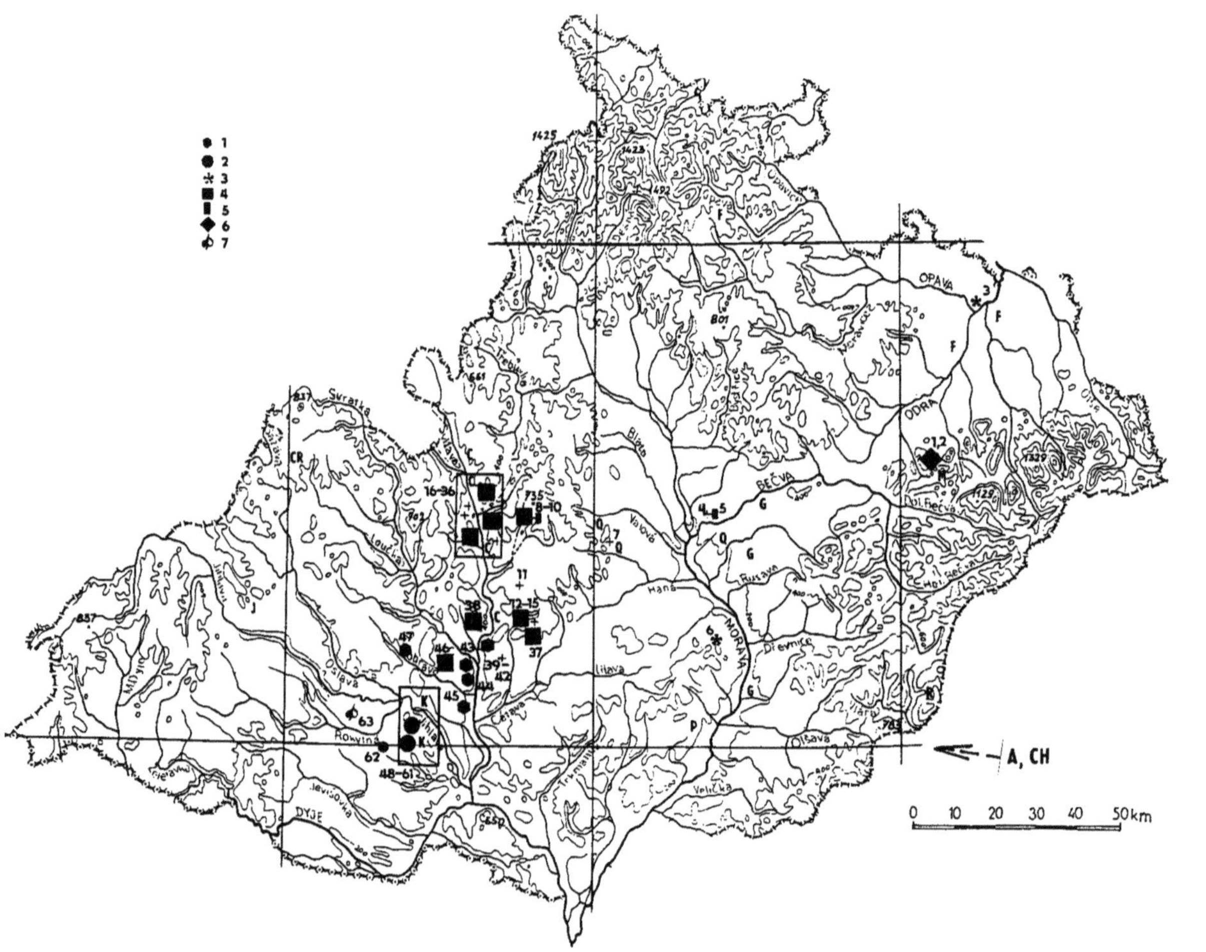

Fig. 1. Paléolithique moyen en Moravie. Matières premières: silex corné (hornstein): K de type Krumlovsky les, C crétacé (spongolite), M de type Šipka, G diverses variétés de graviers; CR cristal de roche, R radiolarite, F silex erratique, P porcelanite, Q quartzite, A andésite, CH calcédoine. Signes: 1 galets aménagés, 2 'Krumlovien', 3 industries aux bifaces plats, 4 Micoquien, 5 Taubachien, 6 Moustérien à denticulés, 7 ind. levalloides, sans types caractéristiques. Sites: 1 Šipka, 2 Čertova díra, 3 Ostrava, Polanka, Bohuslavice, 4-5 Předmosti I et II, 6 Karolin I, 7 Ondratice II, 8-10 grotte Kůlna, différentes cultures, 11 grotte Dratenicka, 12 grotte Pekarna, 13-14 grotte Švédův stůl, couches 14 et 11, 15 grotte Křižova, 16-36 Bořitov, Doubravice, Raječko et autres gisements dans le Bassin de Lysice, 37 Horakov I, 38-43 Brno, trouvailles isolées dans les briquetteries, 44 Modřice V, 45 Popovice V, 46 Troubsko II, 47 Tetčice I, 48-61 Kupařovice II, Maršovice, Vedrovice, Dolni Kounice, Trboušany et autres gisements autour de la Forêt de Krumlov, 62 Moravský Krumlov, 63 Jamolice I.

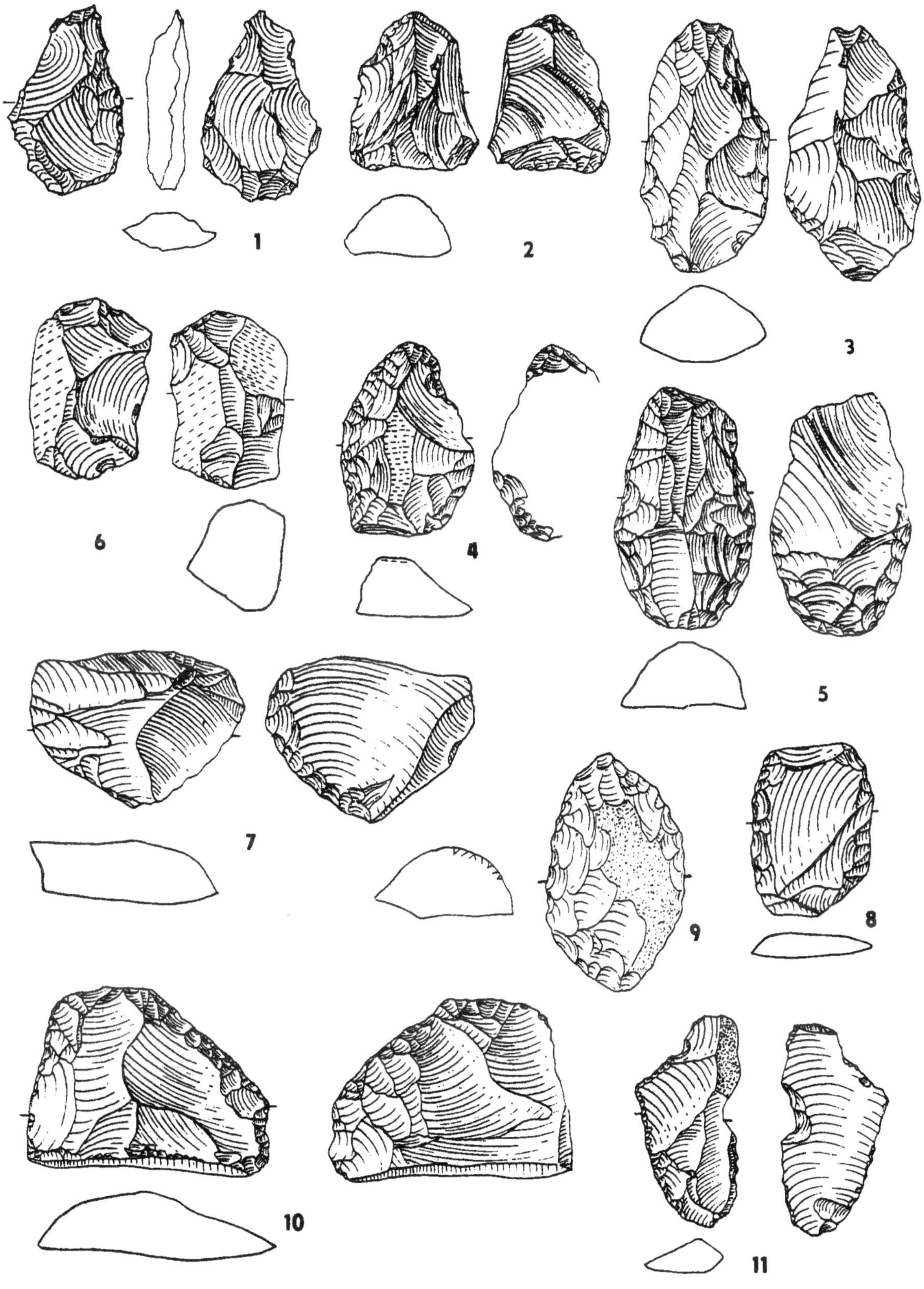

Fig. 2. Kůlna, couche 11 (Taubachien): 1-8 porcelanite, 9 cristal de roche, 10-11 silex erratique. D'après Valoch (1988).

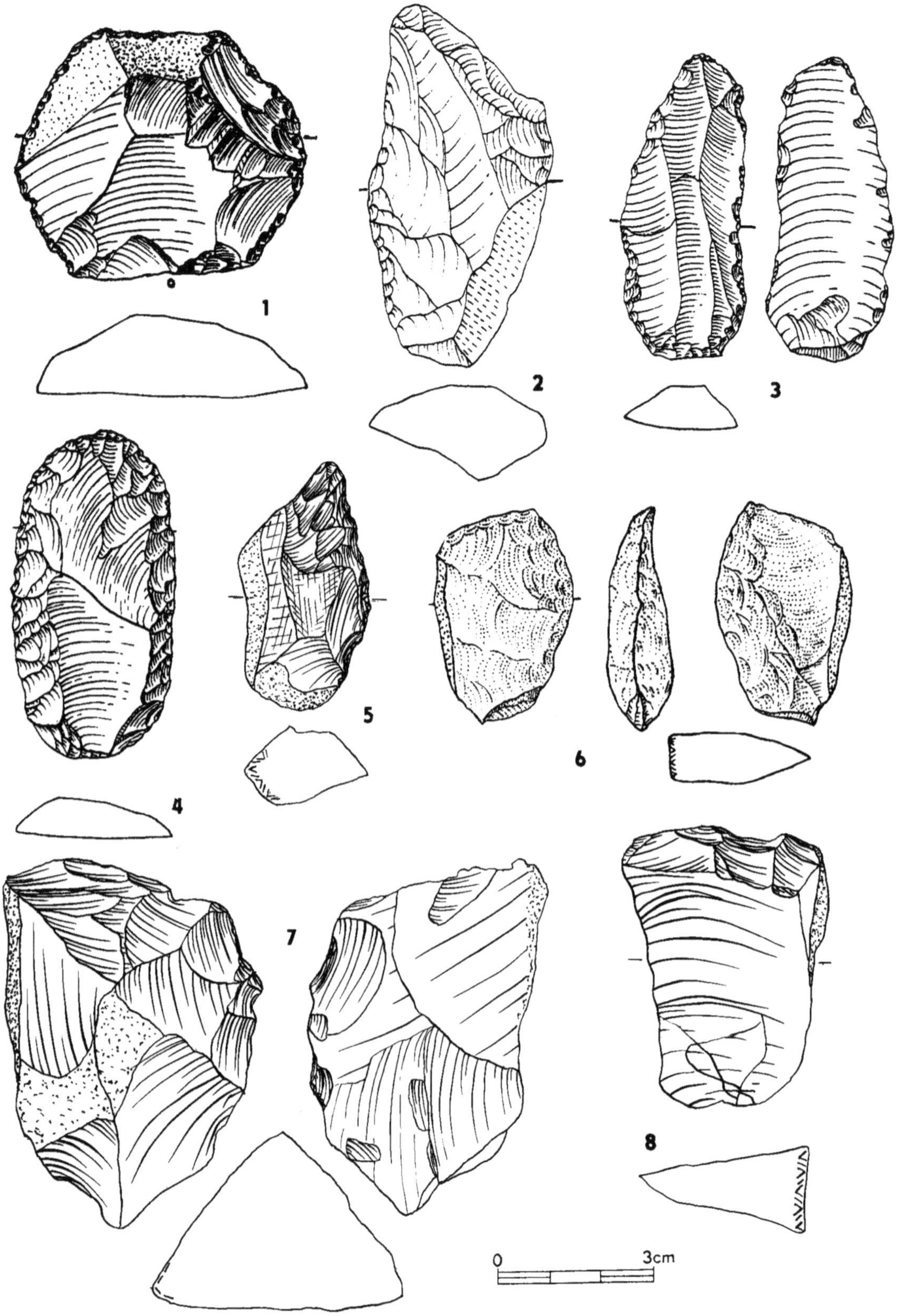

Fig. 3. Kůlna, Micoquien, 1-5 couche 6a, 6 couche 8a, 7 diverticule axiale, 8 couche 7c; 1-2, 6 cristal de roche, 3 radiolarite, 4 silex erratique, 5, 7 calcédoine, 8 roche volcanique. En partie d'après K.Valoch.

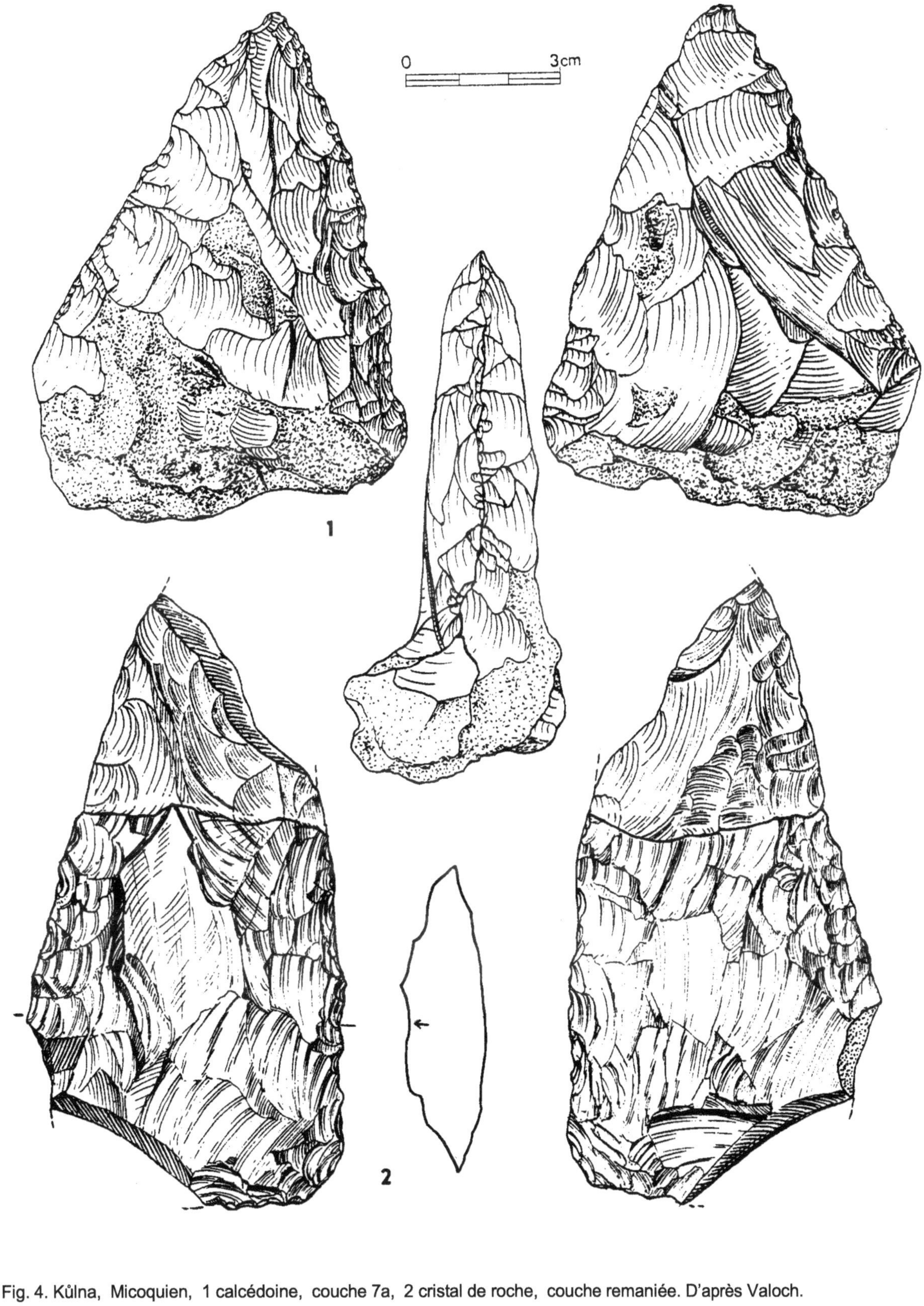

Fig. 4. Kůlna, Micoquien, 1 calcédoine, couche 7a, 2 cristal de roche, couche remaniée. D'après Valoch.

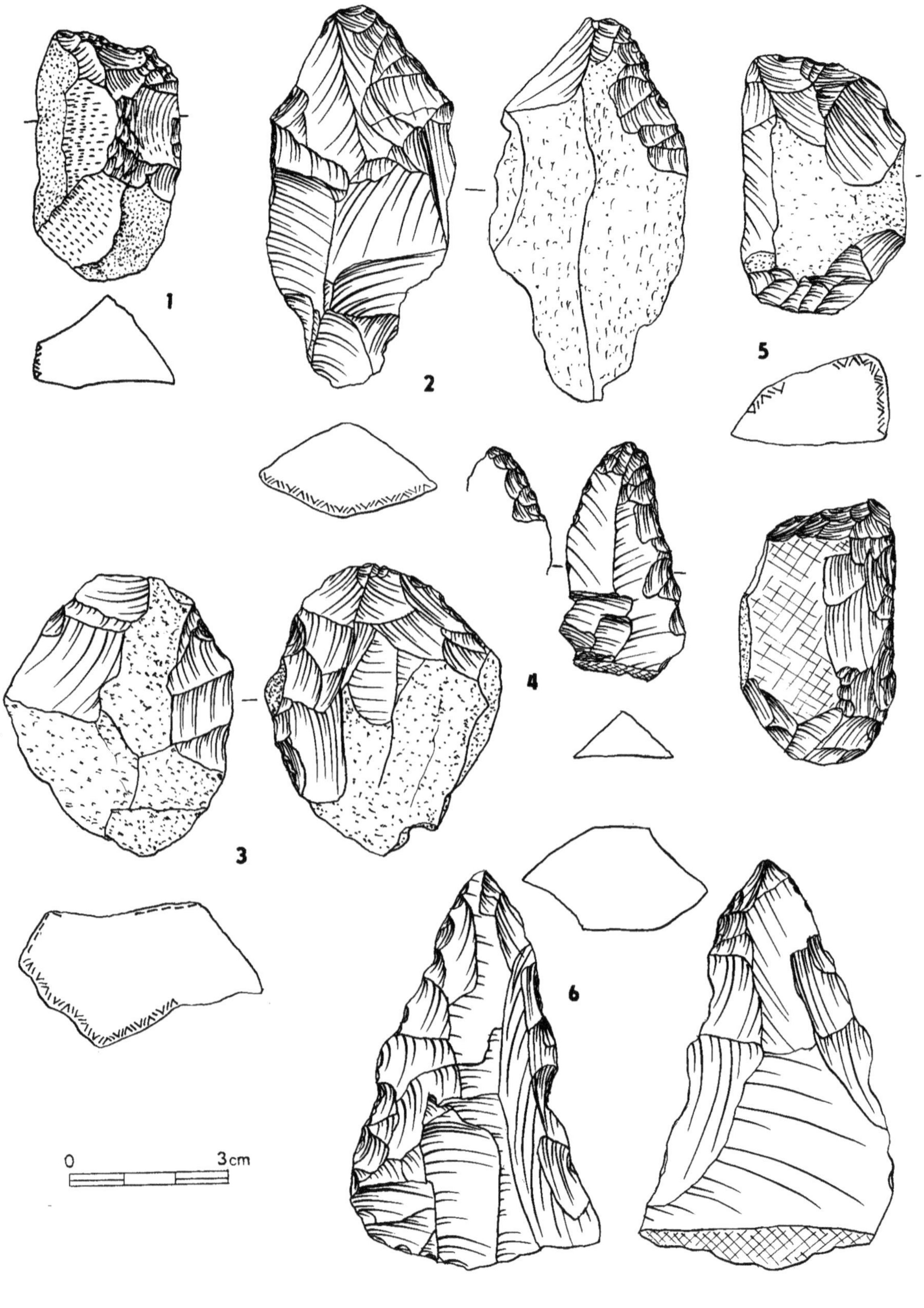

Fig. 5. Kůlna, Micoquien, couche 7a: 1-6 calcédoine

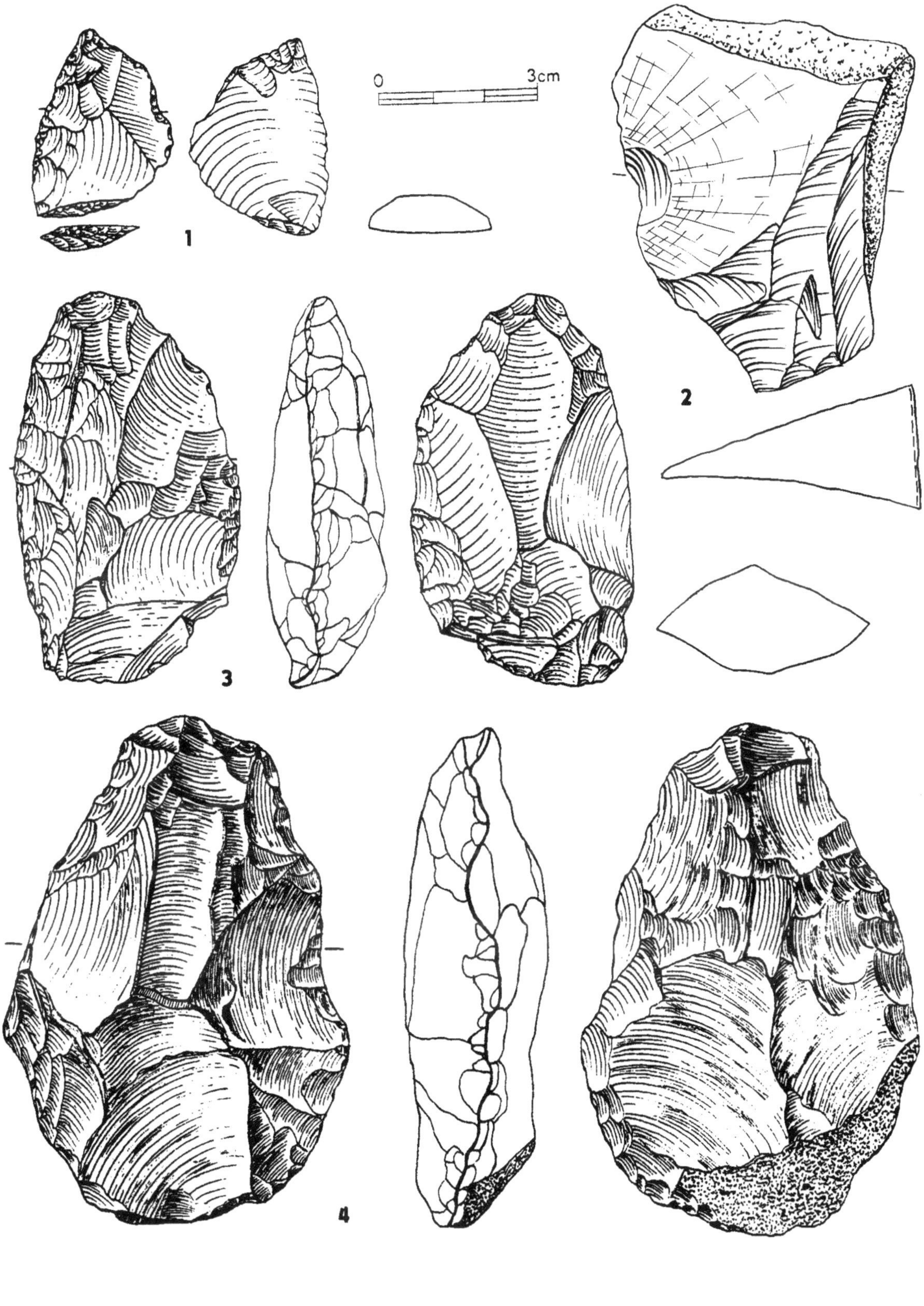

Fig. 6. Kůlna, Micoquien, 1, 3 couche 7c, 2, 4 couche 7a; 1 silex erratique, 2 quartzite limonitique, 3 quartzite fine, 4 andésite.

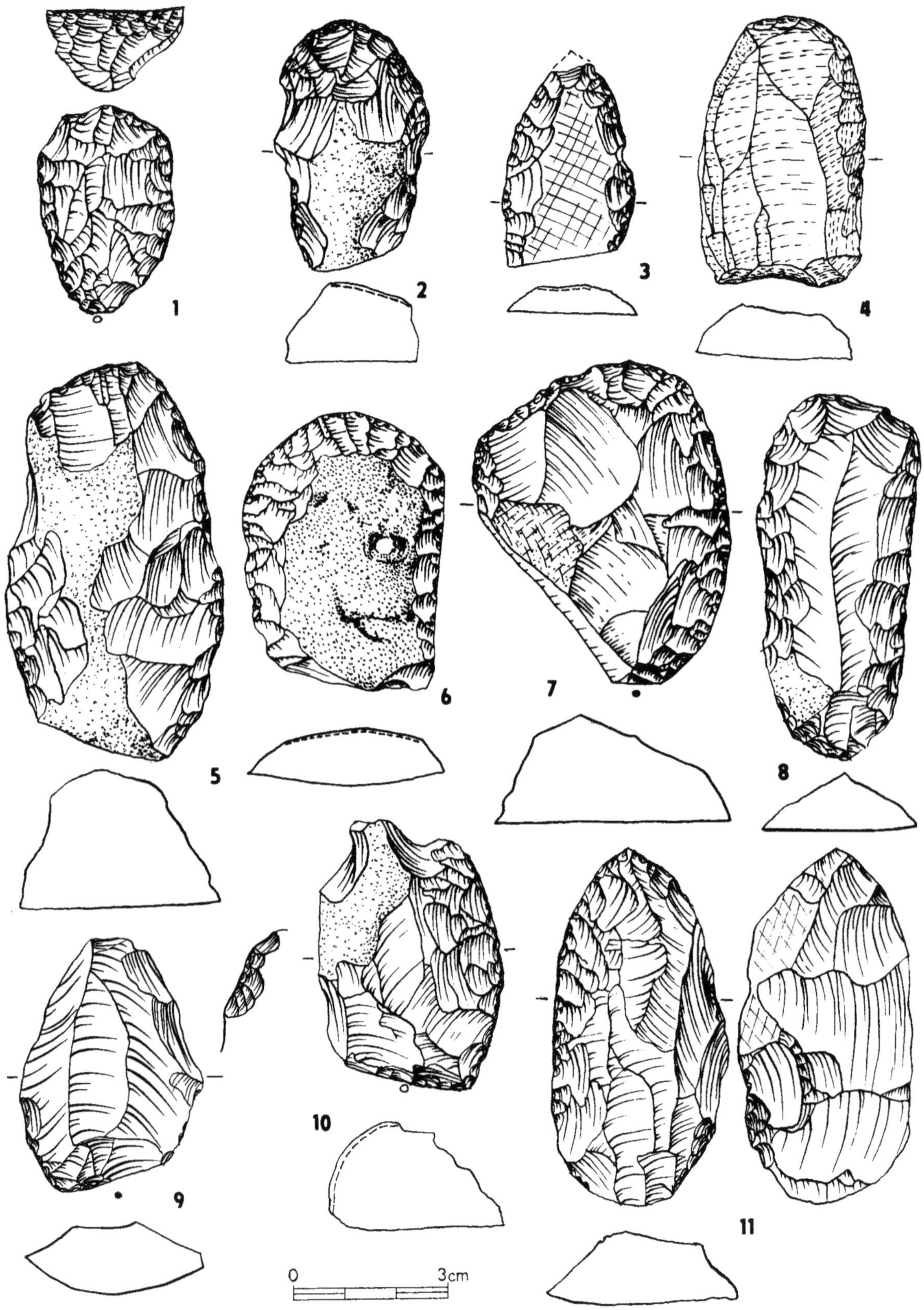

Fig. 7. Bořitov V - Horky, Micoquien évolué, 1-4 radiolarite, 5-6 silex erratique, 7 silex corné jurassique, 8 spongolite zonée, 9 opale, 10 silex corné de Krumlovský les, 11 spongolite opaque.

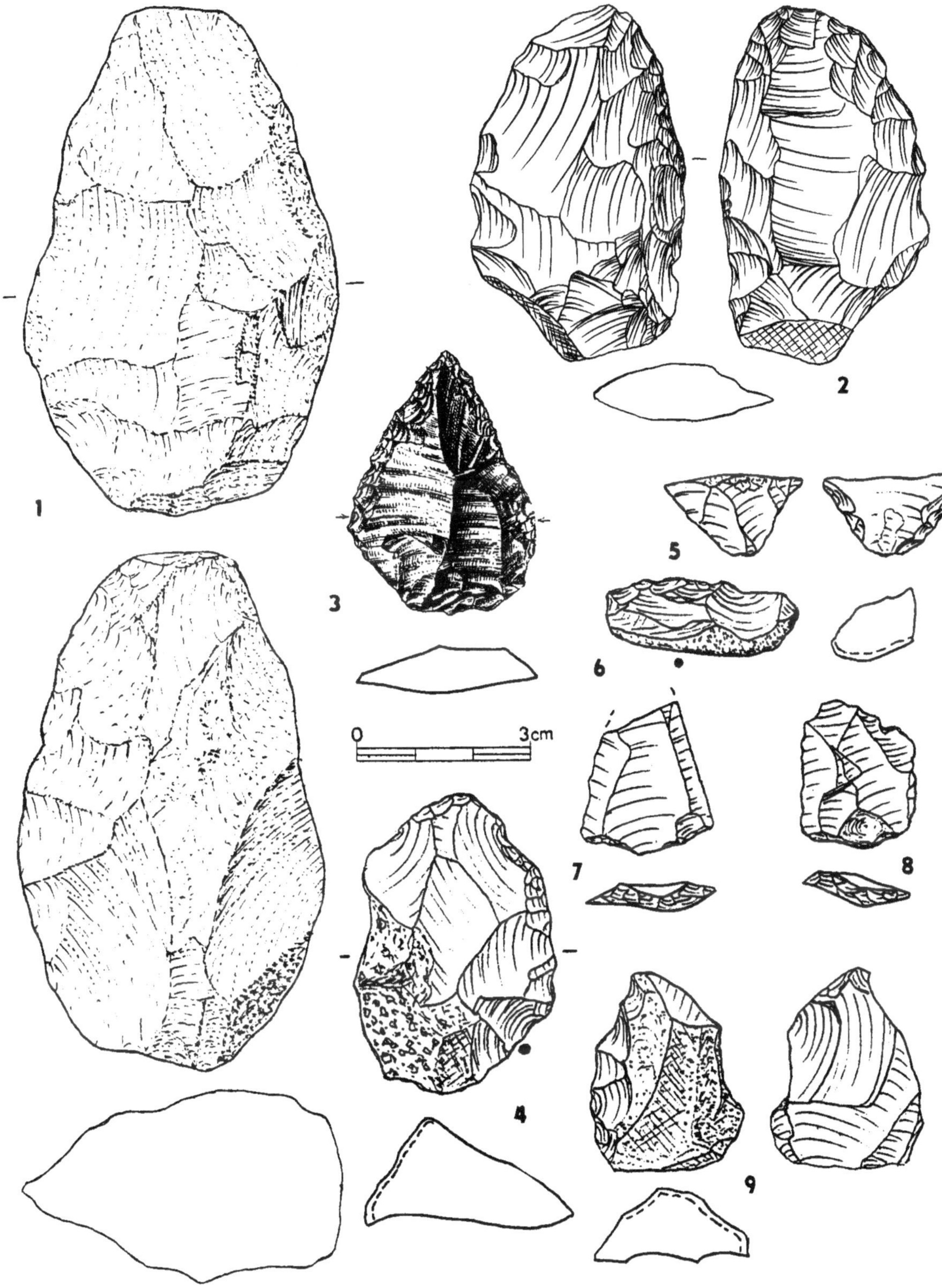

Fig. 8. 1 Maršovice I, andésite, 2 Lubná, silex erratique, 3 Čertova díra, radiolarite, 4-9 Vedrovice Ia, silex corné.

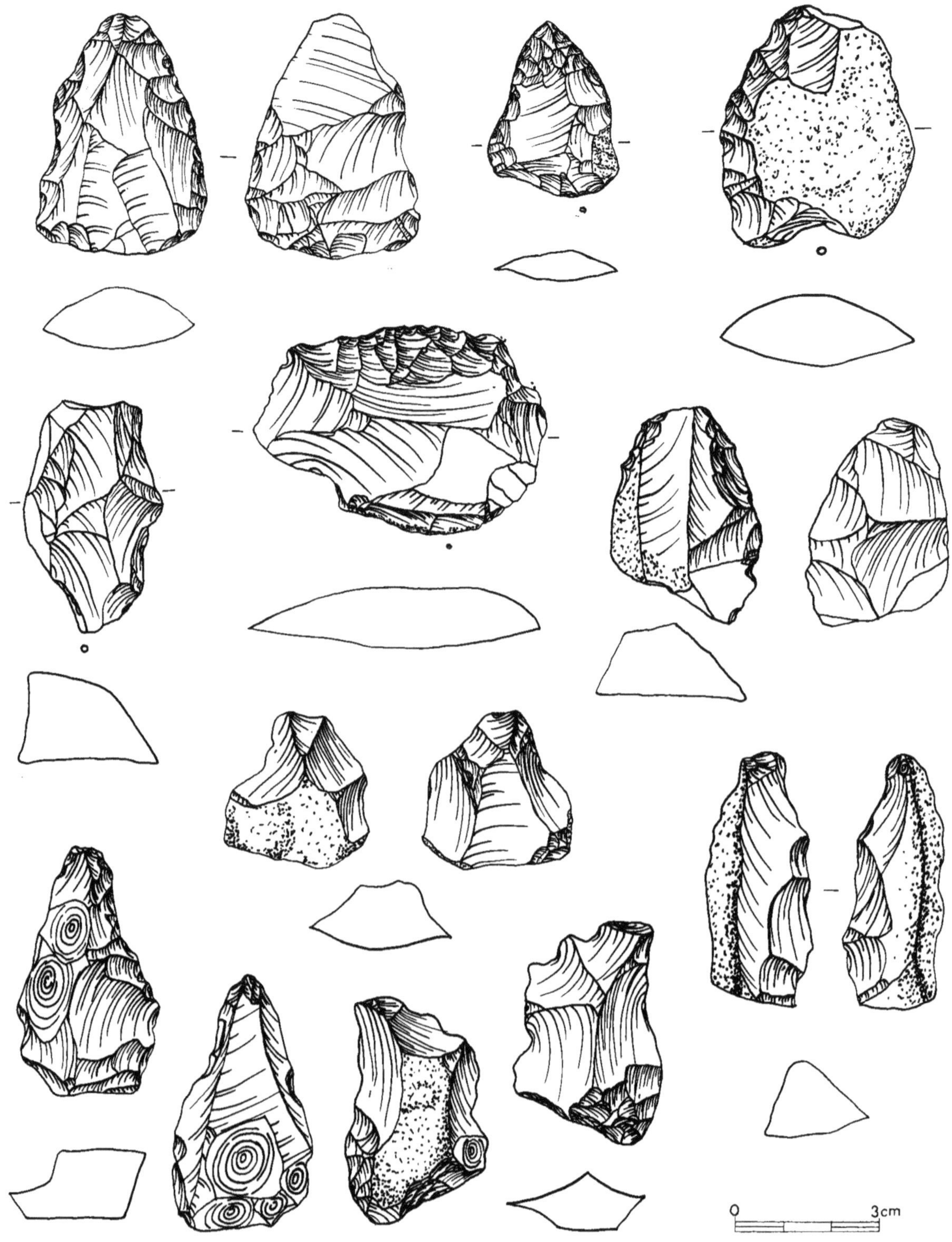

Fig. 9. Kupařovice II, silex corné de Krumlovsky les.

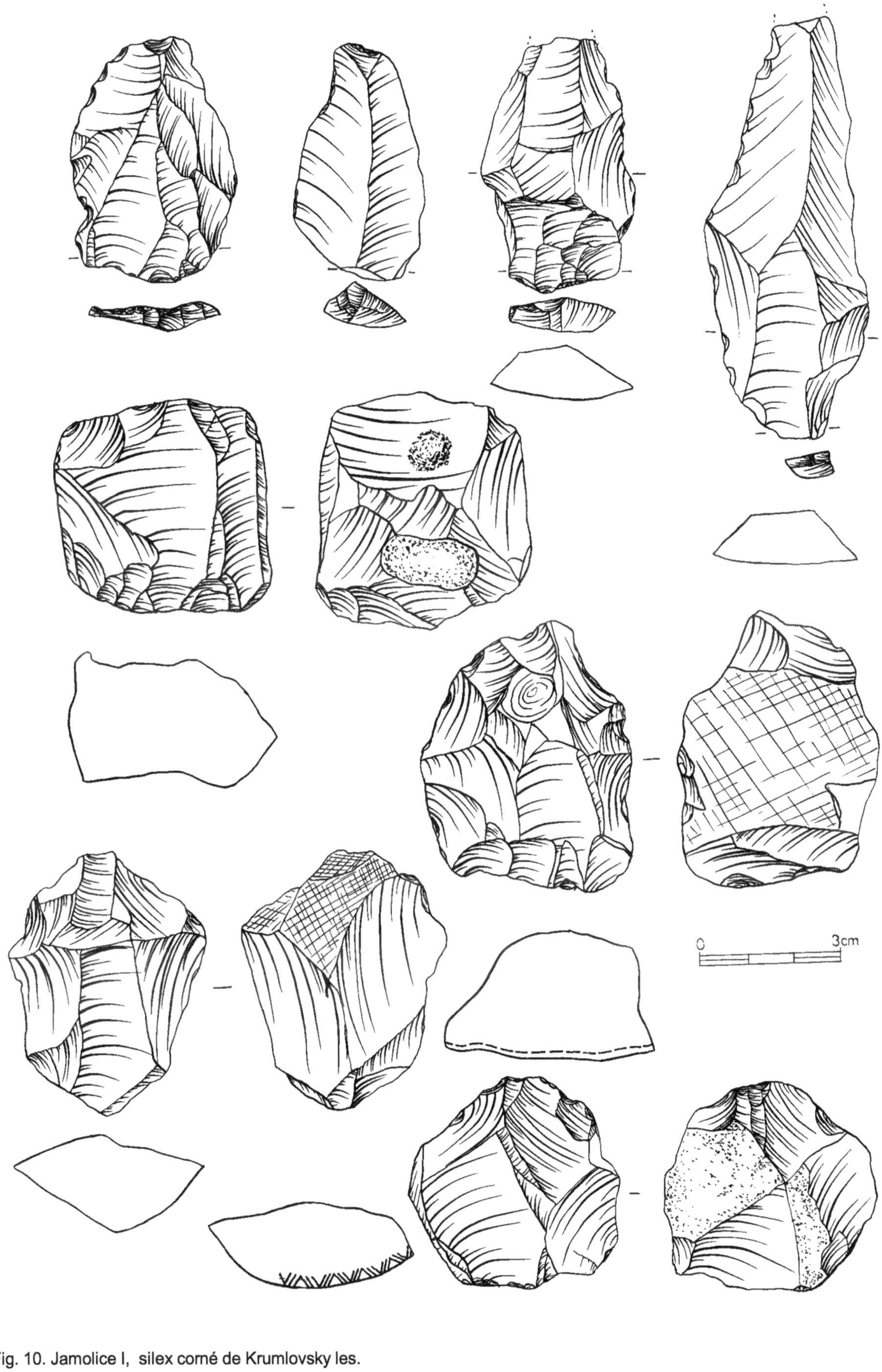

Fig. 10. Jamolice I, silex corné de Krumlovsky les.

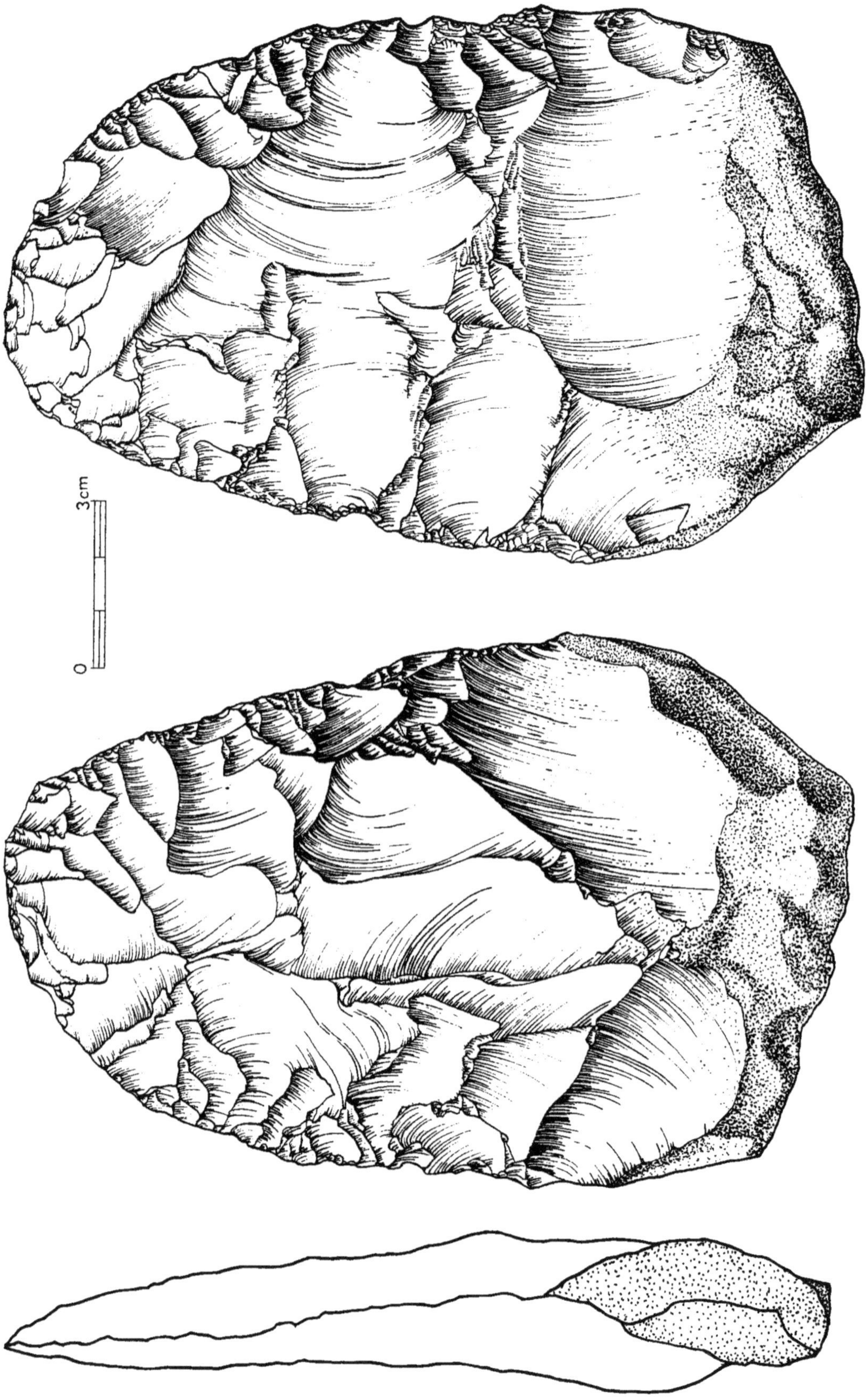

Fig. 11. Karolín I, radiolarite brune

Bibliographie

Boëda, E. 1995: Caractéristiques techniques des chaînes opératoires lithiques des niveaux micoquiens de Kůlna (Tchécoslovaquie). *Paléo,* supplément No.1, 7-72.

Klima, B.- Musil, R.- Jelinek, J.- Pelišek, J. 1961: Die Erforschung der Höhle Švéduv stůl 1953-1955. *Anthropos* N.S.5, Brno.

Oliva, M. 1981: Acheulian Finds from Karolin, distr. of Kromeřiž (Czechoslovakia). *Anthropologie* (Brno) 19, 27-32.

Oliva. M. 1987: Vyvinutý micoquien z návrši Horky u Bořitova. *Acta Musei Moraviae, sci.soc.* 72, 21-44.

Oliva, M. 1991: The Micoquian Open-Air Site of Ráječko I. The land use in the Moravian Middle Paleolithic. *Anthropologie* (Brno) 29, 45-61.

Oliva, M. 1993: Zahajeni výyzkumu paleolitické lokality Vedrovice Ia (okr. Znojmo). *Acta Musei Moraviae, sci. soc.* 78, 11-22.

Oliva, M. 1996: Prehistoric exploitation and utilisation of the Krumlovský les hornstone. In: J. Svoboda, ed., 49-66.

Oliva, M. - Štrof, A. 1985: Přehled paleolitického osidleni Lysické sniženiny a blizkého okoli. *Přehled vyzkumů AÚ ČSAV za rok 1983,* 10-17. Brno.

Ringer, 1983: *Babonyien, eine mittelpalaolithische Blattwerkzeugindustrie in Nordostungarn.* Diss. Arch. Serie II11, Budapest.

Schäfer, J. 1996: Die Wertschätzung aussergewöhnlicher Gegenstände (non-utilitarian objects) im Alt- und Mittelpaläolithikum. *Ethnogr.-Archäol.Z.* 36, 173-190.

Svoboda, J. - Macoun, J - Přichystal, A. 1991: Acheulian finds from Silesia. *Archeologické rozhledy* 43, 371-375.

Svoboda, J. - Škrdla, P. - Ložek, V. - Svobodova, H. - Frechen, M. 1996: Předmosti II, excavations 1989-1992. In: J. Svoboda, ed., 147-190.

Svoboda, J., ed. 1996: *Paleolithic in the Middle Danube Region.* Brno.

Valoch, K. 1965: Die Höhlen Šipka und Čertova dira u Štramberku. *Anthropos* N.S.9, Brno.

Valoch, K. 1971: Eine mittelpaläolithische Industrie von Maršovice in Südmähren. *Anthropologie* (Brno) 9, 29-47.

Valoch, K. 1980: Ein Faustkeil aus Mittelmähren. *Anthropologie* (Brno) 18, 287-289.

Valoch, K. 1988: Die Erforschung der Kůlna - Höhle 1961-1976. *Anthropos* N.S.16, Brno.

Vértes, L., ed. 1964: *Tata, eine mittelpaläolithische Travertinsiedlung in Ungarn.* Budapest.

Vértes, L. 1965: *Az oskokor és az atmenti kokor emlékei Magyarorszgon.* Budapest.

TOWARD MODERN HUMANS: YABRUDIAN AND MICOQUIAN, 400 – 50 kyears ago.
Edited by A. Ronen and M. Weinstein-Evron. BAR S850, 2000.

THE PALAEOLITHIC OF THE LATE MIDDLE PLEISTOCENE IN CENTRAL ASIA, 400 - 100 ka ago

Vadim A. Ranov and Joachim Schäfer

Introduction (J. Schäfer)

Tadjikistan is located in the middle part of central Asia. The mountain ridges of the Hissar-Karategin in the north, the Pamir in the east and the Hindukush in the south surround the so-called Afghano-Tadjikian Depression, which is open to the deserts to the west (Fig. 1). On one hand, then, it is an isolated area. On the other hand it is connecting different geographical regions: the west Eurasian, the Chinese-Siberian and the Indian. As a result of a deflation process which took place in the deserts and several river valleys, aeolian dust was transported to the east and sedimented as loess in the piedmont. These Upper- and Middle Pleistocene loess sediments had accumulated to form huge mantle layers more than 100 metres thick, the highest in the world. From the Matuyama/Brunhes palaeomagnetic boundary up to the Holocene, nine interglacial palaeosol complexes (PC), divided by loess and interstadial sediments, could be counted. Up to the present, 17 archaeological find horizons were distinguished, stratified between palaeosol complexes 8 and 1, spanning the time range between 700,000 and 70,000 years (Fig. 2).

The southern Tadjikian loess-sections enable the reconstruction of the 'cultural evolution' of hominids against the background of climatic and environmental changes during the last 700,000 years. Furthermore, the combination of the Tadjikian loess-chronostratigraphy, the embedded archaeological assemblages and the biostratigraphy enable the dating of many other central Asian sites, whose ages were uncertain. The first results show that the traditional view of the development of the central Asian Palaeolithic has to be changed. The beginning of the Middle Palaeolithic is much earlier than hitherto assumed, and the early appearance of Blade-Industries shed a new light on the Middle- to Upper Palaeolithic transition.

Loess-Palaeosol Stratigraphy (J. Schäfer)

Following TL-measurements by Shelkopljas (1977), the Tadjikian loess-chronostratigraphy had been first worked out by Dodonov (1986, 1991) and Lazarenko (1984). During the last years new investigations took place by several international and interdisciplinary work-groups. A new loess-chronostratigraphy has been established, which dramatically changes the traditional view. According to Dodonov's former view, PC 5 was correlated with the last interglacial. According to Lazarenko, it is PC 4 (on some outcrops PC 3) which had to be correlated with the last interglacial. Aside for some different opinions about the stratigraphical interpretation of some sections, there is a general agreement by the new work groups to correlate PC 1 with the last interglacial, Oxygen Isotope Stage 5 (Forster and Heller, 1994; Bronger et al., 1995; Schäfer et al., 1996b; Shackleton et al., 1995).

The first results were published by Forster and Heller (1994). On the basis of palaeomagnetic susceptibility measurements at the outcrop of Karamaydan, a correlation with the Chinese loess sequence and the oxygen isotope record had been presented. The Matuyama/Brunhes boundary was found between PC 9 and 10, confirming earlier results (Gamov and Penkov, 1977). Nine interglacial palaeosol-complexes have been identified and correlated with the oxygen isotope stages and the Chinese loess sequence of Xifeng: Stage 19 (Chinese S7) = Tadjik PC 9; Stage 18 (L7) = PC 8; Stage 17 (S6) = PC 7; Stage 15 (S5) = PC 6; Stage 13 (S5) = PC 5; Stage 11 (S4) = PC 4; Stage 9 (S3) = PC 3; Stage 7 (S2) = PC 2 and, finally, Stage 5 (S1) = PC 1. Shackleton et al. (1995) had also analysed the outcrop of Karamaydan and came to the same conclusions. Bronger et al. (1995) carried out pedological investigations of the Karamaydan section and the lower part of the Chashmanigar outcrop, concluding that PC 1 had to be correlated with the last interglacial.

Studying several loess-palaeosol outcrops in the south Tadjikian Khovaling district, the present authors came to a similar conclusion by correlating the loess-palaeosol-sequences with the isotope record and the Chinese loess-stratigraphy. An hypothesis was formulated in a first publication (Dodonov et al., 1995), subsequently verified in two additional studies (Schäfer et al., 1996a; 1996b). A new Tadjik loess-chronostratigraphy was presented on the basis of observations in the Khovaling district and their comparison with the Chinese Luochuan loess record, the deep-sea isotope record and the archaeological occurrences (Fig. 2).

The assemblages of the well known sites of Lachuti and Karatau had been dated in the past to 130,000 and 200,000 years, respectively. There has been some discussion about such a recent age for these two Lower Palaeolithic assemblages, because in other parts of the world the Lower Palaeolithic was pushed back by Middle Palaeolithic industries to prior than 200,000 years (we define Middle Palaeolithic on the basis of prepared core technology and true scraper typology). The doubt as to the age of these archaeological assemblages initiated a closer look at the Tadjikian loess stratigraphy. The loess-palaeosol sequences of the south-Tadjikian outcrops show some regularities. PC 5 and 6 were divided by only a thin loessic layer but they were separated from PC 4 and 7 by thicker loess sediments. Following observations by Kukla (1987) and Shackleton (1978), Oxygen Isotope Stage 14 was a milder glaciation whereas Oxygen Isotope Stages 12 and 16 were extremely cold. Consequently, we correlate Oxygen Isotope Stage 14 with the thin, often slightly weathered loess layer between PC 5 and 6 and the very thick layers above PC 5 and below PC 6, with Oxygen Isotope Stages 14 and 16 respectively (Dodonov et al., 1995). Comparing these observations with the Chinese loess stratigraphy, we correlate PC 5 and 6 with the Chinese S 5. We place the Tadjikian sequence from PC 7 downward to the Matuyama/Brunhes boundary, thus confirming the conclusions reached by Forster and Heller (1994) and by Shackleton et al. (1995).

Some additional observations on the Tadjik loess stratigraphy are relevant to the palaeosol complexes. PC 1, 2 and 3 are generally composed of two or three well-developed soils (Bt or Bm in Rosanov's (1883) nomenclature), divided by colder and/or drier events (loess, interstadial sediments, carbonated horizons). PC 4, on the other hand, was not subdivided, consisting of a single well-expressed soil (Schäfer et al., 1996a; 1996b). Looking at the oxygen isotope deep-sea record, there were some interesting similarities: Oxygen Isotope Stages 5, 7 and 9 show a higher differentiated curve while Oxygen Isotope Stage 11 shows only a simple oscillation (Fig. 2).

The new loess chronostratigraphy has far reaching consequences for the exact dating of both the old and the more recently discovered archaeological sites. The problem is not only the replacement of dates in the former loess stratigraphy by Dodonov or Lazerenko, but also the different interpretation offered by our work group. (Schäfer et al., 1996a; 1996b). In sum, we may point out now that the archaeological site Obi-Mazar is not to be placed in PC 3 (contra Lazarenko, 1984), and the youngest Lower Palaeolithic site of Tadjikistan is not embedded in this horizon, with an age estimate of about 300,000 years (Shackleton et al., 1995). Instead, Obi-Mazar has to be placed in PC 4 with an age of about 400,000 years. The question whether the assemblage of this site is Lower or Early Middle Palaeolithic will be discussed below. Our research also shows that the archaeological site of Lakhuti has to be placed within PC 5, about 500,000 years ago. Lakhuti is one interglacial older than Obi-Mazar (Schäfer et al., 1996a). Our recent work has confirmed the stratigrapy of Karamaydan site. The well-known site Karatau, on the other hand, cannot be correctly placed without additional geological investigations. Karatau was formerly thought to be stratified in PC 5 (Lazarenko, 1994), but the present authors will not exclude a position in PC 6.

The Lower Palaeolithic Industries
(V. A. Ranov)

The main region for studying the Lower Palaeolithic of central Asia is South Tadjikistan, where these industries are exposed in a good stratigraphical context. They are tied with the loess-palaeosol sequence and are therefore termed 'loess-Palaeolithic' sites. This loess-Palaeolithic sequence begins with the site of Kuldara which, although containing a few archaeological remains, is nevertheless very important for a geochronological view. Kuldara is located in the Khovaling region not far from the village Lakhuti (Fig. 1). The Lower Palaeolithic artefacts have been found at a depth of 120 m in the loess section, in pedocomplexes (PC) 11 and 12, within the Matuyama Chron. The generalized date of the site is 900 ka BP. As at most loess-Palaeolithic sites in South Tadjikistan, the Kuldara artefacts were found in the morphologically developed and intensive red coloured palaeosols, representing the climatic optimum. Of the 90 objects found at the site, 40 were without doubt artefacts including cores, flakes, slices (citron), perforators, scrapers, pebbles with scars, waste and debris. The artefacts are of a small size; only a quarter of them are in the size range of 4 cm or more. The Kuldara artefacts are typologically non-standardized. Cores are not systematically worked and regular retouch is absent. All the tools are of an individual form and even the best worked pieces were only partly retouched. The technique of knapping used is entirely within the framework of the pebble industry. Generally, the artefacts show a very archaic character (Ranov et al., 1987; Ranov, 1995).

In the Brunhes Chron, there is a series of archaeological finds in different PC and, very rarely, in different parts of the loess. There are only isolated finds in the lower soils, but in PC 6, 5 and 4 they were frequent and in good concentrations, so they could be considered as sites. Altogether, 32 find localities are known in South Tadjikistan. The main sites are Karatau, Lakhuti, Obi-Mazar and Chonako III.

The Karatau site is situated on the loess covered watershed of the Javanien Karatau ridge, 54 km southeast of Dushanbe (Fig. 1). The site is 1,700 m amsl and 1,100 m above the Vaksh river. It is located at a depth of 64 m in the section, probably in PC 6. The artefacts are distributed in the soil of the climatic optimum, which is about 1.5 m thick. In the sites of the loess-Palaeolithic there are not always clear cultural horizons, but some concentrations of stone knapping suggest a repeated occupation by early humans (Lomov and Ranov, 1985). Most of the artefacts are concentrated in a horizon between 20 and 80 cm above the carbonate crust and, more frequently, between 20 and 40 cm. Such concentrations must reflect a more intensive occupation of the site precisely during the formation of this part of the PC.

In Karatau 931 pieces have been found, 616 of which are certainly artefacts. Among them it is possible to define debris with scars, flakes, wedges and slices (citrons), choppers, scrapers and side scrapers, denticulates, notched tools, points, unifaces, blades, pebbles and hammerstones. Very fragile, low quality, magnetic rock served as the primary source of raw material. Silicified limestone, schist and quartzite were of still lower quality. Only one fifth of the artefacts were made of flint, which is very fragile and pitted. The assemblage can generally be described as a pebble industry, because there were true pebble tools (3.2%) and a true pebble technique was used for stone debitage. Blades form only 0.2%. Prepared cores and appropriate pre-forms of flakes and blades, which could have been retouched into various tool types, are completely absent. Bifacial tools are absent too. The presence of slices and wedges (7.7%) is evidence for the use of a particular stone-knapping technique. The numerous primary and semi-primary flakes result from knapping the working edges of choppers, or from the use of pebble cores. Crushing technique was frequently used. Other cores, beside pebble-types, are not numerous – there were only some of globular and multiplatform types, less than one per cent of the assemblage. True platform cores are represented by only a few examples. Eighteen percent of the flakes are between 2 and 3 cm while 30% are larger than 5 cm. At the same time some very large flakes, 10 cm or more, were produced, some of which are reminiscent of the Clactonian.

Tools with secondary retouch form 16.8% of all artefacts. Scraper-like tools dominate. Only a few of these have good working edges and these are mostly side scrapers. They were worked out from flakes of irregular shape. There is a special unifacial type of tool, with a crude elongated triangular outline, worked only on one side. The choppers can be divided into groups, including typical choppers made from pebbles where the working edge encompassed the entire thickness of the convex-flat edge. There are pieces which resemble a primitive core-like ('carinated') scraper. Some artefacts can be compared with Tayacian points. There are also borers, hollow scrapers, notched tools and knives. Many flakes have accidental breakage resembling 'Siret' burins. The Karatau industry can be considered as pre-Mousterian, or 'Acheulean without bifaces'. The pebble technique is dominant, a prepared core reduction practically absent, and there is no standardized typology. Prehistoric humans living in Karatau did not use the Levallois technique, but there are examples for the use of the Tayacian-Clactonian method of stone working. The Karatau pebble-culture is certainly different from contemporaneous Acheulean bifacial cultures of Europe and the Near East (Ranov, 1995).

The Lakhuti site is situated on the right bank of the Obi-Mazar river in the Khovaling district (Fig. 1). The elevation of the valley is 1400 m, and the 5 PC is 65 m above the river bed with an overlying loess cover 45 m thick. As in all the sites excavated in the palaeosols of South Tadjikistan, here too the artefacts are in a 'suspended' position in the soil of the climatic optimum; living floors can only be statistically reconstructed. As in Karatau, there is no hearth or kitchen debris. The raw material is quite variable and includes, among others, quartzite, cornelian, cornestone and chert. There are 1047 finds, separated into: 388 artefacts, 399 debris, 260 manuports. Among these were flakes, wedges and slices, scrapers, rabots, Tayacian points, chopper tools and cores. Only six blades (1.5%) were present. Generally, the technique used here is the same as that used at Karatau. The main artefacts are pebble cores and the main products were flakes. However, there is also a small group of platform cores which can be compared with Levallois cores. Platforms are covered by cortex with a few dihedrals, but no faceted ones. Flakes with a length range between 2.6 and 5.0 cm are the most frequent. Slices and wedges are more frequent here than at Karatau. The quantity of choppers is close to that of Karatau, but they are subdivided into different categories. Some of the blades, and rarely flakes, could be described as Levallois.

The retouched tools are non-standardized, with an absence of recurring forms. The denticulate-notched group and fine marginal retouched tools dominate over the Mousterian regular step-retouch. The main tools are scraper-like instruments with an irregular retouch. There are also real side scrapers with well retouched edges, but those tools are rare. There are some single interesting tools: a side scraper or knife on a large flake with fine subparallel retouch; a side scraper resembling 'La Quina' type; a unifacial point. In sum, the industry of Lakhuti can be described as a pebble industry with Clactonian-Tayacian techno-typological elements and very faint Levallois technology features. The Lakhuti industry seems to be more developed in comparison with Karatau: The appearance of Levallois technology, different forms of retouch, a decrease in the dimension of choppers and cores, and an increased quantity of blades and scrapers. On the basis of these techno-typological distinctions it is possible that the two sites have an evolutionary relationship (Dodonov et al., 1995; Ranov, 1995). The next stage in the development of the original Karatau Culture is the Obi-Mazar site.

Obi-Mazar – the Late Lower Palaeolithic and the Transition to Middle Palaeolithic in Central Asia

(V. A. Ranov)

The Obi-Mazar site, located 600 m southwest of the Lakhuti site, is correlated with the 4 PC which is some 80 m above the river level and 56 m below the top of the high Obi-Mazar section (Fig. 3). The 4 PC, as emphasized above, is to be correlated with Oxygen Isotope Stage 11, i.e., about 400,000 years ago. This is the time range of the beginning of the Yabrudian culture in the Levant and the end of the Middle Acheulean in Europe. However, the industry found in this PC is very different from both the Levant and Europe. It is quite similar to the Karatau and Lakhuti complexes and, although its technical features and typological background are not so clear as those of the two other industries mentioned, it is possible to say that the Obi-Mazar industry represents a more developed stage in the evolution of the Karatau Culture.

The excavation of the PC 4 assemblage had taken place with a break of 11 years, between 1984, by Ranov and Voloshin and 1995, by Ranov and Schäfer. The results of the first excavation were published by Ranov and Amosova (1993) and Ranov (1995). The first excavation covered an area 12 m long with a maximum penetration of 1.2 m into the soil. There were also 10 soundings (each 2-3 m large and 0.5 m deep), showing that the artefacts in this horizon were distributed over a distance of 40 m. Cleaning the section in 1995 showed that artefacts were actually spread over a distance of 70 m. They were embedded in the soil of the climatic optimum, mainly at a height of 1 m above the calcitic crust.

Among the 219 finds of the 1994 excavation, 75% were debris and unmodified pebbles. Twenty-five percent of the finds show traces of artificial knapping, consisting of 5 cores and core-like pieces (2.3%), 33 flakes (15%), 9 citrons (4.1%) and 7 pieces provisionally identified as tools (3.2%). The largest category of finds were debris, a good number of them have an accidental character and look like gravels, but others have conchoidal fractures and can be regarded as the debris of debitage. The slices (citrons) are small and not as typical as in the assemblages of Karatu and Lachuti. The largest flakes are up to 8 cm long, the smaller ones are around 3 cm. Most of the flakes show a flat or cortical butt and only a few demonstrate dihedral or faceted butt. The dorsal parts of the flakes have no regular faceted surfaces. Some implements look like side scrapers, but without a regular retouch. There are also badly prepared notched tools. The cores found in 1984 were very crude with no prepared platforms and no systematic scars.

As a result of the 1984 season, the Obi-Mazar collection appears to represent a decline relative to Lakhuti; as if the technology took a step backwards, but still within the framework of the Karatau Culture tradition and not a Mousterian one. The explanation we offer is that the Obi-Mazar PC 4 assemblage is not a real occupation site, but part of a large workshop. The new work done by Ranov shows that after 11 years it was impossible to find the exact locality of the former excavations. The new excavation was placed, as the former one, in a south-west orientation. It contains eight small units with a total of 30 square metres. The dimensions of the units are as follows:

unit 1-2:	3.6 x 1.2m = 4.32 m^2
unit 3:	2.4 x 1.0m = 2.42 m^2
unit 4:	5.3 x 1.6m = 13.4 m^2
unit 5:	2.2 x 1.4m = 3.1 m^2
unit 6:	1.8 x 1.2m = 2.2 m^2
unit 7:	3.5 x 1.4m = 4.9 m^2
unit 8:	2.5 x 1.4m = 3.7 m^2
TOTAL:	29.8 m^2

The excavated area includes only the main horizon of the artefact concentration. In excavation units 1 and 2, this horizon was distinguished 78 cm above the carbonate crust. In unit 3, it was 74 cm above the crust; in unit 4, 100 cm; in unit 5, 96 cm; in unit 6, 73 cm; in unit 7, 80 cm and in excavation unit 8 it was 82 cm above the crust. Thus, the main concentration of artefacts is between 75 - 100cm above the carbonate crust. The variation of the position of artefacts is shown in Fig. 4. Isolated finds outside the excavated areas show that artefacts were also spread outside the concentrated layer, from the base of the carbonate crust to the top of the soil of the climatic optimum. Only a few artefacts were found in the overlying (interstadial) soil, transitional to the loess.

As always in sites of the loess-Palaeolithic in South Tadjikistan, this time too we did not find the traits of normal cultural layers. The artefacts were embedded in the soil of the climatic optimum, the genesis of which was often explained as a synsedimentary formation with a simultaneous artefact accumulation. In this case, the presence of prehistoric people is documented several times at the same place (Lomov and Ranov, 1985). However, it is quite possible to explain this situation from the point of view of an epigenetic soil formation process (Ranov and Amosova, 1990, 166-168; Dodonov et al., 1995).

The raw material is the same as at Lakhuti, but the percentage of items made of bad-quality metamorphic and sedimentary rocks is higher. There are 1122 finds in the 1995 excavation. They consist of 72% debris and manuports, and only 28% regular artefacts. It is a lower percentage than at Karatau and Lakhuti. The finds were counted as follows: debris (750) = 66.84%, flakes (206) =18.36%, blades (7) = 0.62%, slices (15) = 1.33%, wedges (19) = 1.69%, cores and core-like pieces (26) = 2.31%, pebbles with scars (9) = 0.80%, manuports (60) = 5.34% and modified tools (30) = 2.67%. The diagram of this distribution compared to that of Lakhuti was published earlier (Schäfer et al., 1996b, 104). The modified tools are composed of notches (11) = 36. 66%, scrapers and scraper-like objects (4) = 13.33%, denticulates (4) = 13.33%, carinated scrapers (2) = 6.66%, notch-denticulates (2) = 6.66%, raclets (2)= 6.66%, burin (1) = 3.33%, bifacial tool ? (1) = 3.33%, point ? (1) = 3.33%, chisel (1) = 3.33%, and a chopper (1) = 3.33%.

All tools are atypical and hard to classify. Nearly all show individual characteristics. Typical Middle Palaeolithic tools such as Quina scrapers or Mousterian and Levallois points are absent. In principle, the technology of the PC 4 assemblage is still a Lower Palaeolihic technology with pebbles, and cores and flakes extracted from the pebbles. Most of the butts (N = 104) are covered by cortex or are smooth. Diedhrals count for only 5.76%, and curved platforms count for 9.92%. Faceted butts are completely absent. In 15. 38% of the cases, the dorsal face is covered, wholly or partly, by cortex. Twenty five percent of the blanks have sub-parallel scars on the dorsal face, while the others show a centripetal or a chaotic pattern of dorsal scars. Between the blades there were none of a regular form. They are mostly flake-blades rather than real blades (Fig. 5).

However, there are in the industry of PC 4 some features which make it different from the Lower Palaeolithic industries previously mentioned. There are cores with a diameter of only 6 - 4 cm. The 22 cores fall into the following size groups: 6.9 - 6.0 cm (2) = 9.09%; 6.0 - 5.0 cm (7) = 31.81%; 5.0 - 4.0 cm (7) = 31.81%; 4.0 - 3.0 cm (6) = 27.27%. Their forms also differ, including globular, sub-prismatic, oval, discoidal, triangular, rectangular and a type of 'boat-shape'. Some cores could be named 'platform cores'. It is possible to say that generally, the cores were pre-determined for the preparation of very short blanks, not exceeding 2 - 3 cm in length. The main technique can be termed centripetal (Fig. 6).

A closer analogy to these types of cores is to be found in the Mousterian site Kara-Bura, located in the Vakhsh valley about 160 km from Obi-Mazar (Ranov, 1965, 70-71). Without doubt we can view in these cores some Middle Palaeolithic technological features, dated by comparison to Oxygen Isotope Stage 11, 390 - 410 ka BP. Hence, Middle Palaeolithic technological affinities in South Tadjikistan are among the earliest in the world. It is possible to consider the PC 4 assemblage as a transitional industry within an overall Lower Palaeolithic technology, with Middle Palaeolithic elements more frequent and better represented than in the preceding Lachuti industry of Oxygen Isotope Stage 13, ca. 500 ka ago.

Khonako III and Middle Palaeolithic Variability in Central Asia
(J. Schäfer)

The huge outcrop of Khonako extends from the mountain ridge between the valleys of the Jakhsu and Obi-Mazar rivers down to the Obi-Mazar valley (Fig. 1). Khonako III is situated between Khonako I and Khonako II, nearly on top of the watershed, on the road from the town of Khovaling to the village Shugnou.

Khonako III was known by surface collections made by Ranov. There were a few artefacts, especially Levallois flakes and elongated thin blades, with no stratigraphical context. In 1994, the first archaeological excavations and geological investigations took place by the present authors, together with P. M. Sosin and Sh. Sharapov. Archaeological remains have been found in PC 2 and 4. In the following years until 1996, additional archaeological finds became known from PC 3, from several horizons in PC 2, from interstadial and loessic sediments between PC 2 and 1, and from the upper part of PC 1 (or perhaps a moderately weathered soil above PC 1). These archaeological occurrences, in excellent stratified geological context and accompanied by molluscs, splinters of charcoal and bones (teeth) of large and small mammals, make this outcrop exceptional among the Tadjikian loess sections (Fig. 7). The climatic and environmental changes from the base of palaeosol 2 up to the top palaeosol 1 could thus be studied.

In a small sounding in PC 4 (Oxygen Isotope Stage 11, 400 ka BP) the horizontal distribution of artefacts could be analysed. Because of the small number of artefacts, only some preliminary results are given here. The finds include (excluding debris smaller than 1.5 cm): 3.7% blades (one bladelet), 29.6% flakes, 44.5% angular debris, 3.7% citrus flakes, 7.4% pebbles with scars, 7.4% cores and 3.7% manuports (unmodified pebbles). Among the flakes is one Levallois flake; the only modified artefact is a pebble fragment (or core) with a marginal retouch on one lateral edge. One piece of angular debris show traces of fire. The high variability of raw materials suggests that most of the finds have been carried to the site independently. When compared to the Obi-Mazar PC 4 and Lachuti PC 5 assemblages (Schäfer et al., 1996b), the assemblage of Khanako III PC 4 is different, though closer to the site of the same stratigraphic context, Obi-Mazar PC 4. In contrast to Lachuti PC 5, both assemblages of the PC 4 have a high frequency of preparation flakes and angular debris and a low frequency of citron flakes, pebble tools (pebbles with scars, pebble fragments), manuports and retouched artefacts. The horizontal distribution of Khanako III PC 4 differs from that of Obi-Mazar PC 4, with the finds less concentrated and spread over a larger distance. Future excavations in PC 4 of Khanako III, and a larger sample, will show if the similarities and dissimilarities to Obi-Mazar PC 4 and Lachuti PC 5 depended on environmental conditions or were of 'cultural' aspect, perhaps 'developmental stages'.

In PC 3 (Oxygen Isotope Stage 8, 300 ka BP) only very few artefacts have been found to date. Hence, we can not form a clear idea of the relations between the late Lower Palaeolithic (or early Middle Palaeolithic) of PC 4 and the Middle Palaeolithic Blade Industries of PC 2. Future research will show whether PC 3 can be linked with the Levallois surface finds previously collected by Ranov.

PC 2 (Oxygen Isotope Stage 7, 220 ka BP) could be divided into 3 well-weathered cinnamon soils (Bmt- and Bt horizons), which were separated by loess (L, Lca), interstadial sediments (LB, B, Bca) or carbonate crusts (Sca). The finds from PC 2 were spread over the whole sequence from the base of the lower PC 3 up to the loess above the PC 1 cinnamon soil. However, the finds also show some horizontal concentrations within the soils (Bmt- and Bm- horizons) and the interstadial sediments. The assemblages can be characterized by a high percentage of blades, especially broken blades, accompanied by some Levallois flakes. The tools consist mainly of lightly or steeply retouched blades. The only conventional Middle Palaeolithic tool is a Mousterian point (convergent side-scraper?). The few cores were heavily reduced and they do not mirror the initial flaking pattern.

Divided by 1 metre of loess (L), PC 2 is overlain by interstadial sediments (LI 2b, Oxygen Isotope Stage 6 or the end of Oxygen Isotope Stage 7, with an age around 190 ka BP). In a colder - or drier - stratigraphic position (LB), just following the climatic optimum (B) of this interstadial, a concentration of artefacts had been found. The most important were three pyramidal blade cores of a similar type, showing a high standardization of a core reduction strategy (Fig. 8). The few tools found in this horizon were comparable to those of PC 2. The cores remind us of those of the Upper Palaeolithic and their high standardization suggests that they were not 'accidentally' produced.

About 1 metre below PC 1 there is a weakly developed interstadial (LI 2a, Oxygen Isotope Stage 6, about 150 ka BP). A small area with exposed artefacts could be excavated here. All the finds are very small preparation flakes and debris smaller than 3 cm, with only one exception of a flake made of some unique raw material. In all, there are 80 pieces of debris smaller than 1.5 cm, 6 pieces of angular debris, 7 flakes or flake fragments and 9 blades/bladelets/fragments. This knapping area reminds more of an Epi-Palaeolithic than a Middle Palaeolithic core reduction. We assume that a core rejuvenation, and/or the production of gracile artefacts, could have taken place here, but verification would require a detailed refitting.

PC 1 (Oxygen Isotope Stage 5a - 5e, about 80 -130 ka ago) can be divided into 3 well-weathered soils, if we count the upper soil as part of PC 1 and not as a separate interstadial of the last glaciation. This upper soil (Bm), to be discussed below, is separated by two metres of loess from the underlying well-developed soil 2 (Bmt), which is separated, in turn, by a carbonate layer (Bca) from soil 3 (Bmt). In the upper part of soil 2, a lens of burnt red loess and charcoal was found. In a first publication (Schemer et al., 1996b), this lens has been interpreted as a fireplace. Further observations have shown, however, that there were many lenses of dark brown sediments and charcoal, similar to the first one. They were all situated in the same stratigraphic position at the Khonako III outcrop, and furthermore, similar features have been identified a few kilometers downslope towards the Obi-Mazar valley at Khonako IV. We now assume that these lenses are relics of one or more natural fires, which resemble man-made fireplaces. In the uppermost soil of PC 1 and the adjacent overlying sediments, some scattered artefacts have been found which do not form clear horizons. These few types of tools are common to the Middle- and Upper Palaeolithic of central Asia.

In Khanako IV, some artefacts have been found in the upper soil of a double soil complex. Following geomorphological investigations, we assume that the finds were located in the well-developed soil 2 of PC 1 (Oxygen Isotope Stage 5, about 110 ka BP). The artefacts can easily be characterized as Mousterian with heavy, crude Levallois products. The stratigraphic setting indicates that this industry is younger than the gracile Blade Assemblages from Khanako III - LI 2b and PC 2.

The results of the Khonako excavations give an idea of Middle Palaeolithic assemblage variability in the last 250 ka. A Blade Industry based on pyramidal/prismatic core reduction and laterally retouched blades is overlain by a Levallois-Mousterian. A variety of laterally retouched blades is common in the Middle Palaeolithic of central Asia. These tool types were sometimes assumed to be late Middle Palaeolithic, or even transitional industries. The Mousterian of Ogzi-kichik, Khudji, Obi-Rakhmat and Teshik Tash also have such types (Ranov, 1988). This author will not exclude the possibility that some tools from Warwarsi, published as heavily reduced scrapers (Dibble and Holdoway, 1993), may also be identified as laterally retouched blades. There are also similarities with the Pre-Aurignacian of Yabrud (Rust's collection in Köln). Such tool types were also part of Middle Palaeolithic Blade-assemblages in northwest Europe. Early non-Levallois, Middle Palaeolithic blade production is a well known phenomenon in northwest Europe, from the last interglacial or the beginning of the last glacial. Blade assemblages occur in Seclin, Rocourt, Saint Germain (Révillion and Tuffreau, 1994), Rheindahlen, Wallertheim and Tönchesberg (Conard et al., 1995). The Pre-Aurignacian/Amudian may even be older. At the site Obi-Rakhmat in Uzbekistan (Suleymanov, 1972), a long sequence of Blade Assemblages was overlain by Mousterian layers, as in Khanako. This means that non-Levallois blade technology and laterally retouched blades are not necessarily linked with a late Middle Palaeolithic, or even with a transitional culture; it has to be accepted as a variant within the range of Middle Palaeolithic variability.

Recent research in western Europe clearly shows that different Middle Palaeolithic cultures (for example, Ferrassie-type Mousterian) were not attached to any specific time period. Considering the assumption of a 'cultural evolution' visible in the techno-typological 'development' of artefact assemblages, one has to acknowledge that there is no constant progress during the Middle Palaeolithic period, from 250 to 50 ka ago. The pyramidal blade core from LI 2b of Khanako III is an additional example that there were innovations in the Middle Palaeolithic already similar to the early Upper Palaeolithic. The Middle Palaeolithic is a period of a high variability, with a lot of innovations concerning artefact production. In this respect, there is no clear boundary between the Middle and the early Upper Palaeolithic.

Conclusions (V. A. Ranov)

Some hypotheses were formulated prior to the excavations in Khonako III PC 2 and LI 2b. In Ranov's opinion, the Mousterian could have originated from PC 4 or PC 3 while an Upper Palaeolithic was believed to appear in PC 2 (Lazarenko, 1996, 59). This was supported by surface finds of Mousterian blades, flakes, cores and a few retouched tools collected on the slope of Khonako and Obi-Mazar outcrops, and by TL dates which, at that time, were 90 - 70 ka BP for PC 4, and 40 ka BP for PC 2. It should be emphasized that the surface collection could be divided into two groups, a 'light' one with thin, elegant and sometimes pointed blades, and a 'heavy' group with thicker and larger pieces. The latter group, made on quartzite, shows similarities with Late Mousterian industries from Khudji and Djarkutan. The first group, made of diverse rocks, is similar to the Upper Palaeolithic blades of Shugnou (Ranov, 1965; 1973; Ranov and Amosova, 1984). These assumptions were not confirmed by the new excavations, because only the 'light' group and not the 'heavy' one has been found in Khanako III, PC 2. Despite the small number of modified tools, the Khanako assemblage can be characterized as Middle Palaeolithic rather than a real Mousterian. In Europe, the Mousterian now seems to be only one special version of the Middle Palaeolithic entity, which has developed in the Würm period. We may have the same situation in central Asia: the well known Mousterian sites such as Teshik-Tash, Khudji or Ogzi-Kichik, may also be special, Upper Pleistocene, 'Würmian' cultures (Ranov, 1988). These sites have very few and no concordant absolute dates: Khudji is ^{14}C dated to 38.9 ka and Ogzi-Kichik to < 30 ka BP (Ranov and Amosova, 1984). The ^{234}U date for Teshik-Tash is 44 ka and for Obi-Rakhmat, 70 and 170 ka BP (Cherdynzev, 1969). On the other hand, the lower limit of the Middle Palaeolithic in Europe is now placed around 200, even 300 ka ago (Leroi-Gourhan, 1988). In the Near East, where 'Mousterian' is synonymous to 'Middle Palaeolithic', the beginning is placed around 170 or even 250 ka BP (Sanlaville, 1994). If the position of PC 2 of Khonako on the oxygen isotope curve, as mentioned above, is accepted, then we have in central Asia the same chronological and cultural sequence as in Europe and the Levant.

Acknowledgments

We use the local (Russian) pedological system after B. G. Rosanov (1983). The pedological data are based on research by P. M. Sosin. We are grateful for his permission to use partly unpublished results from the Obi-Mazar and Khonako sections.

We thank the Deutsche Forschungsgemeinschaft (DFG) and the Deutsches Archäologisches Institut (DAI) for financial support.

Authors' addresses:
V. A. Ranov, Institute of History, Archaeology and Ethnography, Dushanbe, 734025 Tadjikistan.
ranov@ac.tajik.net

J. Schäfer, Humboldt University, Berlin, Germany.

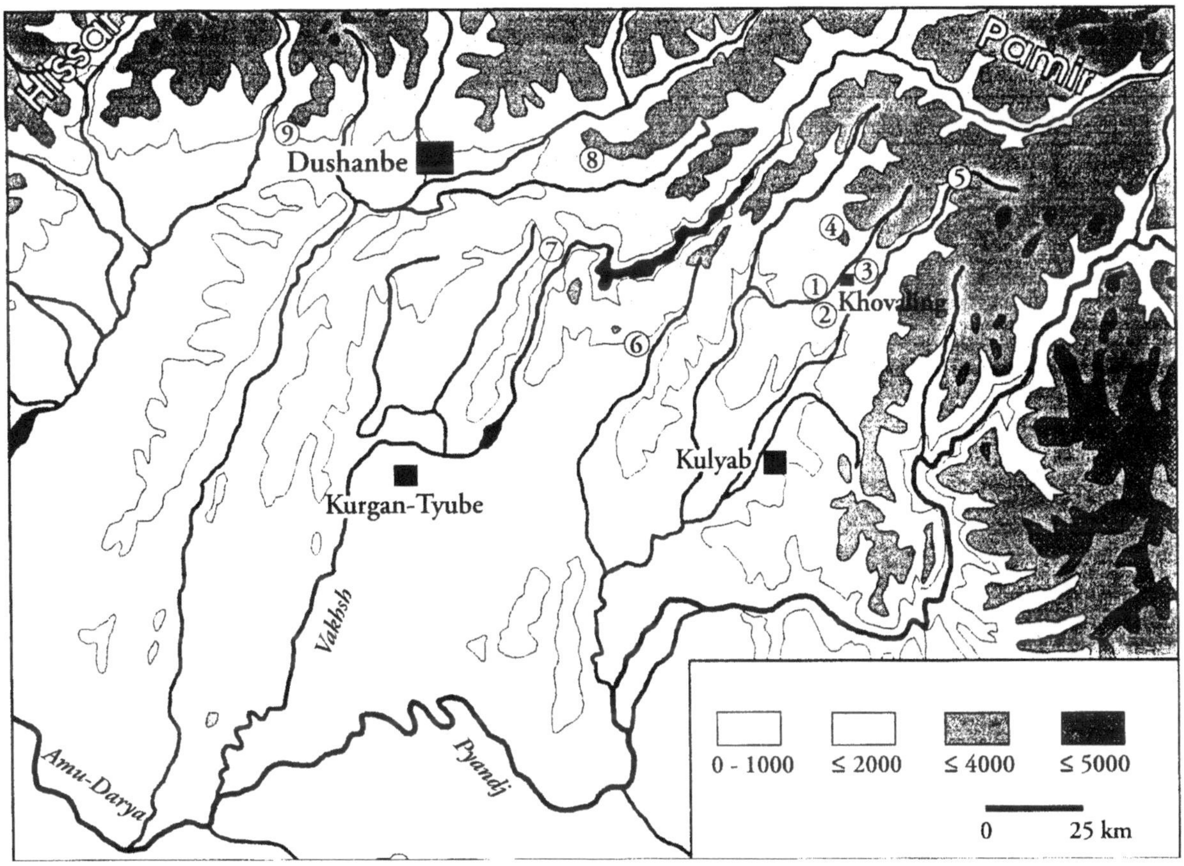

Fig. 1. South-Tadjikistan. 1 - Lakhuti/Obi-Mazar; 2 - Kuldara; 3 - Khonako; 4 - Chashmanigar; 5 - Shugnou; 6 - Ogzi-kichik; 7 - Karatau; 8 - Karamaydan; 9 - Khudji.

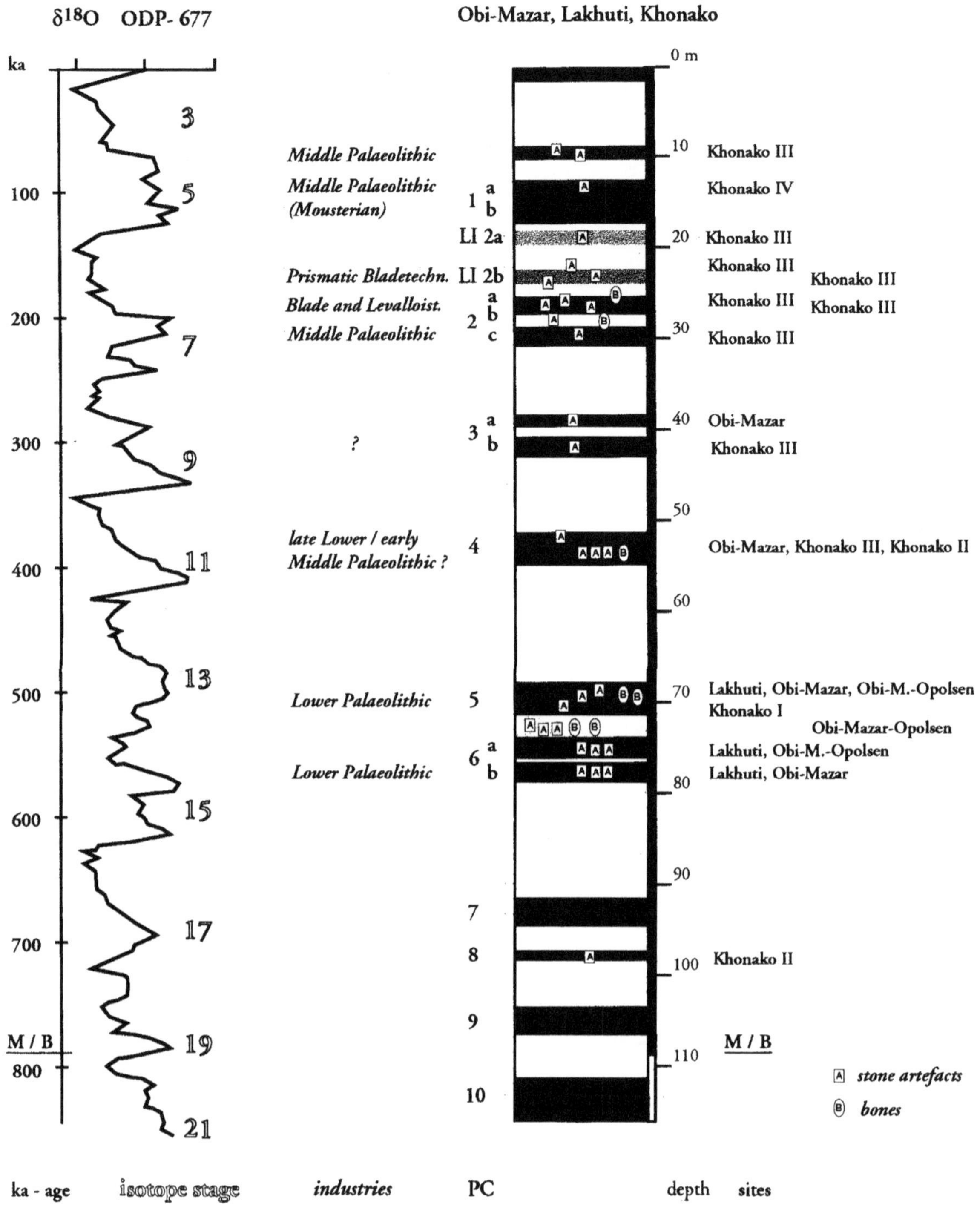

Fig. 2. The Tadjik loess-chronostratigraphy (modified after Schäfer et al., 1996b) and the oxygen isotope record (after Shackleton et al., 1990).

Fig. 3. The Obi-Mazar outcrop, palaeosols and the PC 4 excavation.

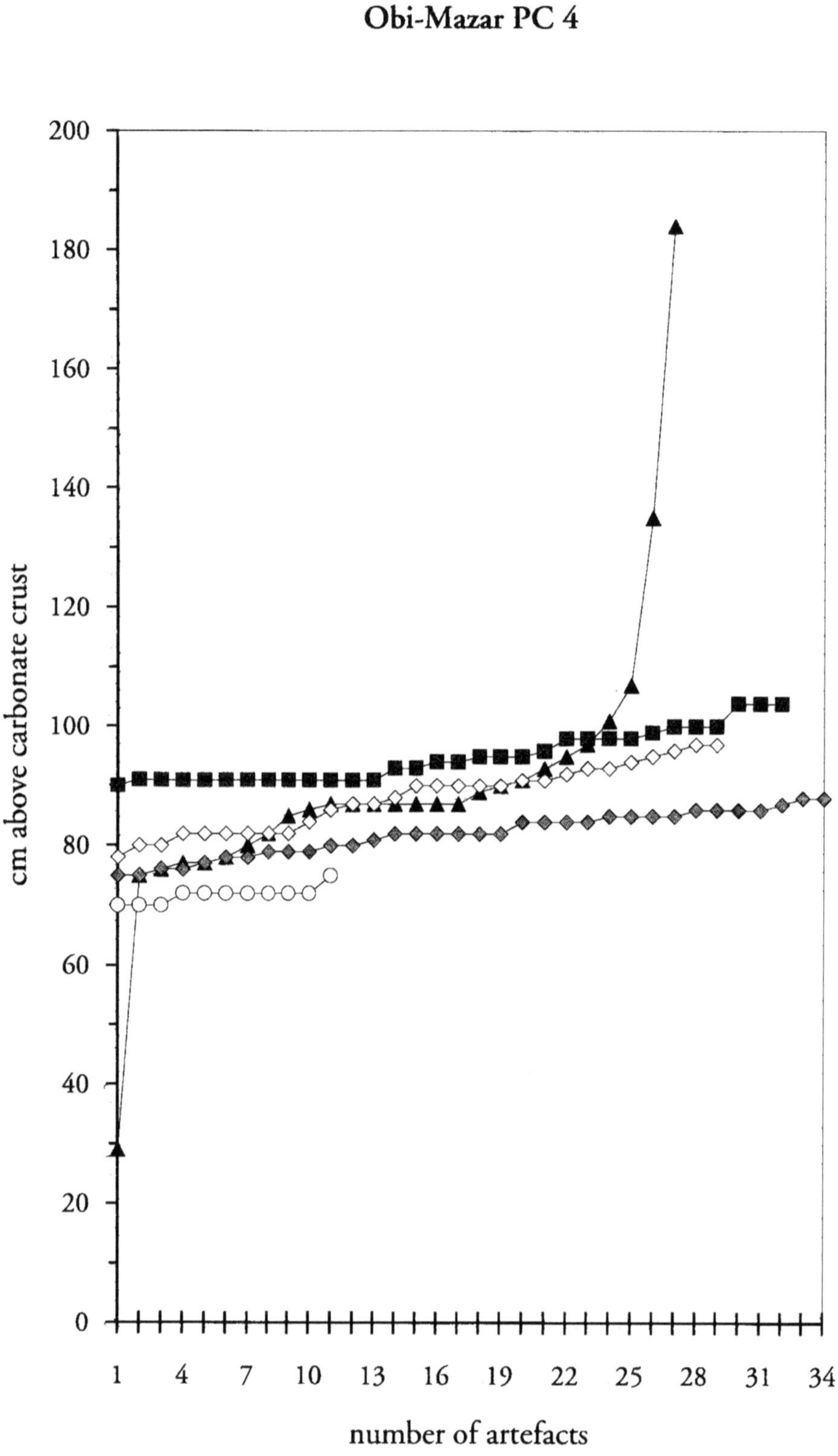

Fig. 4. Vertical distribution of artefacts in different excavation units of PC 4.

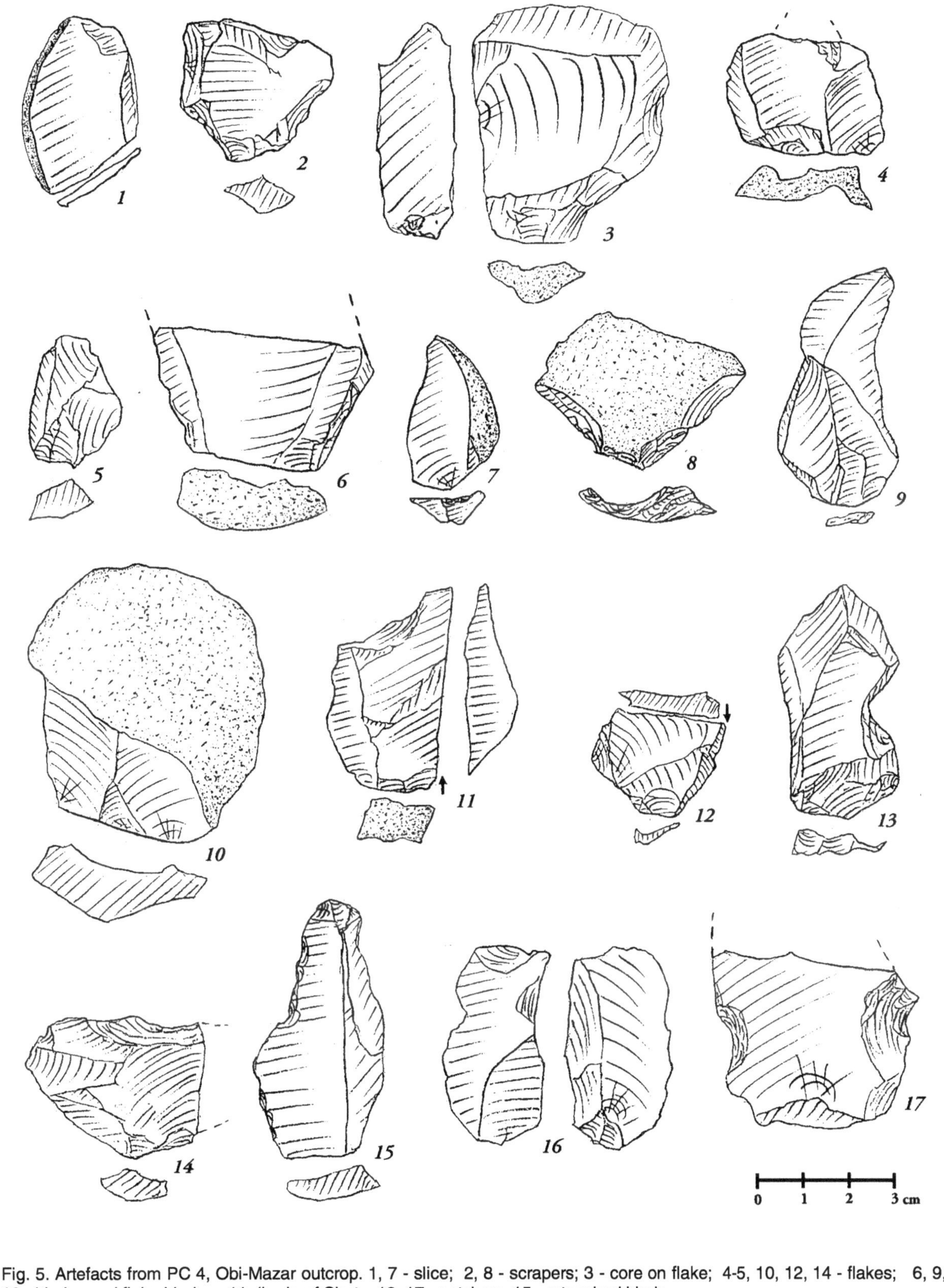

Fig. 5. Artefacts from PC 4, Obi-Mazar outcrop. 1, 7 - slice; 2, 8 - scrapers; 3 - core on flake; 4-5, 10, 12, 14 - flakes; 6, 9, 16 - blades and flake-blades; 11- 'burin of Siret; 13, 17 - notches; 15 - retouched blade.

Fig. 6. Artefacts from PC 4, Obi - Mazar outcrop. Cores: 1 - multidirectional (globular); 2 - shapeless; 3, 7 - single platform; 6 - double platform; 4, 8 - discoid; 5 - pyramidal (round, with a rocking-chair base); 9 - notched blade.

Fig. 7. The Khonako III outcrop - the excavation units of PC 3, PC 2, LI 2b, LI 2a and PC 1a.

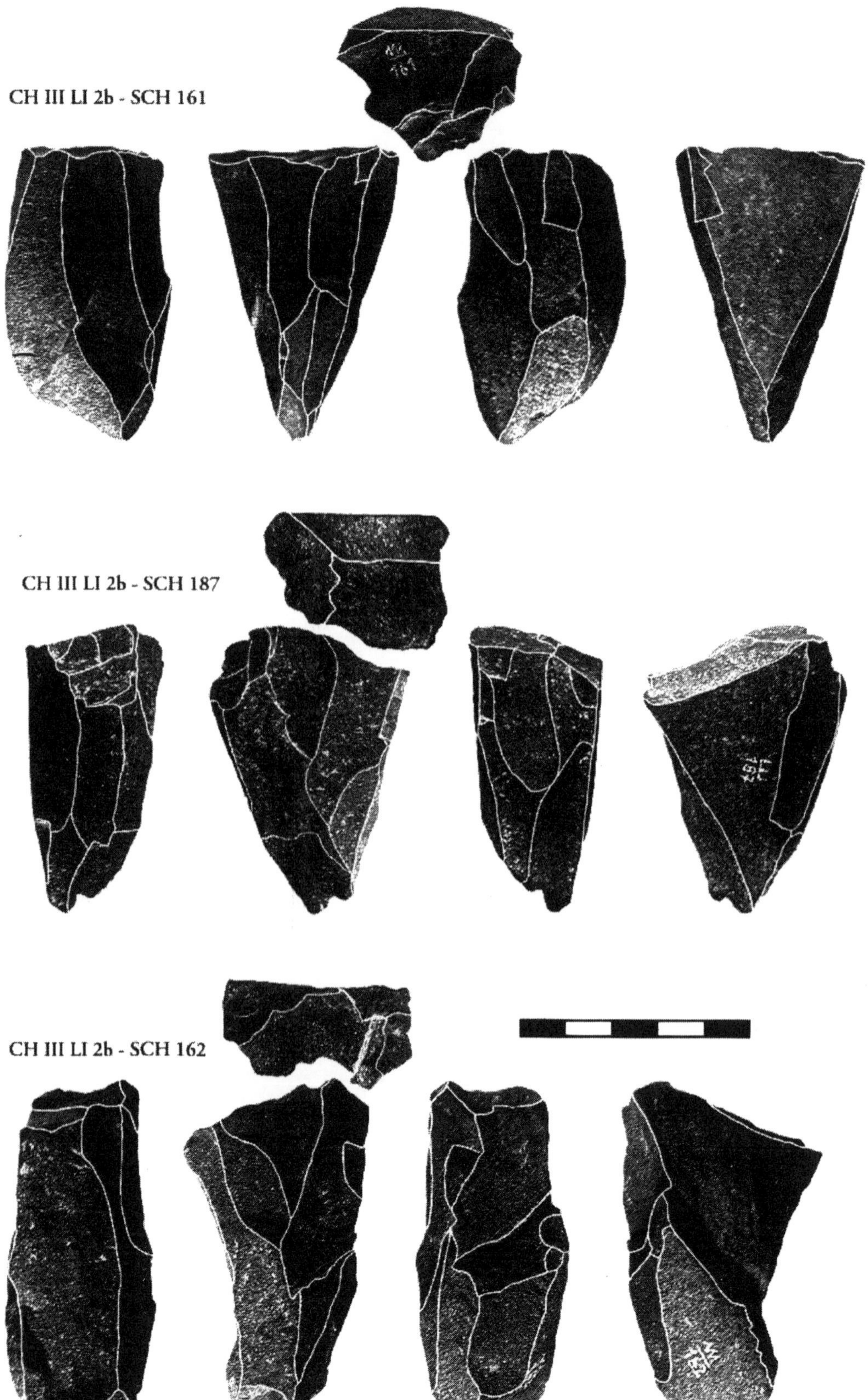

Fig. 8. Khonako III LI 2b artefacts, pyramidal cores.

Bibliography

Bronger, A., Winter, R., Derevyanko, O. and Aldag, S., 1995. Loess-palaeosol-sequences in Tadjikistan as a palaeoclimatic record of the Quaternary in central Asia. *Quaternary Proceedings* 4, 69-81.

Cherdynzev, V. V., 1969. *Uran-243 (Uranium-243).* Moscow (in Russian).

Conard, N. J., Adler, D. S., Forrest, D. T. and Kaszas, P. J., 1995. Preliminary archaeological results from the 1991-1993 excavations in Wallertheim. *Archäologisches Korrespondenzblatt* 25, 13-27.

Dibble, H. L. and Holdoway, S. L., 1993. The Middle Paleolithic Industries of Warwarsi. In: Olszewski, D. I. and Dibble, H. L. (eds.), *The Paleolithic Prehistory of the Zagros-Taurus.* University Museum, Pensylvania.

Dodonov, A. E., 1986. *Antropogen jujnogo Tadjikistana (Quaternary of the Southern Tajikistan).* Moscow (in Russian).

Dodonov, A. E., 1991. Loess of central Asia. *Geo-Journal* 24/2, 185-194.

Dodonov, A. E., Melamed, Ja. R. and Nikiforova, K. V. (eds.), 1977 *Granitsa neogena i chetvertichnoj sistemy (The Boundary between Neogene and Quaternary systems).* Moscow (in Russian).

Dodonov, A. E., Ranov, V. A. and Schäfer, J., 1995. Das Lösspaläolithikum am Obi-Mazar (Tadshikistan). *Jahrbuch des Römisch-Germanischen Zentral-museums Mainz* 39, 209-243.

Forster, Th. and Heller, F., 1994. Loess deposits from the Tajik depression (central Asia): magnetic properties and paleoclimate. *Earth and Planetary Science Letters* 128, 501-512.

Gamov, L. N. and Penkov, A. V., 1977. Paleomagnetizm verchnekajnozoiskikh otlojenij (Paleomagnetism of the Upper Cenozoic deposits). In: Dodonov, A. E., Melamed, Ja. R. and Nikiforova, K. V. (eds.), *Granica neogena i chetvertichnoj sistemy.* Moscow, pp. 15-17 (in Russian).

Kukla, G. J., 1987. Correlation of Chinese, European and American loess series with deep sea sediments. In: Liu Tungsheng (ed.), *Aspects of Loess Research,* pp. 27-38. (China Ocean Press, Bejing).

Lazarenko, A. A., 1984. The loess of central Asia. In: A. A. Velichko (ed.), *Late Quaternary Environments of the Soviet Union,* pp. 125-131. London.

Lazarenko, A. A., 1996. Osnovnie cherti paleo-ecologicheskoi modeli drevnego lessovogo paleolita Tsentral'noj Asii (Key paleoecological features of the oldest Palaeolithic in loesses of central Asia). *Drevnosti. Trudi izdavaemye Rossiiskim Arkheologicheskim Obchestvom* 19, 57-61 (in Russian).

Leroi-Gourhan, A., 1988. *Dictionnaire de la Préhistoire.* Paris.

Lomov, S. P. and Ranov, V. A., 1985. The pecularities of the pleistocene palaeosol formations and distribution of embedded Palaeolithic tools. *Current Trends in Geology* 6, 227-240. Climate and geology of Kashmir, New Delhi.

Ranov, V. A., 1965. *Kamennij vek Tadjikistana (The Stone Age of Tajikistan).* Dushanbe (in Russian).

Ranov, V. A., 1973. Shugnou - mnogosloinaja paleoliticheskaja stojanka v verkhovijakh r. Jakhsu (Shugnou - a mulitilayered Palaeolithic site in Upper Jakhsu river). *Arkheologicheskie Raboty v Tadjikistane* 10, 42-61 (in Russian).

Ranov, V. A., 1988. *Kamennij vek Jujnogo Tadjikistana i Pamira (Stone Age of Southern Tadjikistan and Pamir),* Novosibirsk (in Russian).

Ranov, V. A., 1995. The 'loessic Palaeolithic' in South Tadjikistan, central Asia: its industries, chronology and correlation. *Quaternary Science Reviews* 14, 731-745.

Ranov, V. A. and Amosova, A. G., 1984. Raskopki must'erskoi stojanki Khudji v 1978 g. (Excavations of the Mousterian site Khudji in 1978), *Arkheologicheskie Raboti v Tadjikistane* 18, 11-47 (in Russian).

Ranov, V. A. and Amosova, A. G., 1990. Raboti otrjada po isucheniju kamennogo veka v 1982 g. (Activities of the expedition for studying Stone Age sites in 1982). *Arkheologicheskie Raboti v Tadjikistane* 22, 161-194 (in Russian).

Ranov, V. A. and Amosova, A. G., 1993. Raboti otrjada po isucheniju kamennogo veka v 1984 g (Activities of the expedition for studying Stone Age sites in 1984). *Arkheologicheskie Raboti v Tadjikistane* 24, 172-215 (in Russian).

Ranov V. A., Dodonov A. E., Lomov, S. P., Pakhomov M. M. and Pen'kov, A. V., 1987. Kul'dara - novij nijnepaleoliticheskij pamjatnik Tadjikistana (Kul'dara - a new Lower Palaeolithic site in Tadjikistan). *Bull. Kom. po isuch. chetv. perioda* 56, 65-74 (in Russian).

Révillion, S., and Tuffreau, A. (eds.), 1994. *Les industries laminaire au Paléolithique Moyen.* Dossier de Documentation Archéologique 18. Paris: Éditions CNRS.

Rosanov, B. G., 1983. *Morfologia pochv (Soil Morphology).* Isd. MGU (in Russian).

Sanlaville, P. 1995. Late Quaternary chronology and paleoclimates of the Eastern Mediterranean. *Paléorient* 21, 145-148.

Schäfer, J., Ranov, V. A. and Sosin, P. M., 1996a. Obi-Mazar und Lachuti - zur Stratigraphie und Archäologie des Lösspaläolithikums in Süd-Tadjikistan. In: K. W. Beinhauer et al. (ed.), *Homo erectus heidelbergensis* von Mauer, Kolloquium I, 133-140. Jan Thorbecke Verlag Sigmaringen.

Schäfer J., Sosin, P. M. and Ranov, V. A., 1996b. Neue Untersuchungen zum Lösspaläolithikum am Obi-Mazar, Tadjikistan. *Archäologisches Korrespondenzblatt* 26, 97-109.

Shackleton, N. J., 1978. Oxygen isotopes, ice volume and sea level. *Quaternary Science Reviews* 6, 183-190.

Shackleton, N. J., Berger, A. and Peltier, W. R. 1990. An alternative astronomical calibration of the Lower Pleistocene timescale based on ODP Site 677. In: The Late Cenozoic Ice Age. *Transactions of the Royal Society of Edinburgh: Earth Sciences,* 81: 251-261.

Shackleton, N. J., An, Z., Dodonov, A. E., Gavin, J., Kukla, G. J., Ranov, V. A. and Zhou, L. P., 1995. Accumulation rate of loess in Tadjikistan and

China: relationship with global ice volume cycles. *Quaternary Proceedings* 4, 1-6.

Shelkopljas, V. N., 1977. Datirovanie subaeralnykh plejstocenovykh otlojenij Tadjikistana termoljuminescentnym metodom (Thermoluminiscence dating of subaerial Pleistocene deposits in Tadjikistan). In: Dodonov, A. E., Melamed, Ja. R. and Nikiforova, K. V. (eds.), *Granica neogena i chetvertichnoj sistemy*. Moscow. pp. 17-19 (in Russian).

Suleijmanov, R. Ch., 1972. *Staticheskoe isuchenie kul'turj grota Obi-Rakhmat (Statistical Study of Cultures in the Obi-Rakhmat Rock Shelter)*. Tashkent (in Russian).

DISCUSSION

Avraham Ronen: The oldest fireplace you showed is at the top of soil 1. Do you have an older indication of fire?

Joachim Schäfer: I am not sure that it is a fireplace. The sediments do not clearly show the presence of fireplaces. We have some charcoal and some reddish loess in many places, but may be this is not a fireplace. There may have been a natural fire in the vicinity, and a tree lying there may have burnt so heavily that the loess became red. We have signs of fire in the second soil, with burnt flint. Small pieces of charcoal are found in every horizon.

Nicholas Conard: I would also recommend caution in situations like that. So often one finds, within naturally iron stained horizons, sediments that look very red and may contain charcoal. I think that to be able to argue that it is definitely a fireplace you need further analytical methods such as made by Tsatskin at Tabun.

Joachim Schäfer: Still, I think that it is fire. May be a natural one, but nevertheless fire.

Nicholas Conard: I still think that even if it looks like a fireplace, it should be examined using whatever sedimentological and mineralogical methods that are available. As we know from Steve Weiner and others, there are a lot of ways to look at it, and a lot of interesting results can come out of these studies. It strikes me as unusual that there are no artefacts there. If it is a fireplace which is a part of an archaeological site, one would expect cultural material to be with it.

Joachim Schäfer: I think it is fire because the loess is burnt. This is clearly visible. Maybe it was a natural fire, which means it is not a fireplace.

Leonid Vishnyatsky: Could you further explain why the new datings are more reliable than the old ones?

Joachim Schäfer: It is the correlation with the deep-sea records and the Chinese loess sections. We have no absolute dates. Due to the old system by Lazarenko, the third soil is Last Interglacial while for Dodonov, it is the fifth soil.

Alexander Tsatskin: Do you distinguish between pedosediments and soils *in situ?*

Joachim Schäfer: We have the idea that everything is *in situ*. We have only a thin layer at Obi-Mazar, about 5-6 cm, which has no matrix, with some erosion. But the soils were *in situ*.

Alexander Tsatskin: How do you explain the thickness there?

Joachim Schäfer: We assume that it is a sinsedimentary soil development.

Alexander Tsatskin: How do you distinguish between a sinsedimentary soil which can not be interglacial from properly interglacial soils?

Joachim Schäfer: Why can sinsedimentary soil not be interglacial?

Alexander Tsatskin: By definition.

Joachim Schäfer: We can see present-day sedimentation and we are living in an interglacial.

Alexander Tsatskin: In the slides we saw dipping and tilting of sediments, and truncation of different layers, so they cannot be *in situ*.

Joachim Schäfer: The truncation was in the Obi-Mazar section, in the first and second soils.

Alexander Tsatskin: I am very much impressed by your data about the location of artefacts within the soils. In the palaeorelief of a catena, where are the artefacts concentrated: on top, on the slope or at the bottom?

Joachim Schäfer: The artefacts were spread from the base of the soil up to the loess. Some layers had more dense concentrations. But, we have no real sites in Tadjikistan. We have a geomorphological situation where people came, made some activities and left, then came again, ten years later, and had some other activities. We have no real camp sites.

Gerhard Bosinski: Please allow me to say some personal words. The things you developed in the last twenty years are fantastic. I had the chance to be there and to see for myself. To me it looks unique and this is a real step towards something in the future. There are high loess sections in China, but almost without sites.
I do not understand why did you hesitate to accept Middle Palaeolithic finds in the second and third soil. It is commonly accepted that Middle Palaeolithic, in the same character you have shown, starts at least at about Stage 8. Concerning the cave Selungur, there is at least one typical biface.

Vadim Ranov: Even if it is a biface, it is unique.

Viola Dobosi: The Karatau culture seems to me just the same as Vértesszőlős. Chronologically and typologically, these pebble cultures are the same. Could you tell me what was the raw material of these tools?

Vadim Ranov: There is one main difference between the Karatau and the Vértesszőlős industries: in Vértesszőlős you have two types of tools, one made on flint, with good retouch, and then larger pebbles and choppers. The biggest tools have some similarities with Karatau. The material of Karatau is not of good quality. There are metamorphic rocks and quartzites. I see also some similarities with Érd.

The assemblage from Karatau is nearly one thousand pieces, of which 11% are retouched.

Viola Dobosi: I am confused because of the short and the long chronology. In Vértesszőlős we have a Middle Pleistocene fauna and the chronology is no more than 350,000 years. In Hungary it is considered real Lower Palaeolithic. Yet, you wrote that it is a Middle Palaeolithic technology of central Europe.

Vadim Ranov: The problem is, when does the Middle Palaeolithic begin? The beginning in Europe is now placed at 250,000 years, I think perhaps it is 300,000 years. it is very difficult to say.

Avraham Ronen: This bulge of new dates is perplexing and one has to get used to it. Do not forget that only some thirty years ago, we had to get used to the idea that humans did not appear half a million years ago, but two million. We have to get constantly used to longer chronologies.
About central Asia, we have only seen Middle Palaeolithic up to the top of the section. Is there Upper Palaeolithic or Epipalaeolithic in any of the soils?

Vadim Ranov: We have very little loess above the first palaeosol. In this loess, Joachim found an interstadial palaeosol. In this interstadial palaeosol we found only two or three artefacts. It may be Upper Palaeolithic. Where is the Epipalaeolithic? At the top of the section we have the Holocene soil with Neolithic. May be below this Holocene soil, or in it, we will find the Epipalaeolithic. And all of this accumulation is above soil number one.

Avraham Ronen: How much accumulation do you have above soil 1?

Vadim Ranov: About twenty metres.

Joachim Schäfer: I would like to answer Avraham. Before we began our joint work we knew that there were definitely finds in the fourth and fifth soils. Now, we are very happy to have finds up to the second soil. We also have a few finds in the first soil. In the future we may have more finds in the first soil, but this first soil is very difficult to reach. It is very steep.
For example, At Lachuti and Obi-Mazar the first soil was so far away from the river bed that may be people were not present there. They were present in the fourth soil, 60 metres above the water. Closer to the river, the first soil was eroded. We can not correlate the Upper Palaeolithic site Shugnou with our section, because that site was destroyed by gold mining. That is the only site with real Upper Palaeolithic sequence that we had.

Alexander Tsatskin: It was a pleasure to see the presentation of these data. I remember ten years ago, when we heard in the Geological Institute in Moscow the report of the Quaternary Commission on Central Asia, there was a feeling that central Asia was not part of the loess world. It seems to me that we are on the right road to reassess in this meeting the dates of archaeological cultures and actually, the loess stratigraphy, including that of central Asia.
You make global correlations using the Shackleton curve and the magnetic susceptibility measurements of Heller. I think that Heller's measurements were not used properly, because in China Heller's maximum magnetic susceptibility is in the fifth soil, while your maximum is located very high in the section. May be your high peak is connected with the fires you mentioned. Magnetic susceptibility is more sensible to fires than the siliceous aggregates.

Vadim Ranov: I understand that the palaeomagnetic situation in central Asia is the same as in China.

Joachim Schäfer: The amplitude in the fifth and sixth soils is very low in Tadjikistan.

Nicolas Rolland: You have a series of loess deposits in that region. Usually when there is loess deposition there is no human occupation. Am I correct? Do you have major gaps in human occupations in Central Asia, whether it is indicated by loess deposition or by other features like palynology, or do you have a continuous occupation?

Vadim Ranov: For the palaeoecology and palynology of our sites in south Tadjikistan we do not have good results. We have some palynological diagrams for the Mousterian and the Upper Palaeolithic, but there is a need to correlate them with the oxygen isotope stages.

Nicolas Rolland: Do you think that human occupation was continuous?

Vadim Ranov: According to my old position, there is a local evolution of the Karatau or Pebble Culture of the lowest Palaeolithic. I think that there was a Mousterian immigration, may be from the Near East, of Neanderthals with Levallois culture. Later the Mousterian industries evolved into Upper Palaeolithic types. But now it is much more difficult to say anything about periodization because we have Mousterian features in the second palaeosol.

Nicholas Conard: It is beautiful to see the section, with soils 1-10 and their subdivisions, matched with the isotope curve. But I think, still, it is an hypothesis that needs to be checked independently. It has to be confirmed by biological arguments and dating techniques. And another point. It is clear that you have an enormous loess record, 100 metres. But, based on my experience, it is not necessarily a complete record. There could be missing phases.

Joachim Schäfer: Of course, the correlation is only hypothetical. How can we correlate using TL dating which goes back 100,000 years, when our first soil is 130,000 years old?

PART II

YABRUDIAN

YABRUDIAN AND RELATED INDUSTRIES: THE STATE OF RESEARCH IN 1996

Lorraine Copeland

Description and History of the Sites Yabrud I, Tabun, Zuttiyeh and Abu Sif

Yabrud I

The term 'Yabrudian' (sometimes 'Yabroudian'; 'Jabrudian' in German orthography) derives from the name of the summer resort town of Yabrud, which is located in central Syria on the eastern flanks of the Anti-Lebanon mountain chain. Here, between 1931 and 1933, A. Rust excavated three rockshelters situated in a valley behind the town (Yabrud I, II and III). In Rockshelter I, he unearthed a series of layers (25 - 11). Some of these contained a facies unknown in the Near East. This facies consisted mainly of large non-Levallois flakes fashioned into side-scrapers (hereafter racloirs) through use of steep, scalar retouch reminiscent of that seen in the Quina Mousterian of France. Few other tool-types occurred in such layers. Rust named these layers 'Jabrudian' (Rust, 1950). Jabrudian layers were interspersed with other layers which, using an original nomenclature, he referred to the Acheulean or (when only a few bifaces were present) to an Acheuleo-Yabrudian. Also interspersed were two layers containing assemblages of blades and blade tools which he called Pre-Aurignacian (at that time it was customary to name Near Eastern Upper Palaeolithic, i.e., blade industries, after those of France). According to Rust, the sequence began with the Yabrudian in Level 25. Above Level 11, in Levels 10 – 1, evidence of a series of Levalloiso-Mousterian occupations was found. The stratigraphy was taken at face value and much discussed (e.g., Copeland and Hours, 1981; Hours 1979, Table 1). It was not until much later that new work at the shelter suggested that the chronological sequence might not be as indicated in Plate 4 of Rust's schematic section (Solecki and Solecki, 1986; see below).

Tabun

In the same period (1929-1934), Professor D. Garrod had been excavating at the vast cave of Tabun in Mount Carmel. Here a long sequence beginning with Tayacian and Acheulean layers was overlain in Layer E by a deep deposit containing bifaces, racloirs and (near the top) blade lenses, which were referred to the Upper Acheulean/Micoquian. This was followed in layers D, C and B by Levalloiso-Mousterian occupations (Garrod and Bate, 1937). After Garrod and Rust had examined each other's material at Ahrensberg and Cambridge, they agreed that the blade facies at the two sites were strikingly similar. Garrod also recognized the broad similarity between the Tabun E and Yabrud I, 25-11 sequences. In consequence, she changed the name of the industries of E to 'Acheuleo-Jabroudien' (Garrod, 1956: 40-42). Rust, however, questioned the lack of discrete facies at Tabun, and suggested that the 'melange' was due to faulty excavation or recording methods (Rust, 1933; 1950, 143-146). As we shall see later, the new work of Jelinek and his team at Tabun largely confirmed Garrod's findings on this point (Jelinek, 1975; Jelinek et al., 1973).

Zuttiyeh

Another sizable cave, Zuttiyeh, had already been excavated during 1926-1927 by Turville Petre (1927). The cave opens from a cliff onto the Wadi Amud valley in the Galilee. As Garrod described later, it contained the same mixture of bifaces, Quina racloirs and limited blade tools as in Tabun E. However, as at Zuttiyeh, Mousterian types were also present, she concluded that there had once been a Tabun E-like occupation followed by a Mousterian one (Garrod and Bate, 1937: 115). Of great anthropological interest was Turville Petre's finding of part of a human skull underneath the lowest deposit, called 'Galilee Man'. Kieth (McCown and Kieth, 1939) referred this fossil to the Neanderthals. Since the skull had lain below the probable Yabrudian layer, it has always been regarded as representing the people of the Mugharan Tradition. Its racial affinities have been differently interpreted by various authors (see Vandermeersch, 1989, 1995). U series dates of between 164 and 95 kyr BP have been questioned (Schwarcz, 1994).

Abu Sif

With hindsight it may be suggested that the rockshelter of Abu Sif in Judea, excavated by Neuville (1951), also contained some kind of Acheuleo-Yabrudian-like assemblage in its lowest layer E ('Micoquien'). It appears that assemblages in Layer E may be separated by a sterile layer (D) from an early form of Levalloiso-Mousterian in Layers C and B, which resembles that of Tabun D. Another site which may contain Yabrudian material is Ain Musa (Neuville, 1931).

Early Interpretation Problems

Before the next wave of discoveries, the waters had been considerably muddied by Bordes' interpretation of the Pre-Aurignacian at Yabrud I. Bordes referred these assemblages 'sans discussion possible' to the Upper Palaeolithic, despite their undisputed position underlying the Middle Palaeolithic. He went on to suggest that the Near Eastern Lower Palaeolithic of Yabrud (and therefore of Mount Carmel) took place in Würm II or perhaps I and II, the Pre-Aurignacian in a Würm II-III interstadial, and the Mousterian in Würm III (Bordes, 1955). At this time, the rare ^{14}C dates available could only suggest that Tabun dated to 'before 34,000 B.C.' (Garrod, 1956: 56-58).

This view clashed with the marine shoreline chronology of the Eastern Mediterranean, where it was known that 'Levalloisian' assemblages were associated with fossilized Last Interglacial beaches containing the warm sea mollusc *Strombus bubonius*. Thus, they could hardly refer to Würm III (Garrod, 1958). Yabrudian-like material had also been reported at Adlun on a marine platform about 13m a.m.s.1. (Zumoffen, 1900). In order to advance from the impasse, Garrod decided to 'work with the sea', as this promised to be the key to understanding the relationships of the industries to the marine chronology.

Excavations at Adlun, Masloukh, Jerf Ajla and El-Kowm

Bezez Cave, Adlun

The results of Garrod's two excavations at Adlun may be summarized as follows: In 1963, Bezez Cave produced the earliest material at Adlun. This large cave is situated near the coast between the cities of Tyre and Sidon. At the rear of the cave, an impoverished Acheuleo-Yabrudian-like assemblage was found in Trench M. The assemblage was embedded in, and overlay a pebble beach deposit on the bedrock floor at a height of c. 16m a.m.s.l. Acheuleo-Yabrudian assemblages were found in the centre and mouth areas (Trenches G, K and D). These lay on the surface of a fossil beach (in some areas overlying lenses of small 'Tayacian' flakes). These layers were referred to Level C. Level C was overlain by sterile layers in some exposures, and by Levalloiso-Mousterian (Level B) in others. In a central trench (S), these two levels disappeared into a large central swallow-hole, their relative positions intact. It should be noted that Garrod had by now dropped the term, 'Acheuleo-Yabrudian' in favour of 'Yabrudian' (Garrod, 1966; Garrod and Kirkbride, 1961; Kirkbride, 1983).

Abri Zumoffen, Adlun

The nearby rockshelter of Abri Zumoffen opened from a dead sea cliff onto a marine terrace at about 12 - 13.8 m a.m.s.l. This rockshelter contained a small cave (Zumoffen Cave). In 1958, three soundings (Trenches A, B and C) around 1 m deep (which could not reach bedrock) produced sequences entirely different from those in Bezez Cave. At the base of all three exposures was a marine conglomerate at around 11 – 12 m a.m.s.l. In Trenches A and B, this conglomerate contained the 'Beach Industry' in the upper 30 cm. This industry consisted of cores and choppers on beach pebbles, as well as blade tools. In Trench A, the Beach Industry was overlain by the blade industry which Garrod was later to name 'Amudian'. These occupation layers contained intact hearths, flint and bone in weathering soils separated by sterile calcrete bands. In Trench B, the Amudian was overlain by a grey breccia containing Yabrudian material of a lighter aspect than that in Bezez Cave. This material included some blade tools. In Trench C, the Amudian was found in a red clay soil sandwiched between two beach deposits. The lower one (at c. 11 cm a.m.s.l.) was a pebble conglomerate. The higher one was a pink sandstone which also contained unabraded Amudian tools. No Mousterian occupations were noted, and no *Strombus bubonius* were found in the soundings (Garrod, 1962, 241; 1966; Garrod and Kirkbride, 1961).

Garrod (1966) regarded the occupations at Adlun as having occurred successively as the sea retreated from the Last Interglacial sea-level height of 15 m. This conclusion was subsequently supported in part by Sanlaville, who referred the beaches to oscillations occurring during the transgression stages Enfean I (in Bezez) and Enfean II (at Abri Zumoffen). These transgressions correspond to the Riss/Würm, or Oxygen Isotope Stage 5 (Sanlaville, 1977, 702-706).

Other interpretations and debates concerning the Adlun sequence (including Garrod's suggestion of a symbiosis) are discussed later in this report (see also Copeland, 1983a, 76-81). It had become apparent that the altimetric closeness of the beaches at Adlun to each other was misleading, and that (as just suggested for Tabun) more than one interglacial cycle could be involved (Bar-Yosef, 1989, 600; Sanlaville, 1981; 1996).

Masloukh

A new Yabrudian site was subsequently discovered near Batroun by P. Sanlaville and J. Remiro (Sanlaville, 1977, 442-443; Fleisch et al., 1969). There, a small cave opened onto a marine platform at about 40m a.m.s.l. This was one of many such instances which occurred along the shore. Such features are attributed to the Mindel-Riss Interglacial (or the Jbailian Transgression of Sanlaville, 1977).

Skinner carried out a brief excavation in which he found that the Yabrudian artefacts and animal bones were contained in a rose breccia which clearly post-dated a marine deposit found on the cave floor (Skinner, 1970). He noted that the bifaces, of Late Acheulean type, were rather rare in the upper levels. The racloirs consisted of 55% Quina and offset types. A small blade component was present throughout, although mainly in the upper levels. Among the faunal remains were *Rhinoceros*, *Bos*, *Equus* and *Dama*. Skinner dated the site to a retreat post-dating the Jbailian Transgression, specifically, 'after the onset of the Riss Glacial' (Skinner, 1970, 164). As we shall see, his dating was closer to the mark than he may have realized.

El Kowm

The most recently found Yabrudian sites are at the El-Kowm oasis in the Syrian Desert, half way between Raqqa and Palmyra. These sites came to light during the surveys of the numerous spring mounds carried out by J. Cauvin and his team (Cauvin et al., 1979). These mounds, formed of accumulations of sand and travertines, surrounded the vents. In an effort to find water for irrigation, the peasants living in the area had dug well-shafts into many mounds, throwing up debris which contained Palaeolithic artefacts (Besançon et al., 1982). The most surprising aspects of this situation are: (1) The number of Yabrudian assemblages, which is greater than in the rest of the Near East (although of the nine identified so far, two may be Acheulean); (2) Their concentration within an area of ca. 50 sq km, which contrasts with the isolated positions of each of the Levant sites; and (3) Their entirely different ecological situation. Instead of occurring in caves and rockshelters, they occur in the open air around springs in an artesian basin (Besançon and Sanlaville, 1988). A list of the sites and the artefacts found is given in Copeland and Hours (1983, 23-25 and Table 1).

The Yabrudian material excavated at the Hummal well site at El-Kowm was found at the bottom of the 25 ft deep well shaft in a lm-thick pebble conglomerate. This conglomerate produced 703 artefacts consisting of typical Quina and other racloirs (both large and small), only three bifaces and only 3.5% blades on non-Levallois flakes, as well as cores and debitage (Copeland and Hours, 1983). Contrary to what was at first believed, the Yabrudian (Level Ib) preceeded the Hummalian blade industry, which was contained in a higher, sandy layer. Dosimeters were installed and gave a TL date of 160 kyr BP (Hennig and Hours, 1982). This is broadly comparable to the other El-Kowm dates (e.g, Umm Qubeiba) and the dates for Zuttiyeh and Yabrud I.

Azraq Oasis

Material of Yabrudian type has been noted at spring sites in the Azraq Oasis, located east of Amman in the Jordanian desert. Unlike el-Kowm, the sweet water in these springs (deriving from rainfall on the Jebel Druze basalts) does not give rise to mounds. Thus, the Palaeolithic level is only 3 – 4 m below the present land surfaces of the central endoreic basin, or playa, where the springs emerge to form an oasis (Besançon and Sanlaville, 1988). As at El-Kowm, the artefacts were discovered among the debris when irrigation projects involved the deepening of several springs. Therefore, in all but two cases (where *in situ* levels could be excavated), the artefacts are mixed with Acheulian and Mousterian tools. At Lion Spring (Ain el-Assad), the earlier excavations of D. Kirkbride and later excavations of G. Rollefson failed to produce any distinctly Yabrudian layers, although a few heavy Quina racloirs were present among the Late Acheulean finds (Kirkbride, 1989; Rollefson 1983). In the sounding at C-Spring (Garrard et al., 1987) no clearly Yabrudian types appeared in Levels P or Q, which overlay the Late Acheulean in Level R (Copeland, 1989a). However, an assemblage of typical Yabrudian types (again mixed with large bifaces, and Mousterian and Upper Palaeolithic types) was strewn around a spring at Ain el-Beidha (Copeland, 1989b, Fig. 4). The presence of Yabrudian at Azraq needs to be confirmed by further work there (G. Rollefson, personal communication 1996).

'Old' Sites Revisited: Tabun, Yabrud I and Zuttiyeh

By the 1970s it was accepted that the Yabrudian preceeded the Levalloiso-Mousterian flake complex of the Levant. Where the two are found together, the Yabrudian is always underneath (Copeland, 1975). Its distribution in space was limited (no other sites having been found in spite of intensive surveying) to the central Levant, in a 'slice' running from the Mediterranean coast eastwards to the desert (Fig. 1). The occurrence of the Yabrudian in caves, shelters and open-air sites demonstrated that various ecological niches could be exploited. However, the absolute dating of the phase and the meaning of the lithic variability was still hotly debated (see an account of the various positions in Solecki, 1970).

During the 1970s and 1980s, a number of important re-excavations took place, as follows:

Tabun

In a major project, A. Jelinek worked at Tabun from 1967 to 1971. Study of his results continues to this day. Jelinek excavated a 10 m deep section in the south wall of Garrod's excavation. He managed to broadly correlate Garrod's layers with his many subunits (beds). The beds concerning us here are Garrod's Ea - Ed: Units X - XIII (Beds 70 - 85), i.e., the Yabrudian (Jelinek, 1981, 268, and Table 1). The position of the Amudian in concentration near the top of Ea confirmed Garrod's findings, as did the fluctuation through time of proportions of bifaces against racloirs (Jelinek, 1981, Fig. 2). Jelinek concluded that 'what we are observing here are gradual changes in a single industry.' He therefore proposed to redefine the Yabrudian/Acheuleo-Yabrudian/Amudian material in E as representing a unified 'Mugharan Tradition', in which several variants such as an 'Acheulean Facies' or a 'Yabrudian Facies' could be seen (Jelinek, 1981, 271-274). It must be noted that although we know how the variants developed in the vertical column, we have no information yet as to the spatial distribution of these 'facies' across the cave floor. Jelinek also correlated the sequence at Tabun with the oxygen isotope deep-sea core chronology, placing the Mugharan Tradition layers within the warmer/cooler oscillations of Oxygen Isotope Stage 5, around 110 – 80 kyr BP (Jelinek, 1981, Fig. 3).

Yabrud I

From 1963 to 1965 a Columbia University team under R. S. Solecki investigated along the length of Shelter I. This shelter, as mentioned in the first part of this paper, contained a complex succession of cultural layers where Acheulean, Yabrudian and other assemblages seemed to interdigitate at a depth of about 9 m (Solecki, 1970). It became apparent that Rust's profile did not represent the stratigraphy of any one area. It was not only schematic, but a composite of layers found in different parts of the shelter. Sometimes industries attributed to different facies were found at the same depth but in other exposures. For example, Layer 17 (Acheulean) in Room 2 is at the same depth as Layer 16 (Yabrudian) in Room 1. The Pre-Aurignacian of Layer 13 occurred at the same depth as Layer 12 (End Acheulean - Pre Mousterian). Hence, in both cases, the relative chronological position of the assemblages is unknown. Furthermore, it was believed by de Heinzelin (1966) that the lowest layers (below Layer 19) had been washed into the shelter by fluvial action. On the other hand, Farrand (1982) proposed an aeolian origin for at least two of these layers (21 and 'Flugsand'). This view supports the idea that these layers are natural deposits and not occupation layers (Solecki and Solecki, 1986).

Above the Acheuleo-Yabrudian Layer 11, however, the posteriority of the Mousterian occupations seems certain (see reconstructed stratigraphy in Solecki and Solecki, 1986, Fig. 4). Level 18 ('Micoquian') recently produced a TL date of c. 195 kyr BP (Oxford Research Laboratory, 1990). It may be observed that there are intriguing links between these features and those seen at Adlun. This point will be discussed later.

Zuttiyeh

The breccias remaining in Zuttiyeh Cave were excavated by Gissis and Bar-Yosef in 1973. These investigators managed to extract small samples of artefacts from two separate layers: a lower layer containing artefacts of the Mugharan Tradition (a few bifaces, typical Quina and offset racloirs, limaces, a few blades, as well as cores and flakes), and an upper layer containing Levalloiso-Mousterian artefacts (Gissis and Bar-Yosef, 1974). As mentioned earlier, the cave had been tentatively dated to about 148 kyr BP (Schwarcz, 1994).

Jerf Ajla

This cave, located near Palmyra, had been discovered and excavated in 1955 by C. Coon. Coon reported the presence of bifacial tools in the lowest layer of his excavations, Yellow 2, at a depth of 6 m (Coon, 1956). Of the approixmately 67,000 artefacts recovered, some 4,600 items were available for study by B. Schroeder. In 1965, Schroeder made a small cut in the cave in order to verify the stratigraphy (Schroeder, 1966; 1969). In what appeared to correspond to Yellow 2, he found an ovate biface, 4 bifacial scrapers and two cleavers, in addition to many Levallois and other flakes (including points). However, of the flakes illustrated in Schroeder's Plates I and II, one tool appears to be an Amudian blade, and the other a non-Levallois pointed flake with a nibbled 'back'. The bifacial tools would be at home in either a Late Acheulean (cf. a C-Spring-like bifacial cleaver) or an Acheuleo-Yabrudian industry. It is therefore not clear whether basal Jerf Ajla contained a Yabrudian or an Acheulean assemblage. Nevertheless, the overlying layers certainly represented a Tabun D-like facies Levalloiso-Mousterian, with blades and an elongated Mousterian Point.

Mugharan Tradition, Technology and Typology

Many authors have been struck by the very different approaches to flint-knapping seen in the Yabrudian, as compared to those in the Levalloiso-Mousterian industries. A Yabrudian assemblage is instantly recognizable by its angular and often bizarre flake forms and tool shapes (e.g., Fig. 4, 3). This is especially true in the case of the racloirs. It is clear that the tool morphology is affected from the outset by the core reduction technology.

Technology

At most sites the raw material (universally flint or chert) consists of beach pebbles found in the vicinity (as at Adlun) or wadi pebbles and nodules from seams in the limestone at inland sites. Such core-blanks need minimal preparation to form discoid, globular ('Clactonian') or prismatic cores. Using hard hammer or 'stone-on-stone' techniques (smashing the core on an anvil?), these core-blanks produce the short, thick flakes with high-angle butts seen in the Mugharan Tradition (and Acheulean) assemblages. The blanks must be shaped into tools by retouch, often extensive and abrupt. This is a far cry from the Levallois method in which the process is reversed: in that system cores are carefully shaped to produce a desired flake shape which usually needs no, or little, retouch to make it a tool.

In addition to producing large, thick 'Acheulean' flakes, often wider than long, with pronounced bulbs and high (90°+) butt-angles (Fig. 4, 1 and 2), non-Levallois methods can produce blades by using the orthogonal prismatic cores (Figs. 2 and 3). Many of the blades in the Amudian and Pre-Aurignacian are (judged by Mousterian standards) atypical and unsuccessful, however. This can be seen in the illustrated material from Adlun, Yabrud and Tabun. Many such blades seem to represent debitage: they are either struck off the axis or overshot, the angles of the butts are over 90°, they are rarely faceted, and they are often shattered or consist of cortex. The knappers may not have perfected their methods or been able to produce many pointed or triangular forms such as seen in Fig. 6, 1.

Assemblages sometimes include thin flakes with radial dorsal facets. These types are present in the Amudian and Pre-Aurignacian as has been noted, for example, by Jelinek (1981) and myself (Copeland, 1983). Although these flakes may resemble Levallois products, they could have been struck either from bifaces or from discoid or globular cores (Copeland, 1995). However, the occasional use of Levallois-like techniques need not be ruled out, given that these were, according to some authors, used in the Late and Final Acheulean (cf. the Samokian in North Syria; Muhessen, 1992). Bordes interpreted Rust's Acheulean layers at Yabrud I as including Levallois flakes, and found that this helped to differentiate the Acheulean from the Yabrudian layers (It is questionable how this could be if Levels 25 - 19 were not occupation layers, see above).

Complete flakes in the Mugharan and Mousterian have been studied by Jelinek, who found a technological trend toward greater variance in the ratio of flake thickness to flake width across time (Jelinek, 1982). Nevertheless, this idea has not been universally accepted (Jelinek, 1992, 257).

Typology

The typology of the Yabrudian from the excavated sites has been much discussed. So far, the only detailed description of a large assemblage, analyzed by the Bordean method, is that from Bezez C. For my own covenience, this section will be illustrated with reference to the Bezez C tool types (Copeland, 1983b). The outstanding features of Yabrudian tool-kits are: first, the abundance of racloirs, which outnumber all other types of tool. Second, the high number of offset, bifacial and transverse types (which occur in far lower numbers in Mousterian facies). Third, the frequent use of 'staircase' scalar retouch on the working-edges.

These features relate the industries somewhat to those of the Charentian of France (for instance, at La Quina). Garrod used terms such as 'end-bulb' or 'oblique-bulb' to classify the Tabun racloirs, but today most typologists use the Bordean system based on the position(s) of the retouch relative to the blank's axis (Bordes, 1961). Racloir morphology (and hence classification), especially in the case of transverse types, often stems from the use of large, wide flakes with a pronounced bulb. Transverse racloirs will often rock from side to side (Garrod's 'rocking-chair' action) as a result of these bulbs (Copeland and Hours, 1983). These bulbs are often removed (Fig. 4, 1 and 2). Many racloirs have one or two pointed corners, which are often at 90° to the converging retouched edges, and triple working edges are not uncommon. In contrast, Rust identified a special offset type (dèjetè, or 'winkelkratzer') (Fig. 4, 5). In this offset type, which is characteristic of the Yabrudian layers he identified at Yabrud I, the converging angle is acute and sharply pointed. It would seem that tools with very robust edges were needed by the Yabrudians. The extreme edge is usually sharp, and is often absolutely straight or perfectly convex. The edge often has a characteristic 'crushed' appearance. This feature may derive from a special resharpening method. This characteristic is not seen in Mousterian facies.

Quina-type retouch has been studied by Dibble, who suggests that its frequent use at Tabun and La Quina was 'a function of reduction of the tool through continuous resharpening and remodification of the edges' (Dibble, 1988, 49), and that this is an indication of curation. As sharpening was repeated when needed, the blank was reduced in size to a short, thick, transverse or offset piece (Dibble, 1988). Supporting the idea that such practices occurred in the Near East is the finding that in most Yabrudian assemblages, there is a range of very small, abruptly retouched (i.e., re-worked?) racloirs which morphologically duplicate the larger specimens. However, many transverse scrapers have perpendicular sides to the left and right of the cutting-edge, formed (deliberately?) of cortex or natural fracture-planes. This makes it unlikely that such scrapers began as laterally-retouched types. An alternative idea is that the special debitage methods produced flakes which were already thick, wide and broad butted.

The biface component consists mainly of typical Late Acheulean forms (Fig. 5, 1-5). These are amygdaloid/ cordiform, lanceolate/Micoquian, ovate, etc., finished by soft hammer retouch. These grade into bifacial racloirs (Fig. 5, 6) on 'side-bulb' or 'oblique -bulb' flakes, thinned-back racloirs, and (more rarely) cleavers. These cleavers, however, do not resemble the bifacial cleavers (*bout coupè*) of, for example, C-Spring or Tabun F. The rougher specimens often have an archaic character, but may represent rough-outs. On average, the bifaces are of moderate to small size (although large specimens can occur; Fig. 9). No other attributes seem to distinguish Yabrudian from Late Acheulean bifaces. To some extent the same can be said for the racloirs: only the proportions of each type seem to differentiate the 'facies' discussed below (Fig. 10).

The other tools include choppers made on pebbles, limaces, denticulates and notches, rare end-scrapers and burins (usually atypical), backed knives with varying amounts of abrupt retouch, and large numbers of small, variously retouched pieces, some of which seem to be racloirs.

In the Amudian, while the technology seems equivalent to that of the Pre-Aurignacian, the tool-types differ. The Amudian is characterized by backed knives (some resembling Chatelperron Points, Fig. 6, 9), nibbled blades and 'Adlun burins' (Fig. 6, 8). The burins are robust pieces made on the indirect truncation of a blank from a prismatic core (Fig. 2). Again, in order to form 'tools', the blades had to be retouched into, for example, end-scrapers (Fig. 6, 5). Such forms resemble Upper Palaeolithic types (Copeland, 1983b, 224-228). A number of racloirs on blades are also present (Fig. 6, 2 and 3) which grade into what Garrod termed 'nibbled blades'. Composite tools also occur (Fig. 6, 7).

The Pre-Aurignacian is characterized by end-scrapers of Aurignacian types and dihedral burins. To me, this seems to justify the use of the two names Amudian and Pre-Aurignacian. The Pre-Aurignacian also contained fragments of bifaces and biface repair-flakes (Rust, 1950, Fig. 34). As for the 'Beach Industry' at Adlun, it could represent, as Garrod suggested, a unique 'beachcomber' variant or just a factory facies utilizing a convenient raw material source (are the choppers incomplete or aborted cores?).

To sum up the defining features of Yabrudian industries, we can do no better than to quote the following attributes listed 26 years ago by Skinner (1970, 163):

- Little or no Levallois technique used
- Faceting of striking platforms uncommon
- Flakes mostly removed by high-angle technique
- Implements on flakes and blades are mostly retouched, Quina retouch is prevalent
- A high percentage of side-scrapers, and a low percentage of points
- Quina and dèjeté scrapers are characteristic and present in large numbers
- A small component of implements made on blades
- Denticulate implements are rare
- Bifaces may or may not be present

Following Jelinek and colleagues (1973), it may be added that the only criteria for distinguishing between the 'facies' is the presence/absence of bifaces, or in the Amudian/Pre-Aurignacian, the predominance of blades.

Interpretation and Debates

One tradition with 'facies' or separate groups ('symbiosis')?

This debate could be seen as dividing the 'lumpers' from the 'splitters', as noted in the concluding chapter of the Adlun volume (Kirkbride et al., 1983, 422).

Jelinek's findings at Tabun, summarized above, led him to propose that, for example, the uniformity of the scraper assemblages throughout the whole Tabun sequence was 'strong evidence for a single cultural tradition'. This was one reason why he adopted the name 'Mugharan Tradition' as a unifying term for the Tabun E sequence (Units XI-XIII, Jelinek, 1981, 271). Jelinek envisaged the presence of variants such as an 'Acheulean facies' or a 'Yabrudian facies' (Jelinek, 1981, Fig. 3). The Amudian, which followed an 'Acheulean' layer and preceeded a 'Yabrudian' layer, occurred mainly in concentrations near the top of the pre-Mousterian sequence. Jelinek therefore regarded the Amudian as a third facies of the Mugharan Tradition, from which the earliest Mousterian of Tabun (Level D, with prismatic blades) could be derived.

When single-layer sites are compared, Masloukh and Hummal somewhat resemble each other techno-typologically. At Masloukh, a Yabrudian with low numbers of bifaces seems to be represented. One hundred and ninety one specimens formed less than 2% of the 1,690 retouched tools (Skinner, 1970). Skinner did not see any significant difference in the proportions of types in Levels A, B and C, except for an increase in blades upwards. He wondered if this was connected to the fact that a chimney had opened above the position where the blades increased. He considered the possibility that these blades (which included Upper Palaeolithic types) might have come from an Amudian occupation of the upper terrace above the cave, something which could no longer be investigated (Skinner, 1970, 151-152). At Hummal, the Yabrudian excavated from Level lb contained even fewer bifaces and very rare blades (Copeland and Hours, 1983). Hummal thus resembled the Yabrudian layers at Yabrud I, although it contained no 'interdigitations' in the 1 m-deep section. This seems to indicate that virtually 'pure' Yabrudian variants can occur alone, in caves (Masloukh) and open sites around springs (Hummal).

At Adlun a somewhat different situation was observed. At the mouth of Bezez Cave in the various units of Trench D (see plan in Kirkbride, 1983, Fig. S.2 and section in Fig. S.14), racloirs amounted to about 65% - 68% (or 70% - 71% essential) and bifaces to 0.8% - 13%. In the central area of Trench G, bifaces reached ca. 21% as opposed to 49% racloirs (Copeland, 1983b, 156-157 and Table C.9). Blades only reached 1.1% - 1.3% in these two exposures.

At the rear of the cave (Trench M, where the industry lay in and on the 15 m fossil beach) there were so few racloirs (22% - 27%) and such a good number of bifaces (11.6% - 19.2%) that the 'Yabrudian-ness' of the assemblage was doubted (Copeland, 1983b, 151). At that spot, blades amounted to 6.7%, recalling the situation at Tabun where the Amudian component was correlated with the highest biface-to-scraper ratio: 18.2% (Jelinek 1981, 272). Unfortunately, the exposures were separated from each other by the swallow holes that interrupted the horizontal stratigraphy.
Unless Trench M artefacts represent a particular activity or living area, the assemblage could be seen as a kind of Acheulean (perhaps older) located on a more ancient beach than the material in the other exposures (as previously suggested by Bar-Yosef, 1989 and Sanlaville, 1998). It may be equivalent to, for example, Bed 80 or Layer G at Tabun. Supporting this idea are the similarities between the cumulative graphs for M.152, Bed 80 and Acheulean Level 23 at Yabrud I, which contrast strongly with the graphs for the Yabrudian assemblages (Fig. 7) (Copeland, 1983b, 133-135 and Figs C.19c, C.19d).

The disposition of the Amudian (with its basal Beach Industry variant) at Abri Zumoffen, and the overlying Yabrudian has already been described above. In an earlier work I studied the Amudian industry kept in Beirut and re-classified it according to the Bordean system (Copeland, 1983b, 209-261). Among approximately 1,000 artefacts examined, there were around 234 tools (Fig. 8). Racloirs amounted to about 10% - 28%. Backed knives (typical and atypical) ranged between 12% - 40%. There was one biface. A few end-scrapers and burins were also present (Copeland, 1983b, Tables Z.7, Z.8 and Fig. Z.6, p. 241). The debitage consisted of numerous unretouched flakes and blades, as well as polyhedric, discoid, prismatic and 'semi-Levallois' core-types (Copeland 1983b, Table Z.2). Some excavation units were not published by Garrod and Kirkbride (1961). Therefore it was hard to attribute a number of units to either a 'Yabrudian with backed blades' or an 'Amudian with racloirs' (Copeland, 1983b, 244).

The Yabrudian units produced 2,300 artefacts. The three bifaces found were small and poorly-made. The racloirs resembled those from Tabun, although the dèjetè type was rare (Garrod and Kirkbride, 1961). Amudian retouched blade types, which often outnumbered the racloirs, were present in all the Yabrudian horizons in varying proportions.

It could be concluded that while at first glance the Abri Zumoffen sequence contains distinct 'facies' and resembles more the Yabrud I than the Tabun model, in the detail this is not entirely the case. Once again, there are two ways of interpreting the situation: the Abri Zumoffen industries can be either 'lumped' broadly with the upper part of the Tabun sequence, or alternatively, they could be viewed as representing variants such as those at Yabrud I.

As mentioned above, Garrod proposed an entirely different model which suggested that the Amudians at Abri Zumoffen were a separate group who lived symbiotically with the inhabitants of Bezez Cave and exploited the raw material on the beach. Garrod explained the presence of blades in the subsequent Yabrudian layers as representing 'a kind of Yabrudian-Amudian symbiosis, which had already been noted at other sites (Tabun, Zuttiyeh) but whose meaning was not yet plain' (Garrod, 1962, 242). Apart from Rust and Bordes, this interpretation does not seem to have been accepted by other workers then or later. Rather, it was suggested that the Amudian tools were made in response to particular resource needs or living conditions (Jelinek, 1981; Copeland, 1983b).

Influenced by Rust's findings at Yabrud I, it was at first thought that one could separate 'Acheulean' from 'Acheuleo-Yabrudian' and 'Yabrudian' assemblages (e.g., Bordes, 1955). The re-evaluation of the Yabrud I stratigraphy, discussed above, has rendered the debates concerning the succession of these facies obsolete. However, it does seem that the entities themselves may have existed more or less as Rust believed, even if he was not able to collect all the waste material from any one occurrence.

The evidence presented here demonstrates the extreme variability of the lithic composition at the sites discussed in this paper. Furthermore, the lack of any clear pattern as to the facies occurring at either enclosed or open sites, or in one climatic zone or another, seems to require us to compromise on the question of 'facies' for the time being. We can regard the facies as belonging to one cultural tradition in which different tool-kits were used from time to time. What is clear is that this cultural tradition was a local phenomenon, existing in the central Levant (and nowhere else in spite of extensive surveying) during the period of time between the Late Acheulean and the early Levalloiso-Mousterian such as seen, for example, in Tabun D, Yabrud I.10 and probably Bezez B.

The Hummalian of Hummal Level 1a: Mugharan or Mousterian?

So far, we have not discussed this Hummalian industry because of its equivocal techno-typological composition. It is neither a typical Levalloiso-Mousterian, nor quite Mugharan, yet it is found between an almost 'pure' Yabrudian and a series of typical Mousterian layers (2 - 6). The blade material of Layer la was first discovered at the bottom of the well shaft and at first thought to underlie the Yabrudian (Besançon et al., 1983 and erroneous section in Fig. 4). It was later found that these artefacts had fallen from an *in situ* sandy layer overlying the Yabrudian (Le Tensorer and Hours, 1989).

The very original technology of the Hummalian has been discussed (Copeland, 1981; 1983c), suggesting that this technology was of a Levallois series - blade type. Bergman and Ohnuma (1983) questioned the Levallois definitions, as did Boëda (1995). Boëda found that although some Levallois-like shapes were present (such as elongated points and blades), they were made by 'the Hummal Type Volumetric Construction'. The goal of core construction was the production of a continuous series of blades, and clearly distinct from the Levallois conception (Boëda, 1995, 63).

Like the Yabrudian, the Hummalian tools are fashioned by non-Levallois debitage methods, but the blanks are blades (many being long and well made) instead of thick flakes. Also like the Yabrudian but unlike the Amudian, the tools consist predominently of racloirs, although the edges are formed by flat or semi-abrupt scalar retouch. As Bergman and Ohnuma (1983, 179) point out, the tools might better be classified as 'retouched blades'. Similarly to the Amudian, a characteristic type of Hummalian tool is a kind of backed knife where the backing retouch, instead of being nibbled, is abrupt and greatest near the sharply-pointed tip. This is referred to as the 'Hummalian Point'. These tools grade into elongated Mousterian Points (Copeland, 1985). Bifaces are absent.

Similar assemblages are not known anywhere else in the Near East except for traces at other El-Kowm sites such as Nadaouiyeh and Ain Juwal. However, at Abu Sif C and B (Neuville, 1951) racloirs and 'backed' points were made on non-Levallois prismatic blades. These were found overlying a possibly Yabrudian industry. Similarly, at Tabun the earliest Mousterian in Level D contains prismatic blades (Jelinek, 1982). An unpublished assemblage from the north cave at Nahr Ibrahim may also be relevant (Solecki, 1970, Figs. 14 and 15).

The evidence suggests that the Hummalian may be viewed as a variant technologically related to the Amudian in the Mugharan Tradition range. Alternatively, from a typological point of view, it may represent the earliest form of (non-Levalloiso-) Mousterian. As such, it would have a bearing on the question of 'transitions' (see below). Is the Hummalian part of what Jelinek (1982, 72) sees as the gradual transition to the Middle Palaeolithic? As the industry is not yet dated, this question remains unanswered.

The Mugharan Tradition: Lower or Middle Palaeolithic?

The terminological placement of the Mugharan Tradition assemblages within the Palaeolithic has changed backwards and forwards during the last decades. It has been placed in Lower to Middle Palaeolithic by Garrod (1962) and in the Middle to Lower Palaeolithic by Bar-Yosef (1989, 599). Many authors regarded the question as unimportant, a classificatory 'game' (see contributions to Ronen, 1982). Bar-Yosef (1982, 32) for instance, suggests retention of the terms merely as a convenience in oral discussions.

The conventional view was that, since the Yabrudian was an industry made on flakes, and at first thought to date to the early Würm Glacial, it (and by association the Acheuleo-Yabrudian) belonged in the Middle Palaeolithic (Garrod, 1956). The presence of bifaces was no bar since they were known to occur in the M.A.T. of France, and in any case closely resembled the Middle Palaeolithic Charentian (Quina) industries there (Jelinek, 1981, 272). More recently it has been suggested that a Lower Palaeolithic placement would be more realistic, bearing in mind the much earlier dating now proposed for the industries in Tabun (Bar-Yosef, 1996, 10). In fact, Tabun G - E industries are all referred to the 'Late Acheulean' by Goren-Inbar (1994). The reader may make his choice!

The Mugharan tradition: environmentally or functionally affected?

Efforts were made by Jelinek to correlate the variability of the Mugharan Tradition facies with environmental factors based on what was known of the deep-sea core sequence and its dating (Jelinek, 1982, 71 and Fig. 5), and later taking into account the results of the pollen studies from Tabun (Jelinek, 1992). The industries were placed within the transgressive and regressive oscillations, cooler-to-warmer-to-cooler (but generally towards cooler) conditions of Oxygen Isotope Stage 5, between about 120 to 80 kyr BP (this dating has since been revised, see below). According to Jelinek's view, the Acheulean of F correlated with a warm phase, when olives virtually disappeared. This was followed by Yabrudian facies during minor transgressions, and then by the Amudian during a minor retreat phase in which the pollen indicated cooler conditions (Jelinek, 1992, 262).

Interestingly, a pollen sample from Bezez level C at Adlun (apparently taken from deposits on or near the 15 m beach) contained Mediterranean tree species (olive, carob, and *Quercus calliprinos*). This indicates warm conditions for the Acheulean or Acheuleo-Yabrudian. In contrast, at Abri Zumoffen a Yabrudian pollen sample contained mountain oaks, pines and cedars. This strongly suggests much colder and more humid conditions for the later part of the phase (Leroi-Gourhan, 1980, 87. For the pollen list at Zumoffen see Roe, 1983, 413).

Garrard's study of the Adlun fauna, which consisted predominently of woodland species, tends to support the proposal of Leroi-Gourhan. Garrard suggested (based on the numerous hyaena remains in Bezez B) that the sites were visited on a seasonal basis (Garrard, 1983). Judging by the pollen sample from the probable Yabrudian layer described above, this may have been either during winters or during cold or wet spells. Nevertheless, it should be stressed that the sequences should be re-dated to an earlier cycle, as Bar-Yosef has suggested (1994). In spite of the above indications, Bar-Yosef holds that, 'correlations between changes of environmental conditions and the emergence of different kinds of stone tool assemblages have never been demonstrated to be real....' (Bar-Yosef, 1994, 254).

A functional explanation was suggested for the Amudian and Beach Industry occurrences at Abri Zumoffen (Copeland, 1983b, 243). However, no use-wear studies have been reported on Amudian assemblages. Furthermore, apart from the obvious suggestions that the tools were used in cutting and slicing jobs, it is not clear just what activity was involved. It is also unclear why similar tool-kits were needed both on the Mediterranean seashore and at Yabrud in an inland valley 1,400 m a.m.s.l. which was subject to communication-blocking snowfalls during the winter (of which I have had personal experience!).

For the rest, only slight differences can be seen between the Acheulean and the Mugharan in terms of site type, lithic types and their production method, or game preferences. The only exception is the more limited geographic distribution of the Mugharan. There are also no great socioeconomic differences between the Mugharan and the Mousterian, although we can see a wider distribution and a distinct shift towards a different reduction technology in the Mousterian. In addition, different human types may have been involved, but this is a subject beyond the scope of this paper. Thus it is hard to propose any functional, environmental or ecological reasons to explain the Mugharan phenomenon.

The Mugharan tradition: a continuum or an industry of transition?

Jelinek's interpretation of this question has already been mentioned above. In the Tabun Palaeolithic sequence he saw 'strong evidence for a continuous sequence without evidence of intrusive elements' (Jelinek, 1982, 99). In a later article he describes the results of sediment studies. These showed that there were four episodes of karstic collapse in Layer E, some disturbance in D, and a long gap in occupation between D and C (Jelinek, 1992, 258; Mercier, et al., 1995).

Bar-Yosef proposed an archaeological interpretation of Tabun. He suggested that the 'transitional' levels (Ea to D) are the result of 'a slow, cumulative mixing between the old and the new industries within a basin-shaped area in the cave' (Bar-Yosef, 1994, 254). However, in general he sees a continuum in the Near East and no clear cultural break. He reminds us that human evolution appears to occur in small steps, with innovations and technical improvements along the way (Bar-Yosef, 1982, 30).

At El-Kowm the Yabrudian and Hummalian are regarded as representing, in a purely chronological sense, a transitional phase between the Acheulean and the Mousterian. This phase is contemporary with different industries found in northern Syria, such as the Samokian and Defaian (Muhesen, 1992).

So far Tabun is the only site in the Near East to provide a 'complete' sequence of all the relevant industries in which Mugharan developments through time can be demonstrated. Although some information about the vertical column has been published, it would be interesting to have spatial information as well (would the 'facies' change laterally as they do in Bezez?). In a final word on this subject, the 'butterfly' diagram (Copeland and Hours, 1981, 234, Fig. 4) demonstrates the various techno-typological relationships of the Mugharan industries.

Dating the Yabrudian

The two main sources for dating Palaeolithic artefact assemblages to be discussed here are the radiometric determinations and the Pleistocene marine evidence.

Radiometric Determinations

Mention has been made above of the changing ideas about the dating and re-dating of the Mugharan over the years as new data emerged and new dating methods came on line.

The resulting confusion is well illustrated by the number of authors who published tables comparing the different interpretations (e.g., Schwarcz, 1994, Fig. 2) or proposing alternative hypotheses (Bar-Yosef, 1989, Fig. 30.2). The dates have gradually become older. However, since the material and the methods (U-series, TL and ESR) have been much aired, the stages in this back-dating process need not be recounted here in detail, nor will the various methods be described.

A significant advance came with the publication of ESR dates based on bovid teeth from Tabun, which gave average dates of about 200 kyr BP (Oxygen Isotope Stage 7) for Tabun Ed, and up to 150 kyr BP (Oxygen Isotope Stage 6) for Eb (Grün et al., 1991). At this time a TL date of about 195 kyr BP had just been obtained for Yabrud I, Level 18 (Oxford Research Laboratory, 1990). Criticism was not lacking regarding these interpretations, based on the manifest discrepancies between assemblages dated differently by different methods, as well as other inconsistencies (e.g., Farrand, 1994).

The situation drastically changed when articles appeared giving still older TL dates on 37 burned flints from Jelinek's excavation (Mercier and Valladas, 1994; Mercier et al., 1995). The calculated mean ages were given as 331 kyr BP for Unit XIII (Layer Ed, on 4 flints), 350 for Unit XII (Ec-Eb, on 4 flints), 306 kyr BP for Unit XI (on 10 flints) and 270 for Unit X (Layer Ea - D, on 2 flints) (Mercier et al., 1995, 501). The actual age of each flint is given in Table 2 of that report. These authors propose that the 'deposition of units XIII-II (Layers Ed to D) took place between 330 and 210 kyr BP spanning the oxygen isotope stages 9 and 8. After a gap of several tens of thousands of years the sediments in unit I (Layer C) were deposited at about 171 ± 17 kyr BP, at the end of Oxygen Isotope Stage 7 or the start of Stage 6' (Mercier et al., 1995, 508). Those authors commented that the transition from the Mugharan to the Mousterian would have taken place around c. 250 - 270 kyr BP (Tabun Unit X). This date is not incompatible with similar, recently conducted datings for the French Mousterian which set the age as much older. Nor is it incompatible with recent dating of the Amudian of Unit XI, which has been set for the first time to over 300 kyr BP. This suggests a position for the transition at the end of Oxygen Isotope Stage 9, which is consistent with the evidence from Adlun, where the the Amudian is found in beach deposits formed during a retreat of the sea (Mercier et al., 1995, 506). The principles, methods, tests, necessary estimations, problems of gaps, overlaps etc. are discussed by Mercier et al. (1995) and reasons given for believing in the accuracy of the results.

The present chaotic situation places Tabun 'out of sync' with the other Mugharan sites and suggests the inconceivable: that the Mugharan there ended long before it had begun at the other sites. We find it hard to believe that the Mugharan Tradition industries are not at least broadly contemporary. Therefore, if these new dates are accepted, chronological revisions will surely have to be made. All the other Mugharan Tradition assemblages (still thought to refer to Oxygen Isotope Stages 7 - 6) will have to be redated. Mercier and colleagues (1995, 505-506) maintain that the dates from El-Kowm (Arida, Hummal lb, Umm el-Tleil and Umm Qubeiba), Zuttiyeh and Yabrud I.18 (conducted on travertines and burned flints), as well as the ESR dates from Tabun (on fossil-teeth), must be treated as upper limits. Let us now turn to the question as to how the marine chronology would relate to the new perspectives.

Marine Chronology

The effects of Pleistocene sea-level fluctuations on the impressive limestone coastline of Lebanon and Syria (which is judged to have remained relatively stable in the sense that when it moves, it does so *en bloc*, in relation to glacio-eustatic and tectonic movements) have been much studied in the past (e.g., Wetzel and Haller, 1944). Sanlaville has published a relative chronology for the marine sequence, based on traces of Pleistocene retreats and transgressions in the form of fossil beaches, abrasion platforms, dead-cliffs, etc. (Sanlaville, 1977). The marine transgressions which concern us here are Sanlaville's Jbailian I and II ('Mindel-Riss'), the Pre-Enfean and Enfean I and II, which he had assigned to the Riss-Würm, or Oxygen Isotope Stage 5.

Recently, Sanlaville (1998) has revised his earlier scheme (Sanlaville, 1977; 1981), based on new knowledge concerning the oxygen isotope chronology, as well as further study of the Eastern Mediterranean littoral (Besançon et al., 1994; Sanlaville et al 1995; Sanlaville 1996; 1998). It appears that although eustacy and tectonics play their part in affecting Quaternary sea-level rises and falls, climatically-controlled effects on the Libano-Syrian coastline were (and are) more important than heretofore thought. This is partly due to the proximity of the well-watered mountain chain to the ('drowned') coastline, which has resulted in quantities of alluvium being transported by east-west streams flowing to the seashore. This material is then propelled northwards along the coast by 'longshore drifting' (winds and waves hitting the shore at an angle from the SSW). When augmented by sand, shells and pebbles from shallow seas thrown up by storms, this material can form substantial beaches and bars, giving the false impression of a rise in sea level. It seems that such effects occur cyclically. Sanlaville (1998) suggests a sequence of three phases, as follows:

In Phase 1, torrential erosion during a deterioration of the climate produces the above mentioned disturbances, and storm beaches are deposited at a higher altitude than the actual sea-level. These beaches can contain dead-assemblages of fauna, as in, for example, the famous *Strombus bubonius* conglomerates of the erstwhile Tyrrhenian beaches.

In Phase 2, when equilibrium returns and the vegetation on the slopes re-establishes, erosion diminishes and the volume of beach materials is reduced by aeolian evacuation and tidal usure. The warm sea can then erode the shore, which retreats inland leaving a strip of 'beachrock' at sea level.

In Phase 3 there is a period of relative stability with little erosional activity. The clear waters favour the installation of vermets on the pavements and 'nicks' in the beachrock. This holds until the equilibrium is upset once more by a phase of irregular or violent precipitation which produces another volume of debris.

Sanlaville believes that the stages and sub-stages seen in, for example, the Enfean transgression (Oxygen Isotope Stage 5) consist of several of these thousand-year-long 'climato-sedimentary' cycles. This makes it less certain than he at first thought (based on the absence of any beach section containing two *Strombus* layers) that only one '*Strombus*' episode occurred on the Levantine coast. As a working hypothesis, Sanlaville proposes the following relative chronology for the Upper Pleistocene shorelines of the Levant:

The beaches formerly referred to the Pre-Enfean (Sanlaville, 1977) should be put back to Oxygen Isotope Stage 7. The preceeding beaches (such as those containing the Acheulean at Ras Beirut, for example) should be put back to Oxygen Isotope Stage 9. The *Strombus* beaches could now be referred to Oxygen Isotope Stage 5e (Enfean I), as could the 8 - 15 m beach sites at Ras Beirut containing 'Levalloisian'. The occupation layers at Naame (and probably Ras el-Kelb) could belong to a subsequent regressive phase (5d?), the U /Th dates of c. 93 kyr BP being regarded as too young. If the *Vermet* level at Naame refers to the next rise (5c), then the lowest *Strombus* levels near sea level could represent a later rise (5a?).

Taking all the available evidence into account, Sanlaville suggests that the Enfean of Oxygen Isotope Stage 5 is not the phase responsible for the beaches at Adlun. However, the evidence as to the number of fossil sea levels represented there is conflicting as well as incomplete. This confines us to archaeological interpretations. Two hypotheses can be suggested:

Alternative 1. The most logical placement would be in Oxygen Isotope Stage 7 (Pre-Enfean), since it is hard to see how the occupation deposits could have survived sea-level returns if they referred to Oxygen Isotope Stage 9. Oxygen Isotope Stage 7 accords with the ESR dates, around 230 - 150 kyr BP.

Alternative 2. According to the Tabun TL dates, the Mugharan Tradition industries in Bezez should refer to immediately post Stage 9. Jbailian II should date around 300 kyr BP, with the occupations continuing in Oxygen Isotope Stage 8. The same should hold true for Abri Zumoffen (Sanlaville, 1998).

Conclusions

The available archaeological and other evidence concerning the Mugharan, discussed in this paper, has so far not provided us with the reasons for its peculiar distribution or its lithic variability. With Galilee Man as the only fossil known to be tentatively associated with a Mugharan assemblage, we have only vague ideas as to who the Yabrudians were, from whom they were descended (unless it was the Late Acheuleans) and what happened to them (unless they 'became' the Mousterians). All these subjects are beyond the scope of this paper and will no doubt be dealt with by other contributors.

The problem of the dates remains unsolved for the moment. Until these are more securely established, it will be difficult to comment usefully on the relationships between the Mugharan and like industries elsewhere in the world, such as the Charentian or the African MSA. However, Mercier and colleagues (1995, 507) point out that their dates for Tabun bring the Mugharan into line with the dates for the French Mousterian industries, now known to date (by 'modern radiometric methods' and other evidence) to before the penultimate glacial.

Meanwhile, we seem to be relying heavily on the oxygen isotope stages as they are presently published. What would happen if the next 'bombshell' comes from revisions of this chronology? Prehistorians' imaginations must be prepared to boggle.

Author's address: Chateau de Marouatte, Grand Brassac, 24350 Tocane St. Apre, France

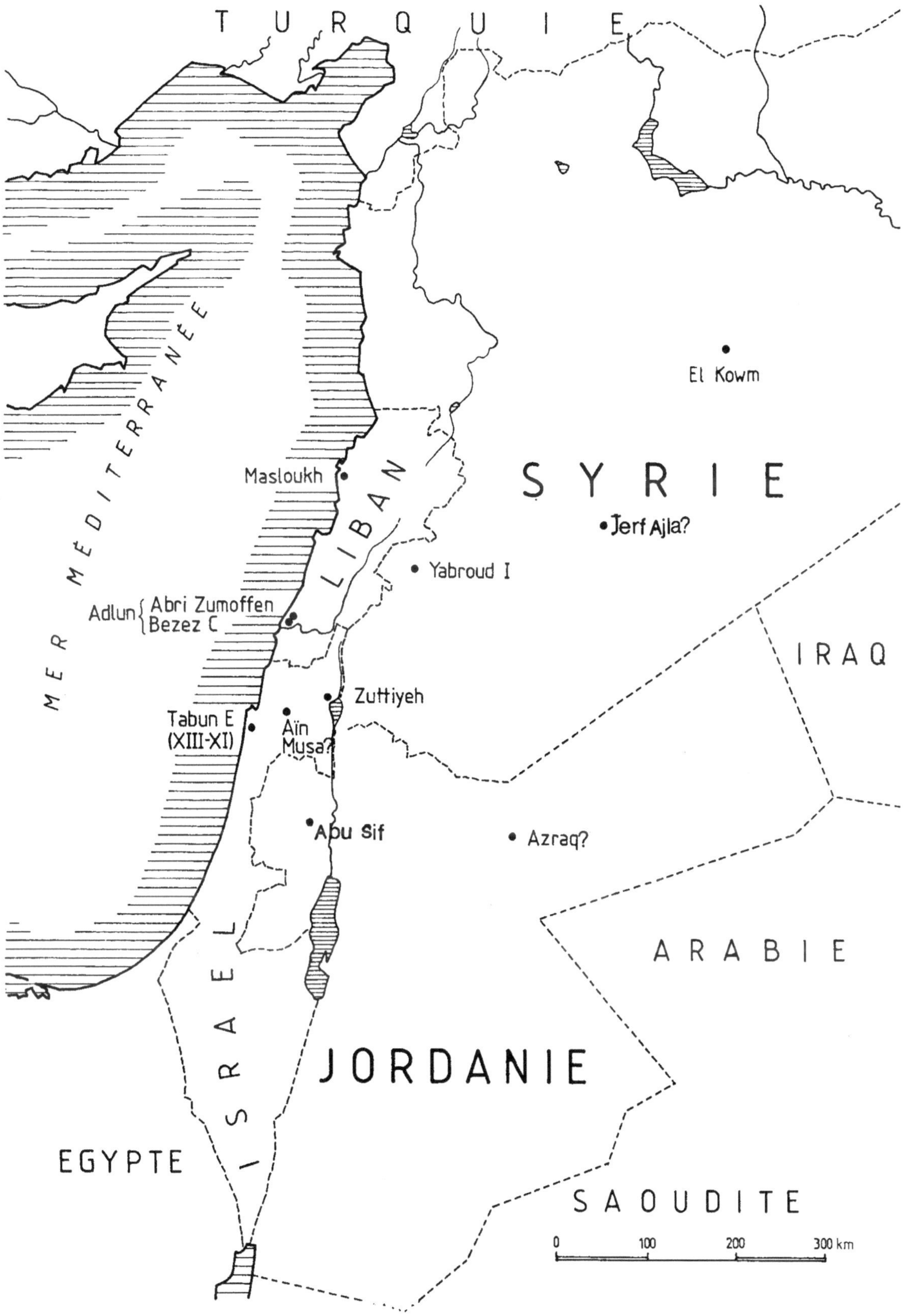

Fig. 1. Mugharan Tradition sites, after Copeland and Hours 1983, Map 1. The El-Kowm sites are Hummal 1b, Arida 10, Ain Juwal 13, Nadaouiyeh 7, Umm el-Tleil 26, Ain Beni Ali, Ain Cheikh Ali, Ain Sebkha and Umm Qubeiba.

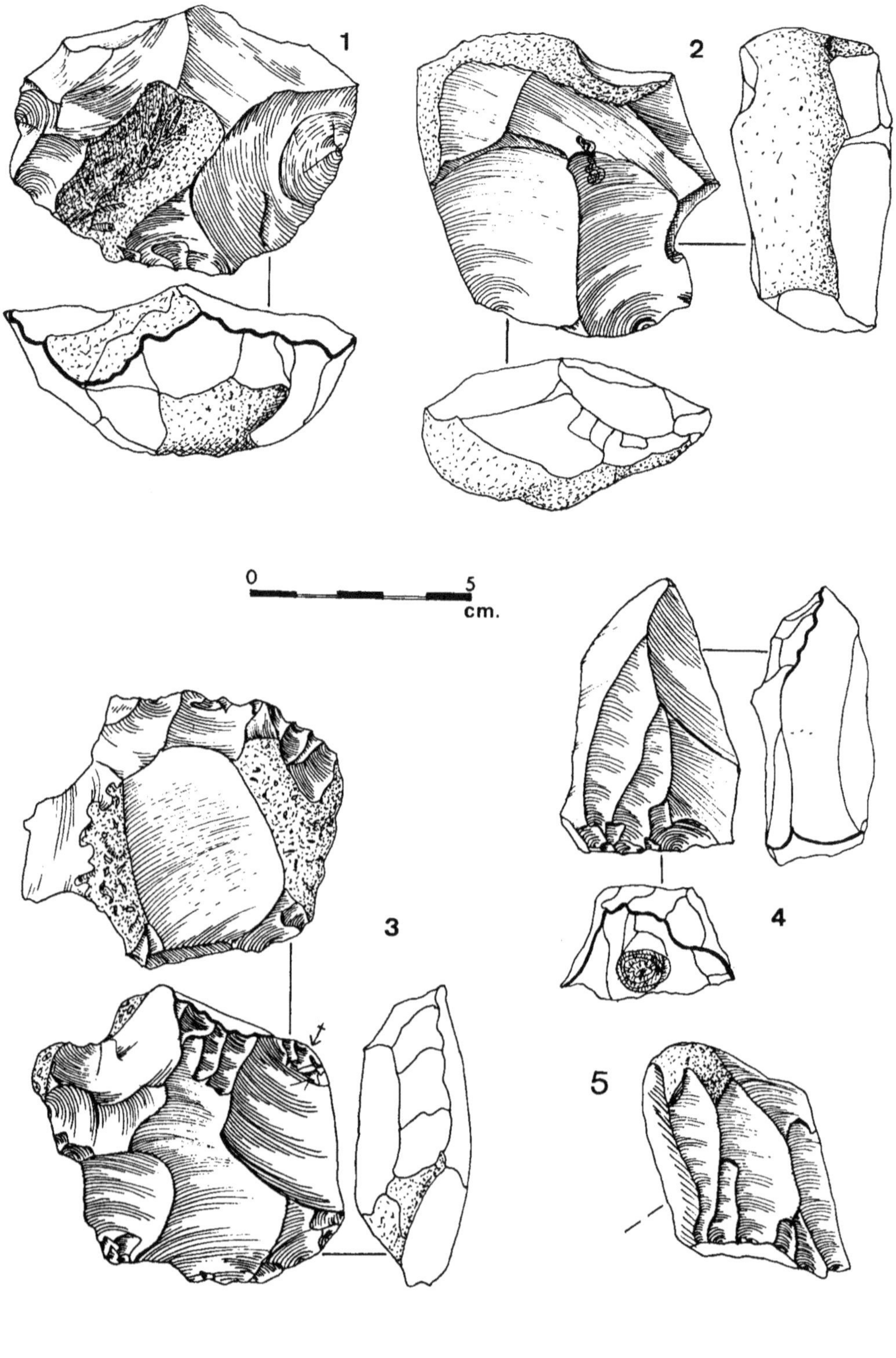

Fig. 2. A selection of cores from Abri Zumoffen. 1-4 - Beach Industry; 5 is a prismatic core from the Amudian of Trench A. After Copeland, 1983b, Plate Z.2.

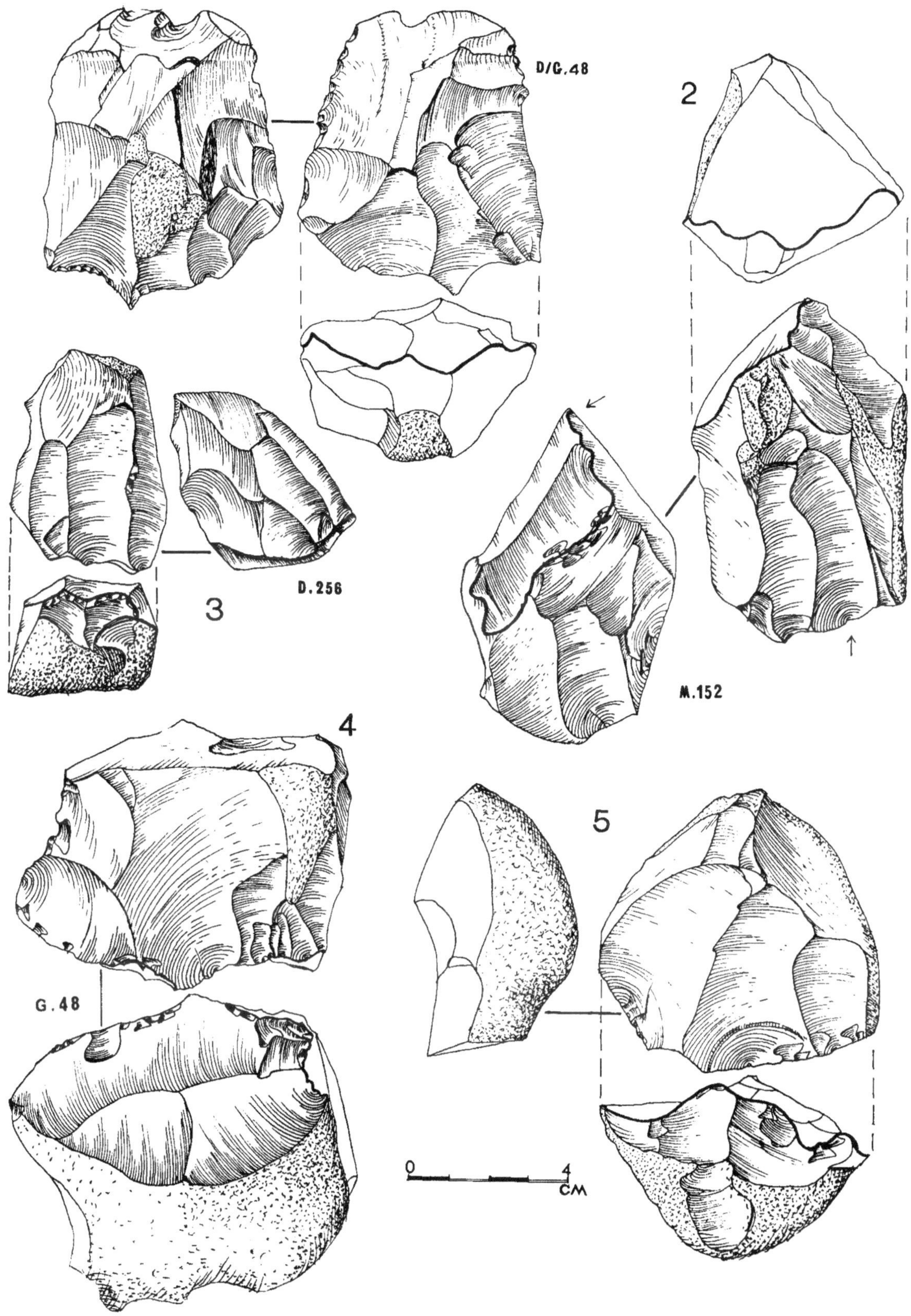

Fig. 3. A selection of the smaller cores from Bezez Level C. After Copeland, 1983b, Plate C.26 (see also the 'chopping tools', Plate C.2).

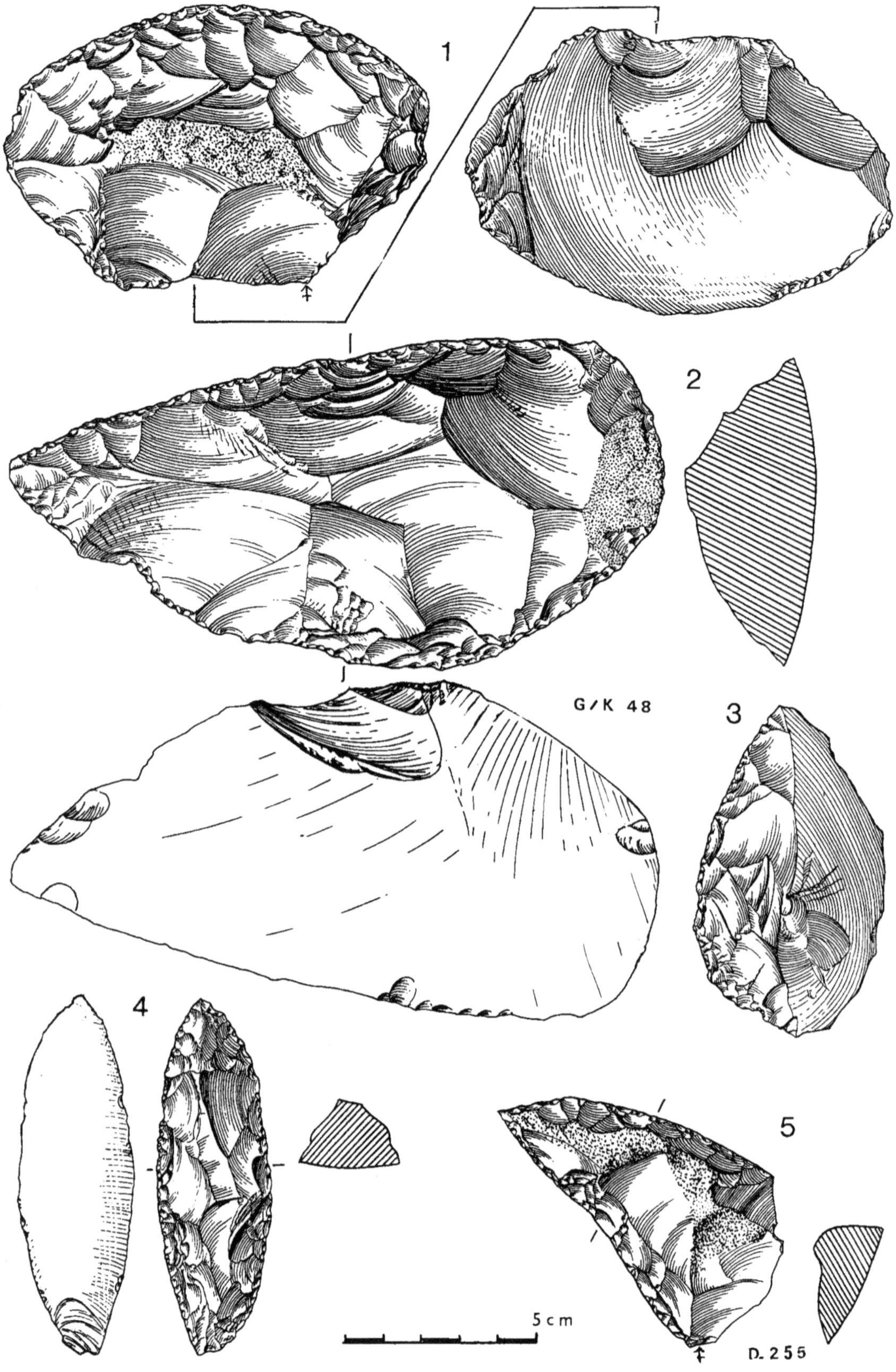

Fig. 4. Racloir types from Bezez Cave level C. After Copeland, 1983b, Plates C.14, C.20 and C.21. 1 - Transverse convex Quina racloir with butt and bulb removed; 2 - transverse straight Quina racloir with butt removed and bulb thinned; 3 - a 'bizarre' specimen, racloir on the ventral surface of a cortex flake; 4 - double-pointed limace; 5 - 'winkelkratzer' type of offset racloir.

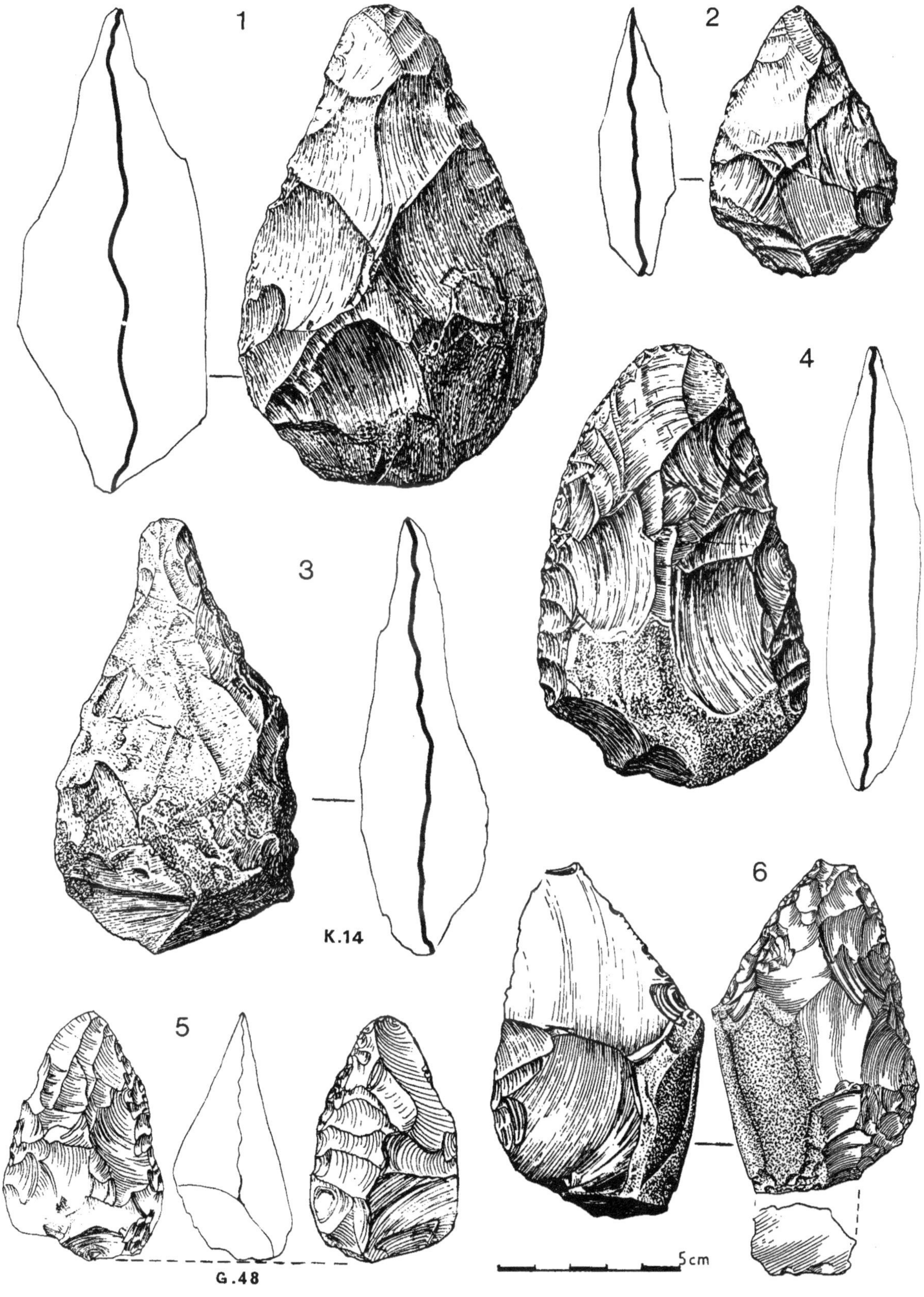

Fig. 5. A selection of medium-sized and small bifaces from Bezez level C. After Copeland, 1983b, Plates C.7, C.8 and C.10. 1 - Amygdaloid; 2 - cordiform; 3 - Micoquian; 4 - elongated cordiform; 5 - piriform, perhaps on a flake; 6 - partial biface or bifacial racloir.

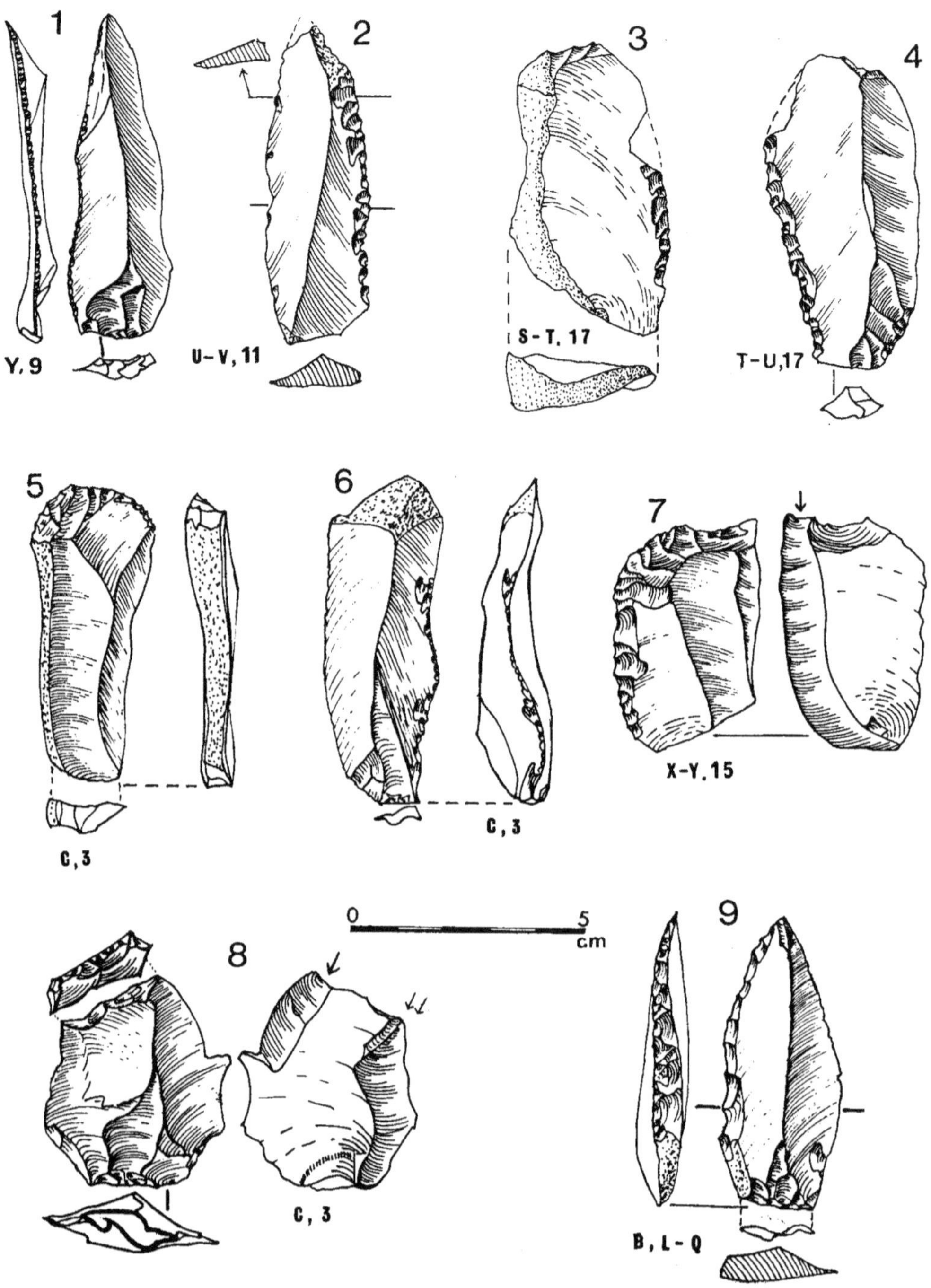

Fig. 6. A selection of Amudian tools from Abri Zumoffen. After Copeland, 1983b, Plates Z.9 and Z.10. 1, Backed knife; 2 -, retouched blade or single racloir; 3 - single racloir on the inverse of an offset cortex flake; 4 - denticulated blade; 5 - end-scraper; 6 - nibbled blade; 7 - composite racloir and burin; 8 - double Adlun burin; 9 - backed knife, cf. Chatelperron Point.

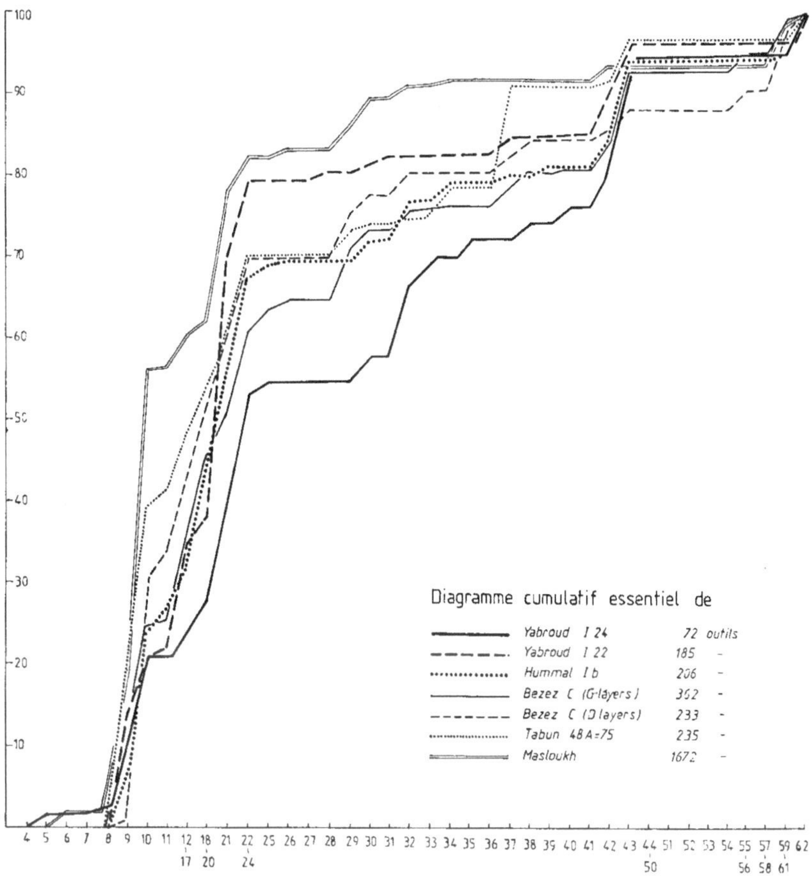

Fig. 7. Bordean (1961) cumulative graph comparing the proportions of flake tools in Yabrudian assemblages. After Copeland and Hours, 1983, Fig. 6.

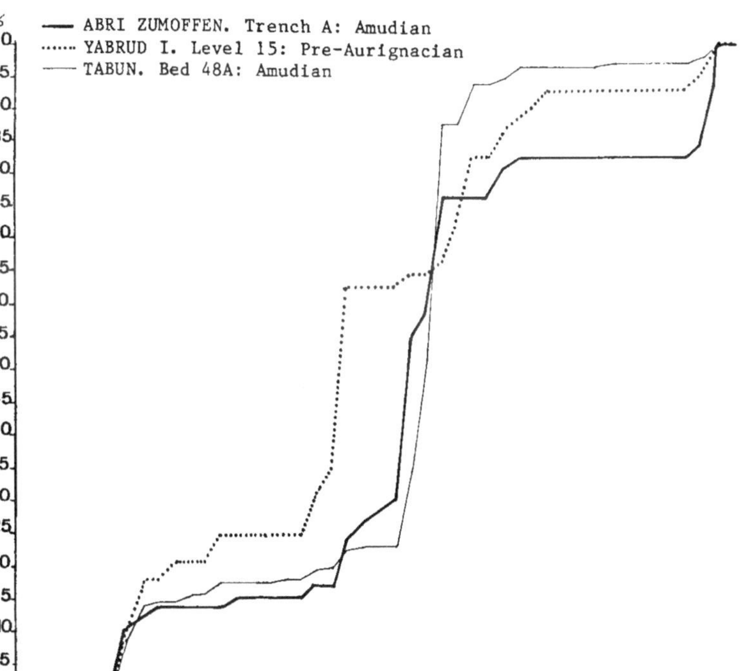

Fig. 8. Cumulative graph comparing the proportions of flake and blade tools in Amudian and Pre-Aurignacian assemblages. After Copeland, 1983b, Fig. Z.7.

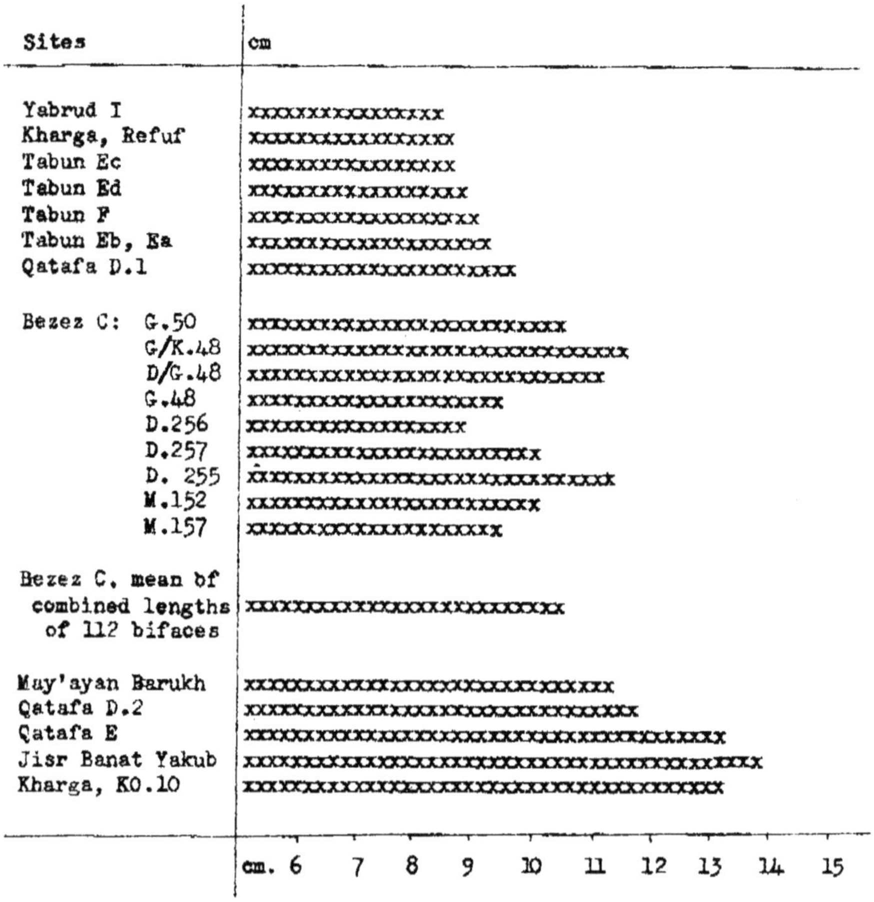

Fig. 9. Comparison of the mean lengths of biface groups. After Gilead, 1970, Fig. 2, with the addition of Bezez bifaces from Copeland, 1983b, Fig. C.21.

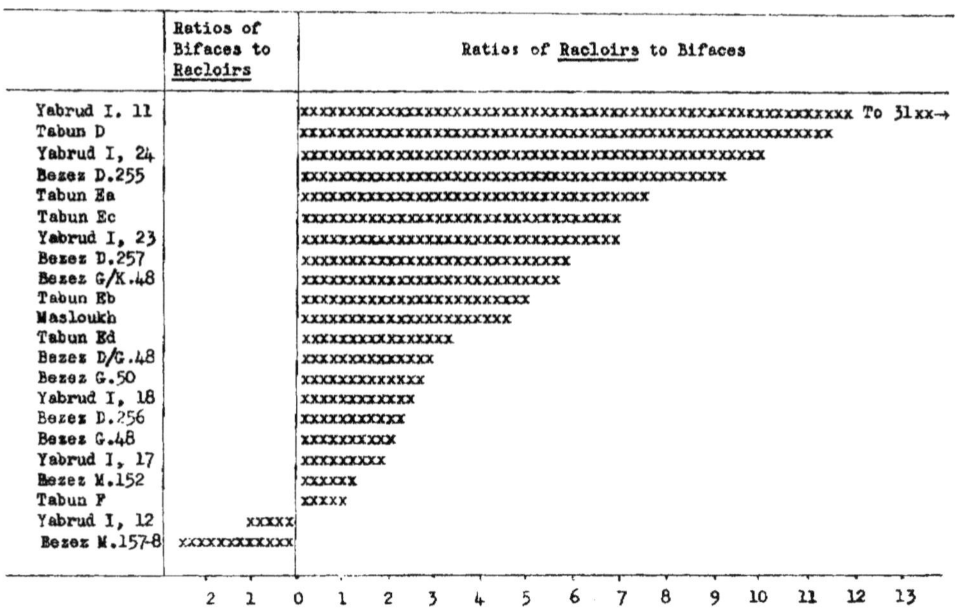

Fig. 10. Comparison of the ratios of racloirs to bifaces at Tabun, Yabrud and Bezez. After Copeland, 1983b, Fig. C.22.

Bibliography

In cases where an author has written several articles on the same subject, usually the latest is the one cited here.

Bar-Yosef O. 1982 Some remarks on the nature of transitions in Prehistory. In: Ronen, A. (ed.), *The Transition from Lower to Middle Palaeolithic and the Origin of Modern Man.* BAR IS151, 29-33

Bar-Yosef O. 1989 Geochronology of the Levantine Middle Paleolithic. In: Mellars, P. and C. Stringer, (eds.), *The Human Revolution.* Edinburgh University Press, 589-610.

Bar-Yosef O. 1994 The Lower Paleolithic in the Near East. J. *World Prehistory* 8, 211-265.

Bergman C. and K. Ohnuma, 1983 Technological Notes on some Blades from Hummal Ia, EI-Kowm, Syria. *Quartär* 33/34, 171-180.

Besançon J., L. Copeland, F. Hours, S. Muhesen and P. Sanlaville 1982 Prospection géographique et préhistorique dans la bassin d'EI-Kowm (Syrie); rapport preliminaire. *Cahiers de l'Euphrate* 3, 9-26.

Besançon J., L. Copeland, F. Hours, S. Muhesen and P. Sanlaville 1994 Prospection géomorphologique et préhistorique dans la region de Tartous (Syrie) *Paléorient* 20/1, 5-19.

Besançon J. and P. Sanlaville 1988 L'evolution géomorphologique du Bassin d'Azraq (Jordanie) depuis le Pleistocene Moyen. *Paléorient* 14, 23-30.

Boëda E. 1995 Levalloiss: un concept volumetrique, des methods, une technique. In: Bar-Yosef, O. and H. Dibble (eds.), *The Definition and Interpretation of Levallois Technique.* Prehistory Press, Philadelphia, 93-116.

Bordes F. 1955 Le Paléolithique inférieur et moyen de Yabrud (Syrie) et la question du Pre-Aurignacien. *L'Anthropologie* 59, 486-587.

Bordes F., 1961 *Typologie du Paléolithique ancien et moyen.* Delmas, Bordeaux.

Cauvin M.-C., J. Cauvin and D. Stordeur, 1979 Recherches préhistoriques à el-Kowm (Syrie). *Cahiers de l'Euphrate* 2, 80-117.

Coon C. 1956 *The Seven Caves.* Knopf, New York.

Copeland L. 1975. The Middle and Upper Paleolithic of Lebanon and Syria in the light of recent research In: Wendorf, F. and A. E. Marks, (eds.), *Problems in Prehistory: North Africa and the Levant.* Southern Methodist Univ. Press, Dallas, 317-350.

Copeland L. 1981 Chronology and distribution of the Middle Paleolithic as known in 1980 in Lebanon and Syria. In: Cauvin, J and P. Sanlaville (eds.), *Préhistoire du Levant.* CNRS, Paris, 239-263.

Copeland. L. 1983a The 1958 Excavation of Abri Zumoffen in Retrospect and the Extension of 1963. In: Roe, D. (ed.), *Adlun in the Stone Age. The excavations of D.A.E. Garrod in the Lebanon 1958-1963*, BAR IS 159, Oxford, 75-86.

Copeland L., 1983b The Palaeolithic Stone Industries. In: Roe, D. (ed.), *Adlun in the Stone Age. The excavations of D.A.E. Garrod in the Lebanon 1958-1963,* BAR IS 159, Oxford, 89-365.

Copeland L., 1983c Levallois/non-Levallois determinatiens in the early Levant Mousterian: Problems and Questions for 1983. *Paléorient* 9, 15-28.

Copeland L. 1985 The Pointed tools of Hummal Ia (El-Kowm, Syria). *Cahiers de l'Euphrate* 4, 177-190.

Copeland L. 1989a Analysis of the Paleolithic artifacts at C-Spring, Azraq, 1985 season. In: Copeland L. and F. Hours (eds.), *The Hammer on the Rock. Contributions on the Early Paleolithic of Jordan.* BAR IS 540, Oxford, 325-390.

Copeland L. 1989b Surface finds from Ain el-Beidha, Wadi Rajil and Qa Jashsha. In: Copelahd, L. and F. Hours (eds.), *The Hammer on the Rock. Contributions on the Early Paleolithic of Jordan.* BAR IS 540, Oxford, 451-468.

Copeland L. 1995 Are Levallois flakes in the Acheulean the Results of Biface preparation? In: Dibble H. and O. Bar-Yosef (eds.), *The Definition and Interpretation of Levallois Technology.* Prehistory Press, Philadelphia, 171-184.

Copeland L. and F. Hours 1981 La fin de l'Acheuléen et l'avénement du Paléolithique moyen en Syrie. In: Cauvin J. and Sanlaville P. (eds.), *Préhistoire du Levant,* C.N.R.S. Paris, 225-238.

Copeland, L. and F. Hours 1983 Le Yabrudieh d'EI-Kowm (Syrie) et sa place dans le Paléolithique du Levant. *Paléorient* 9, 21-37.

Copeland L. and N. Moloney (eds.) 1998 *The Mousterian Cave Site of Ras el-Kelb*, Lebanon. BAR IS 706, Oxford.

Dibble H. 1988 The interpretation of Middle Paleolithic Scraper reduction patterns. In: Otte M. (ed.), *L'Homme de Neanderthal, 4, La Technique,* ERAUL, Liège, 49-58.

Farrand W. R. 1982 Environmental conditions during the Lower/Middle Paleolithic transition in the Levant and the Balkans. In: Ronen A. (ed.), *The Transition from Lower to Middle Palaeolithic and the Origin of Modern Man.* BAR IS 151, Oxford, 105-108.

Farrand W. R. 1994 Confrontation of geological stratigraphy and radiocarbon dates from Upper Pleistocene sites in the Levant. In: Bar-Yosef O. and R. S. Kra (eds.), *Late Quaternary Chronology and Paleoclimates of the Eastern Mediterranean.* Radiocarbon, University of Arizona, Tucson, 33-53.

Fleisch H., J. Remiro and P. Sanlaville 1969 Gisements préhistoriques découverts dans la region de Batroun (Liban). *Bulletin du Musée de Beyrouth* XLV, 1-28

Garrard A. 1983 The Palaeolithic Faunal Remains from Adlun and their Ecological Context. In: D. Roe (ed.), *Adlun in the Stone Age.* BAR IS 159, Oxford, 397-410.

Garrard A., A. Betts, B. Byrd and C. Hunt 1987 Prehistoric environment and settlement in the Azraq Basin: and interim report on the 1985 excavation season. *Levant* XIX, 5-26.

Garrod D. A. E. 1956 'Acheuleo-Jabrudian' et 'Pre-Aurignacian' de la grotte du Taboun (Mont Carmel): étude stratigraphique et chronologique. *Quaternaria* 3, 39-59.

Garrod D. A. E. 1958 The Ancient Shore-line of the Lebanon and the Dating of Mount Carmel Man. In: G. von Konigswald (ed.), *Hundert Jahre Neanderthaler*, Utrecht, 182-184.

Garrod D. A. E. 1962 The Middle Palaeolithic of the Near East and the problem of Mount Carmel Man, *Journal Royal Anthrop. Inst.* 92/2, 232-259.

Garrod D. A. E. 1966 Mugharet el-Bezez, Adlun: interim report, July, 1964. *Bulletin du Musée de Beyrouth* XIX, 5-10.

Garrod D. A. E. and D.M.A. Bate 1937 *The Stone Age of Mount Carmel,* I. Clarendon Press, Oxford.

Garrod D. A. E. and D. Kirkbride 1961 Excavation of the Abri Zumoffen, a Paleolithic Rock-shelter near Adloun, South Lebanon, 1958. *Bulletin du Musée de Beyrouth* XVI, 7-46.

Gilead D. 1970 Handaxe industries in Israel and the Near East. *World Archaeology* 2, 1-11.

Gissis I. and O. Bar-Yosef 1974 New Excavations in Zuttiyeh Cave, Wadi Amud, Israel. *Paléorient* 5, 175-180.

Goren-Inbar N. 1994 The Lower Paleolithic of Israel. In: T. Levy, (ed.), *The Archaeology of Society in the Holy Land.* Leicester University Press, London, 93-109.

Grün R., C. Stringer and H. Schwarcz 1991 ESR dating of teeth from Garrod's Tabun cave collection. *Journal of Human Evolution* 20, 231-248.

Hours F. 1979 La Fin de l'Acheuléen en Syrie du Nord, note préliminaire. *Paléorient* 5, 9-16.

Heinzelin J. de 1966 Revision de site de Yabroud. *Annales archéologiques Arabes Syriennes* XVI/2, 157.

Hennig G. and F. Hours 1982 Dates pour le passage entre l'Acheuléen et le Paléolithique Moyen à El-Kowm (Syrie). *Paléorient* 8/1, 81-83.

Jelinek A. J. 1975 A preliminary report on some Lower and Middle Paleolithic industries from the Tabun Cave, Mount Carmel (Israel). In: Wendorf, F. and A. Marks, (eds.), *Problems in Prehistory: North Africa and the Levant*, Dallas, 279-316.

Jelinek A. J. 1981 The Middle Paleolithic in the Southern Levant from the perspective of the Tabun Cave. In: Cauvin J. and P. Sanlaville, (eds.), *Préhistoire du Levant,* C.N.R.S., Paris, 265-280.

Jelinek A. J. 1982 The Middle Paleolithic in the Southern Levant with comments on the Appearance of modern *Homo Sapiens.* In: A. Ronen, (ed.), *The Transition from Lower to Middle Palaeolithic and the Origin of Modern Man* BAR IS 151, Oxford, 57-104.

Jelinek A. J. 1992 Problems in the chronology of the middle Paleolithic and the first appearance of early modern *Homo sapiens* in Southwest Asia. In: Akazawa T., K. Aoki and T. Kimura (eds.), *The Evolution and Dispersal of Modern Humans in Asia.* Hokusen-sha, Tokyo, 253-275.

Jelinek A. J., W. R. Farrand, G. Haas, A. Horowitz and P. Goldberg 1973 New Excavations at the Tabun Cave, Mount Carmel, Israel, 1967-72: a preliminary report. *Paléorient* 1, 151-183.

Kirkbride D. 1983 The Excavations at Adlun, 1958 and 1963; Section I. The Soundings in Bezez Cave. In: D. Roe, ed. *Adlun in the Stone Age,* BAR IS 159, Oxford, 23-68.

Kirkbride D. 1989 Description of Stone Age Sites at Azraq, Jordan. In: Copeland L. and F. Hours (eds.), *The Hammer on the Rock*, BAR IS 540, Oxford, 153-169.

Kirkbride D., S. de St. Mathurin and L. Copeland 1983 Results, tentative interpretations and suggested chronology. In: Roe, D., (ed.), *Adlun in the Stone Age* BAR i.s. 159, Oxford, 415-431.

Leroi-Gourhan A. 1980 Les analyses pollinique au Moyen Orient. *Paléorient* 6, 71-91.

Le Tensorer J.-M. and F. Hours 1989 L'occupation d'un territoire a la fin du Paléolithique ancien et au Paléolithique moyen à partir de l'example d'El-Kown (Syrie), In: M. Otte, (ed.), *L'Homme de Neanderthal* 6, ERAUL, Liège, 107-114

McCown T. and A. Kieth 1939 *The Stone Age of Mount Carmel II.* Clarendon Press, Oxford.

Mercier N. and H. Valladas 1994 Thermoluminescence dates for the Paleolithic Levant. In: Bar-Yosef O. and R. Kra, (eds.), *Late Quarternary Chronology and Paleoclimates of the Eastern Mediterranean* Radiocarbon, University of Arizona, Tuscon, 13-20.

Mercier N., H. G. Valladas, J.-L. Reyss, A. J. Jelinek, L. Meignen and L. Joron 1995 TL Dates of burnt flints from Jelinek's excavations at Tabun and their Implications. *Jrnl. Archeol. Sci.* 22, 495-509.

Muhesen S. 1992 The Transitional Lower-Middle Paleolithic Industries in Syria. In: Akazawa T., K. Aoki and T. Kimura (eds.), *The Evolution and Dispersal of Modern Humans in Asia.* Hokusen-Sha, Tokyo, 51-65.

Neuville R. 1931 l'Acheuléen supérieure de la grotte d'Oumm Qatafa (Palestine). *L'Anthropologie* 41, 13-51, 249-263.

Neuville R. 1951 *Le Paléolithique et le Mésolithique du desert de Judée.* Archives de l'Inst. Paléont. Humaine, Memoire 24.

Oxford Research Laboratory for Archaeology 1990 Date List 4. *Ancient TL* 8(3), 43.

Roe D., (ed.) 1983 *Adlun in the Stone Age.* BAR IS 159, Oxford.

Rollefson G. 1983 Two Seasons of Excavation at Ain el-Assad, Eastern Jordan, 1980-1981. *BASOR* 252, 25-34.

Ronen A. (ed.) 1982 *The Transition from Lower to Middle Palaeolithic and the Origin of Modern Man* BAR IS 151, Oxford.

Rust A. 1933 Beitrag zur Erkenntnis der Abwicklung der vorgeschichtlichen Kulturperioden in Syrien. *Prehist. Zeitschrift* 24, 205-218.

Rust A. 1950 *Die Hohlenfunde von Jabrud (Syrien).* Neumunster.

Sanlaville P. 1977 *Etude Géomorphologique de la region littorale du Liban.* Université Libanaise, Beirut.

Sanlaville P. 1981 Stratigraphie et chronologie du Quaternaiare marin du Levant In: J. Cauvin and P. Sanlaville, (eds.), *Préhistoire du Levant,* C.N.R.S. Paris, 21-32.

Sanlaville P. 1996 Changements climatiques dans la région levantine à la fin du Pleistocene supérieur et au debut de l'Holocene. Leurs relations avec l'évolution des sociétés humaines. *Paléorient* 22/1, 7-30.

Sanlaville P. 1998 The deposits of Ras el-Kelb in the Framework of the Regional Chronostratigraphy In: Copeland L. and Moloney N., *The Mousterian Cave Site of Ras el-Kelb.* BAR IS 706, Oxford, 37-44.

Sanlaville P., R. Dalongeville et J. Evin 1995 Rythmes et modalites de évolution du littoral syrien à l'Holocene. *Bul. Assoc. Georg. Fr.* 5, 410-427.

Schroeder B. 1966 The Lithic Material from Jerf Ajla: A preliminary report. *Annales. Arch. Arab. Syriennes* XVI/2 201-213.

Schroeder B. 1969 *The Lithic Industries of Jerf Ajla and their Bearing on the Problem of the Middle to Upper Paleolithic Transition.* Unpublished Dissertation, Columbia University.

Schwarcz H. 1994 Chronology of Modern Humans in the Levant. In: Bar-Yosef O. and R. Kra (eds.), *Late Quaternary Chronology and Paleoclimates of the Eastern Mediterranean.* Radiocarbon 14, University of Arizona, Tuscon, 21-32.

Skinner J. 1970 El-Masloukh: a Yabrudian site in Lebanon. *Bul. Mus. Beyrouth* XXIII, 143-172.

Solecki R. S. 1970 A sketch of the Columbia University Archaeological Investigations at Yabroud (Syria). *Fundamenta*, Reihe A. Band 2, 199-211. Cologne.

Solecki R. S. and R.L. Solecki 1986 A Reappraisal of Rust's cultural stratigraphy of Yabroud Shelter I. *Paléorient* 12/1, 53-60.

Turville Petre F. 1927 *Researches in prehistoric Galilee 1925-1926 and a report on the Galilee Skull.* London, British School of Archaeology in Jerusalem.

Vandermeersch B. 1989 The Evolution of modern humans: Recent evidence from southwest Asia. In: Mellars P. and C. Stringer, (eds.), *The Human Revolution.* Edinburgh University Press, Edinburgh, 155-164.

Vandermeersch B. 1995 Le rôle du Levant dans l'évolution de l'humanité au Pleistocene Supérieur. *Paléorient* 21/2, 25-34.

Wetzel R. and J. Haller 1945 Le quaternaire côtier de la region de Tripoli (Liban). *Notes et Mem. de la Section Géologique. Déleg. Génerale de Fr. au Levant,* 4, 1-48.

Zumoffen G. 1900 *La Phènicie avant les Phèniciens.*

OBSERVATIONS ON THE YABRUDIAN OF TABUN CAVE, ISRAEL

Anat Shifroni and Avraham Ronen

The cave of Tabun, on the southern bank of Nahal Hamearot in Mount Carmel, has become a yardstick in world prehistory during Dorothy Garrod's initial excavations in the 1930s (Garrod and Bate 1937). The roughly 25 m thick sequence (Fig. 1) still provides, after some 70 years, the most important reference for the late Lower Palaeolithic and the Middle Palaeolithic of Western Asia (e.g., Bar-Yosef, 1994; Goren-Inbar, 1994).

Garrod's layers do not always conform to the very complex natural stratigraphy in the cave, largely affected by bending and sliding into a central 'swallow-hole'. This entailed some mixture of the cultural content in Garrod's excavations in the upper half of the sediments, and considerably more so in the lower half, which was far more heavily affected by the slidings.

New Excavations were undertaken in Tabun by A. J. Jelinek between 1967 and 1972. Jelinek re-excavated Garrod's layers C, D and the upper part of E (Jelinek, 1977; 1982; Jelinek et al., 1973). Our excavations (A. R.), started in 1975, continue those of Jelinek in re-checking the remainder of Tabun sequence, Garrod's lower E and layers F and G (Ronen and Tsatskin, 1995).

Layer E, some 5-7 m thick, was initially termed Micoquian by Garrod. Following Rust's excavations in the Syrian Rockshelter Yabrud I (Rust, 1950), Garrod recognised the Yabrudian character of Layer E.

Again conforming to Rust's observations at Yabrud, Garrod noticed the presence of blade assemblages close to the top of layer E (Garrod, 1956), with blades approaching Upper Palaeolithic quality. At Tabun, however, contrary to Yabrud, this assemblage did not so much resemble the Aurignacian, and Garrod proposed for it the term Amudian. At Tabun, the Yabrudian is covered by the oldest Middle Palaeolithic, biface-free assemblage of Layer D, with Levallois technique.

The Renewed Excavations

In Layer E Jelinek has distinguished an alternation between biface-rich and scraper-rich beds. Jelinek interpreted the two as climatically determined varieties, suggesting that bifaces were more numerous in colder conditions. Jelinek further suggested that the blade-rich Amudian was also an adaptation to some environmental requirements, and hence need not be considered as a separate culture. Accordingly, he created the term 'Mugharan Tradition' to encompass the three occurrences within layer E (Jelinek, 1982). It should be stressed, however, that the extreme scarcity of any blade form in the Yabrudian stands in a stark contradiction to the abundance of fine blades in the Amudian, manufactured in a quasi-Upper Palaeolithic technique (Ronen, 1992). This discrepancy renders the concept of a single Mugharan tradition problematic.

Dating

Initially, the Tabun sequence was believed to cover the Last Interglacial - Last Glacial periods, Isotope Stages 5 through 2 (Garrod and Bate, 1937; Jelinek et al., 1973). Recent dates place, however, the youngest Middle Palaeolithic occupation around 100 Kyears and the Yabrudian layer E, around 350 Kyears (Table 1).

Lithic Analysis

The series used in the present analyses originate in Squares 47 and 48 in Ronen's excavations (Fig. 2), in the south-eastern corner of the swallow-hole, where sediments bend ca. 40° northward and ca. 10° westward (Fig. 3). The results presented here should be regarded as preliminary, because

1. the sample studied is only a part of the material excavated in the corresponding layers, and
2. minute stratigraphical correlations of mini-faults and mini-slides, still under way, may slightly alter the placement of some artefacts. The overall picture, we believe, is not likely to change to any considerable degree.

Two assemblages were recognized: the Upper assemblage, containing 959 items, belongs to the basal part of Garrod's layer E (our layers 64 through 62) and a Lower assemblage with 1514 items from the top part of layer F (our layers 68 through 65). In the present article, the Upper assemblage will be described. Only the flake component will be discussed here; bifaces and cores are excluded.

The Assemblage

About one third of the series studied here consists of modified tools, including 2.5% handaxes (Table 2). This high ratio points definitely to a selection by the Yabrudians, which in turn indicates a residential-activity area as against a knapping zone. The relatively low frequency of cores (14%) and of debris (8.9%) is in line with the former observation. The high number of unretouched blanks indicate that a choice had been made somewhere (at the knapping place?) and suitable blanks were brought to our study location. A part of these blanks may have been used, but wear-sign studies were not yet undertaken.

Technological aspects

Dimensions. The most significant trait appears to be that modified items are larger and thicker than the unretouched

Tabun Layer	ESR EU	LU	TL	U-SERIES
B	86±11	103±16		50.7±0.2
				97.8±0.4
				101.7±1.4
C	102±17	119±11	171±17	105±2.6
			212±22	
			244±28	
D	122±20	166±20	263±27	110±0.9
			270±22	
				159±1.3
Ea	154±34	188±31		168±2.6
			306±33	
Eb	151±21	168±15		
			350±33	
			315±20	
Ec	176±10	199±7	317±36	
			324±22	
			247±27	
Ed	182±15	213±46	331±30	
			415±27	
F				
G				

Table 1. Dating Tabun Layers by Various Methods
After Grün et al., 1992; McDermott et al., 1993; Mercier et Valladas, 1994; Mercier et al., 1995; Mercier et al., nd.

Type		N.	%
Tools		307	32.01
Bifaces		24	2.50
Cores		134	13.97
Debitage		409	42.65
Debris		85	8.86
	Total	959	
Debitage			
Primary element (complete)		10	2.44
Broken primary element		25	6.11
Flakes (complete)		100	24.45
Broken flakes		206	50.37
Broken blades		21	5.13
Biface spall		9	2.20
	Total	409	
Debris			
Chunks		37	43.53
Pebbles		9	10.59
Edge flakes/spall		39	45.88
	Total	85	

Table 2. Major Elements

	Tools (202)		**Flakes (137)**			
	Mean	**S.D.**	**Mean**	**S.D.**	**T**	**Prob>T**
Length	49.38	12.74	42.74	12.88	-4.6735	0.0001
Width	33.85	8.89	30.32	9.04	-3.5485	0.0005
Thickness	11.03	4.18	9.27	4.23	-3.7938	0.0002

Table 3. Metrical Data of Yabrudian Artefacts

blanks. This implies that the largest blanks (=debitage) were chosen to be retouched. Considering that the size of the tools had decreased by the modification, the original difference was even larger than that reflected in Table 3. It is noteworthy that even the thickness played a role in choosing the appropriate blanks to be retouched.

Dorsal scar pattern (Table 4). The parallel and opposite scar patterns are practically equally represented, together forming ca. 55% of the cases. The radial pattern is weakly represented, together with cortical flakes. Some 20% of the flakes have irregular or unrecognisable scar patterns.

Type	%
Radial	6.63
Sub-Parallel	3.37
Parallel	25.05
Opposite	25.05
Natural back	10.53
Cortical	8.0
Irregular	8.63
Unknown	12.63

Table 4. Dorsal Scar Patterns (N=702)

Butt forms (Table 5). The butts are largely governed by cortical or otherwise unprepared, ca. 38% of the cases. Faceted butts are rare. Surprisingly, in two thirds of the flakes the butt is either unknown, broken or removed.

Type	%
Cortical	7.58
Unprepared	30.03
Dihedral	10.88
Faceted	4.68
Unknown	23.44
Broken	32.51
Removed	10.88

Table 5. Butt Forms (N=729)

At the time we were puzzled by this frequency, given that the fine sandy matrix did not constitute an environment in which a high breakage ratio would be expected. It seems that some other cause must be sought, most probably something related to use. Only later, however, did we notice that some 'removed' butts represented a specific pattern which, until now, has only been observed in the Yabrudian. To our knowledge, this deliberate and systematic pattern of severing a flake has not been described in other Yabrudian or Acheulean assemblages

N	Type	N	%
1	Levallois flake, typical	1	0.33
2	Levallois flake, atypical	1	0.33
8	Limace	1	0.33
9	Racloir, simple straight	4	1.30
10	Racloir, simple convex	48	15.64
11	Racloir, simple concave	4	1.30
12	Racloir double straight	1	0.33
13	Racloir double straight – convex	1	0.33
15	Racloir double straight – concave	5	1.63
17	Racloir double concave – convex	1	0.33
18	Racloir convergent straight	9	2.93
19	Racloir convergent convex	18	5.86
20	Racloir convergent concave	1	0.33
21	Racloir dèjeté	4	1.30
22	Racloir transversal straight	2	0.65
23	Racloir transversal convex	5	1.63
25	Racloir on ventral side	4	1.30
28	Racloir, bifacial	7	2.28
30	Grattoir	17	5.54
31	Grattoir, atypical	7	2.28
32	Burin	23	7.49
33	Burin atypical	3	0.98
34	Awl	3	0.98
35	Awl atypical	6	1.95
36	Backed knife	1	0.33
37	Backed knife, atypical	4	1.30
38	Natural backed knife	42	13.68
39	Raclette	2	0.65
40	Truncation	3	0.98
42	Notch	18	5.86
43	Denticulate	9	2.93
44	Burin, alternating	1	0.33
45	Retouch on ventral side	4	1.30
46	Retouch, alternating	1	0.33
48	Steep retouch, alternating	7	2.28
54	Notch on end	4	1.30
62	Divers	14	4.56
65	Emiroid	1	0.33
67	Used item	16	5.21
Total		**307**	

Table 6. Yabrudian Tool Types (Bordes' List)

(e.g., Copeland, 1983). Hence, we propose to term it 'Tabun Snap' (Ronen and Shifroni, nd).

The Tabun Snap

The snap described here affected the proximal end of flakes, and was conducted by a blow on the dorsal face (Fig. 4, 4-5; Fig. 5, 1-3; Fig. 6, 5). This produced a triangular spall with two bulbs of percussion (Fig. 6, 1-2), which we have considered at first as a form of Combewa flake. But soon it was realised that these spalls differed from Kombewa in their form, method of production and probably their function. The removed proximal end has the flake's original bulb of percussion and the bulb formed by the subsequent snapping. While the original bulb has the flake's original butt, no butt accompanies the later bulb. We have found only one case - probably accidental - of snapping on the ventral face (Fig. 6, 3-4). This treatment was not reserved for a specific tool, nor for any blank form or size. The possible aim of this act was not thinning, unlike the 'dos aminci' of Bordes (1979, 36). In some cases the Tabun snap created, in fact, a thicker proximal end than the removed, original one (illustrated in Fig. 5, 1). The only clear result of the snapping has been to shorten the blank.

The only other example of similar snapping known to us was noted in an end-Pleistocene Mushabian site in the Negev (Kaufman, 1983). As at Tabun, here too the blank was invariably struck from the dorsal face (D. Kaufman, personal communication). But in the Mushabian assemblage, the technique had affected a single tool type, grattoirs, 9 items of 24 were snapped. There too, the snapping was seemingly aimed to shorten the blanks.

Typological Aspects

Tool types The series is characterized by the abundance of racloirs, the diverse types of racloirs and the importance of Yabrudian types (Table 6): convex (Fig. 7; Fig. 5, 2), transversal (Fig. 7, 1), convergent (Fig. 4, 3-5, 7) and bifacial racloirs (Fig. 8, 1-6). Blanks were frequently side-struck.

Retouch types (Table 7) The series is governed by scale retouch (31%), followed by Yabrudian/Quina-type step retouch (19.4%). For comparison, in the underlying Acheulean assemblage from the top part of layer F, step retouch amounts to only 2.2% of the cases.

Type	%
Scale	30.88
Parallel	2.64
Fine	13.53
Coarse	0.59
Steep	2.65
Burin	8.24
Step	19.41
Semi-step	5.0
Clactonian notch	6.47
Other	10.59

Table 7. Retouch types

Summary

This note, the first quantitative study of the Yabrudian of Tabun, shows that the assemblage is non-Levallois and non-laminar, heavily based on flakes. The high frequency of end products and the scarcity of primary elements indicate that the initial steps of the operational procedure were taken elsewhere. Butts are mostly unprepared; the parallel and opposed dorsal scar patterns dominate. The largest flakes were more readily chosen for further modification. Retouched items characteristically have their butt end narrowed or removed by ventral retouch. A special way of shortening the blanks - the Tabun Snap - had been noticed, and seems to be confined to the Yabrudian. Typologically, Yabrudian scraper types dominate with step, or Quina-type retouch.

Acknowledgments

This article is based on a Masters thesis by the senior author. The authors thank Dr. Daniel Kaufman for his help with the statistical analyses.

Authors' address: The Zinman Institute of Archaeology, University of Haifa, Haifa 31905, Israel aronen@research.haifa.ac.il

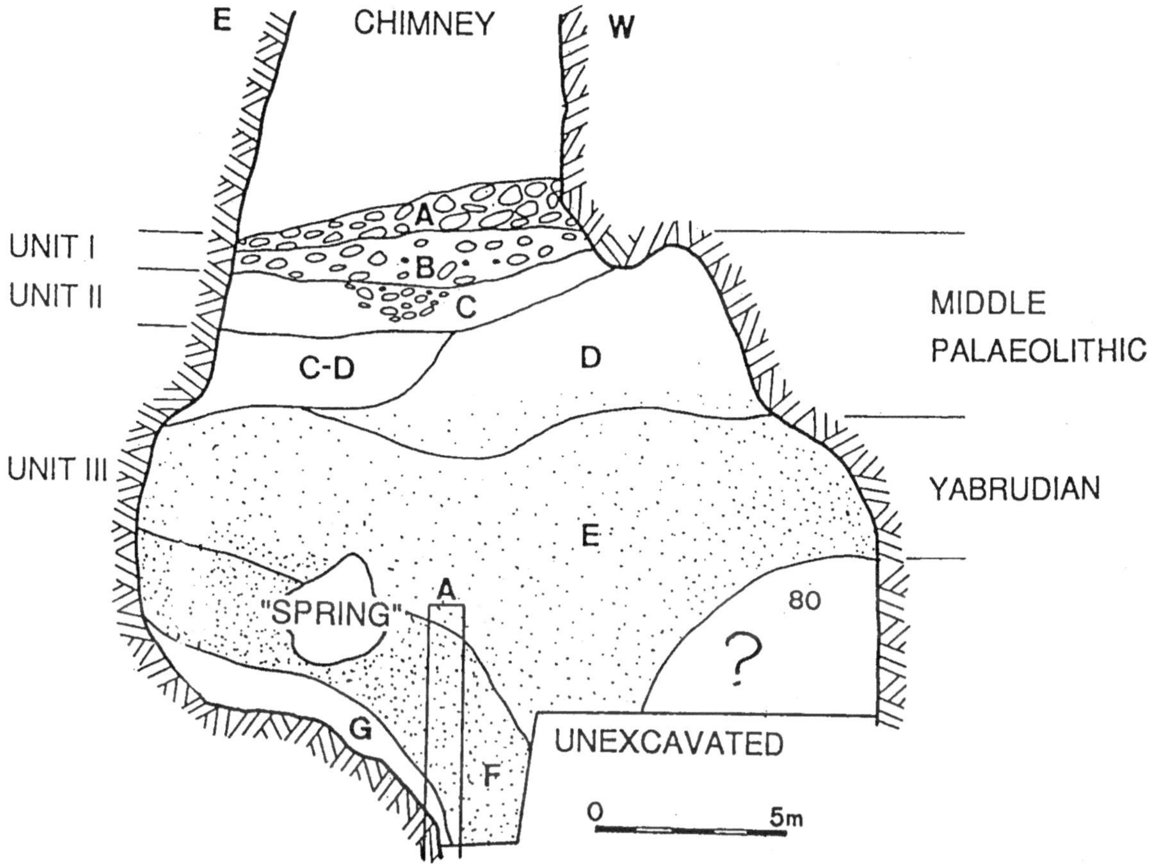

Fig. 1 Schematic section after Garrod and Bate, 1937. Goldberg's geomorphological units on the left (Jelinek et al., 1973). A - location of the section in Fig. 3.

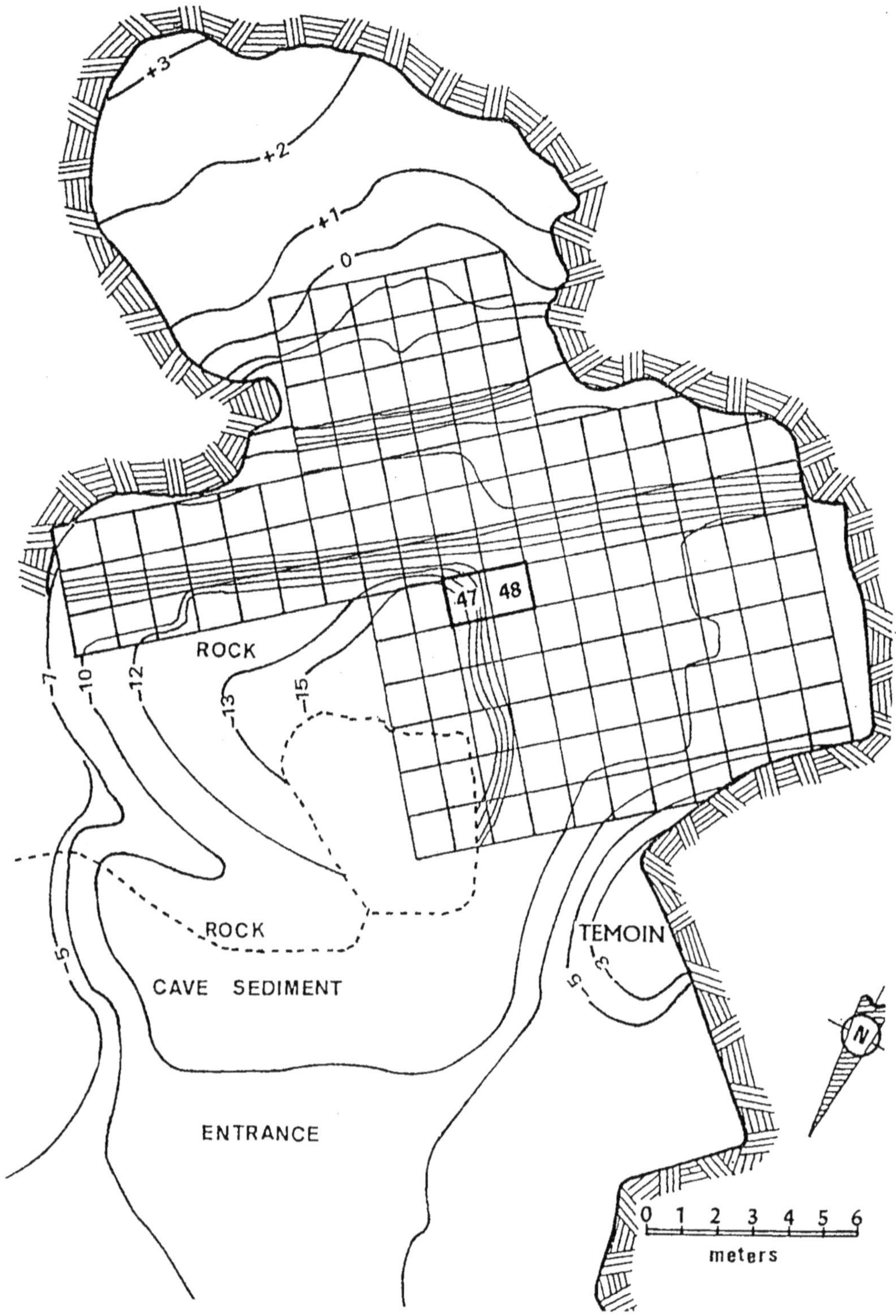

Fig. 2 Plan of Tabun and study squares.

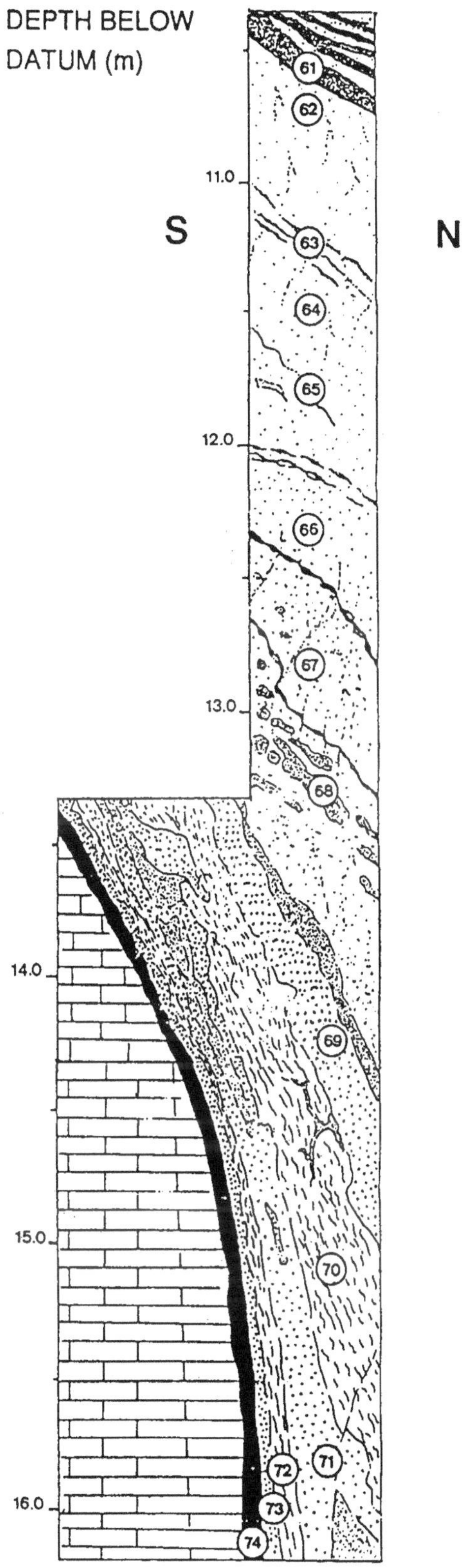

Fig. 3 Section in study area, on edge of swallow hole (by A. Tsatskin). Above layer 64, Yabrudian; 64 and below, non-yabrudian, layer F Acheulean.

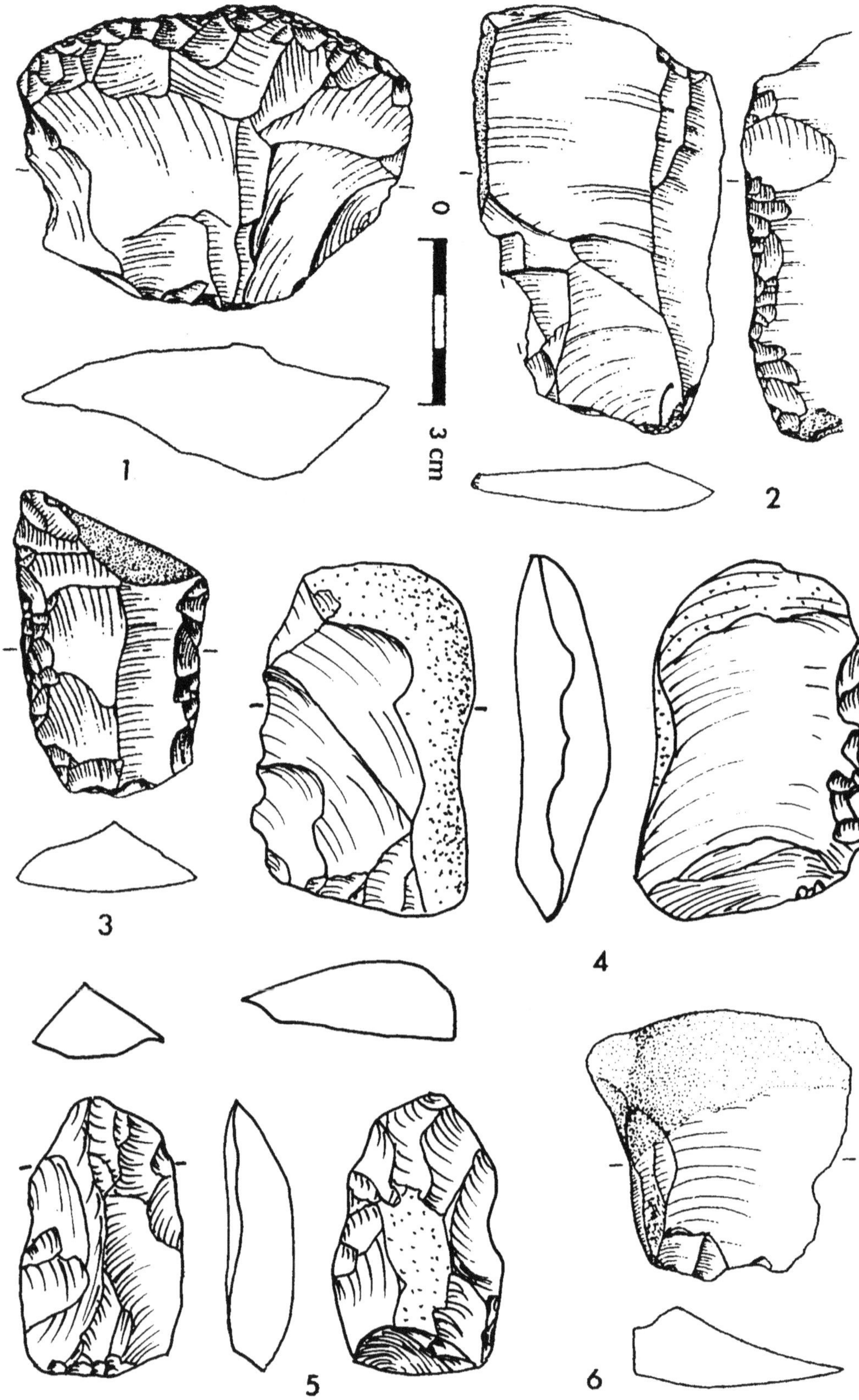

Fig. 4 1 - Transversal racloir; 2 - on ventral face with parallel retouch; 3 - double racloir; 4 - denticulate, base snapped; 5 - bifacial piece, base snapped; 6 - natural backed knife.

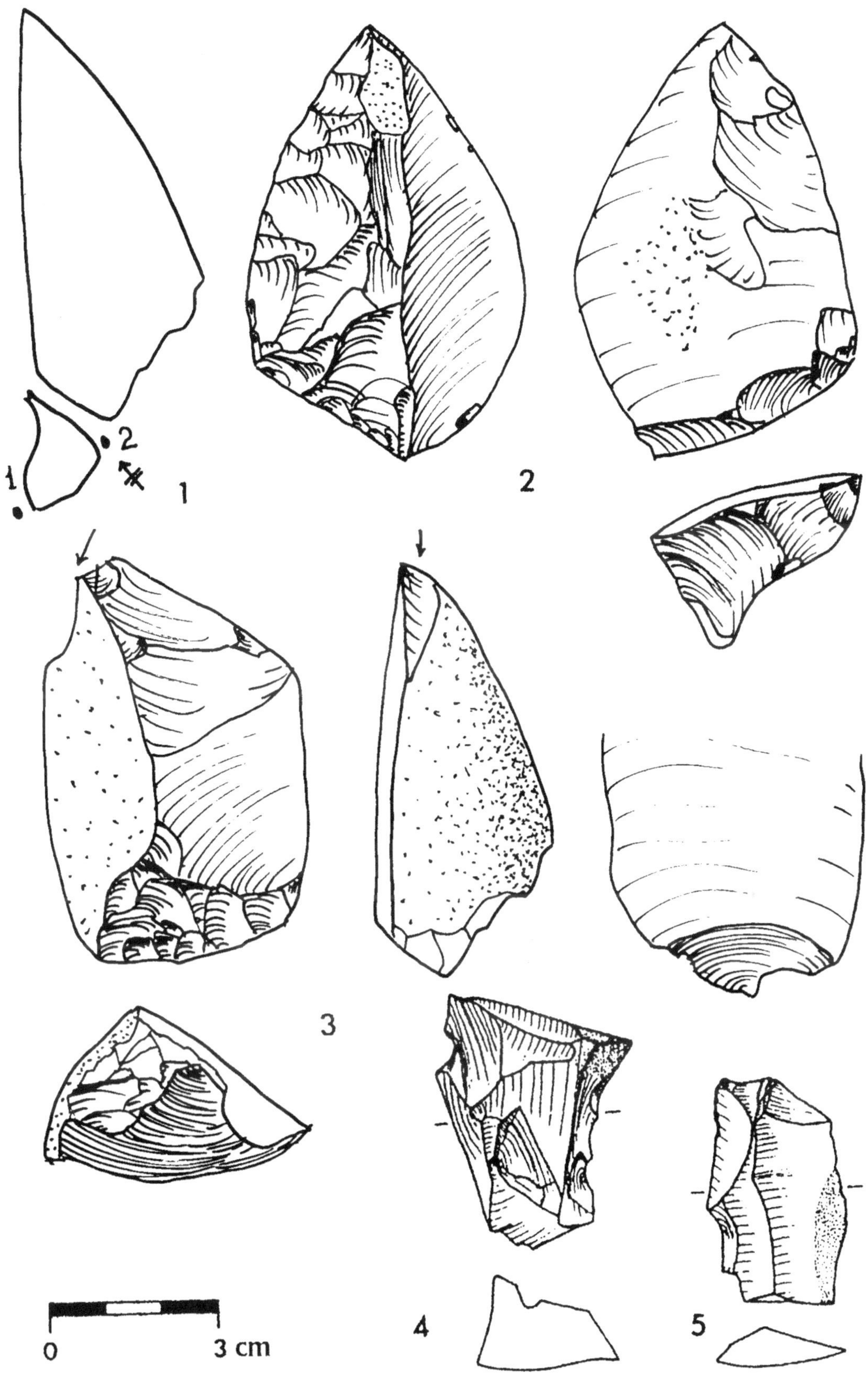

Fig. 5 1- Illustration of Tabun Snap (non related parts). The waste is the piece illustrated in Fig. 6, 1; the blank is that of Fig. 5, 3. 2 - convex racloir, base snapped and subsequently retouched; 3 - burin with base snapped and subsequently retouched; 4 - core-trimming waste; 5 biface trimming flake (?).

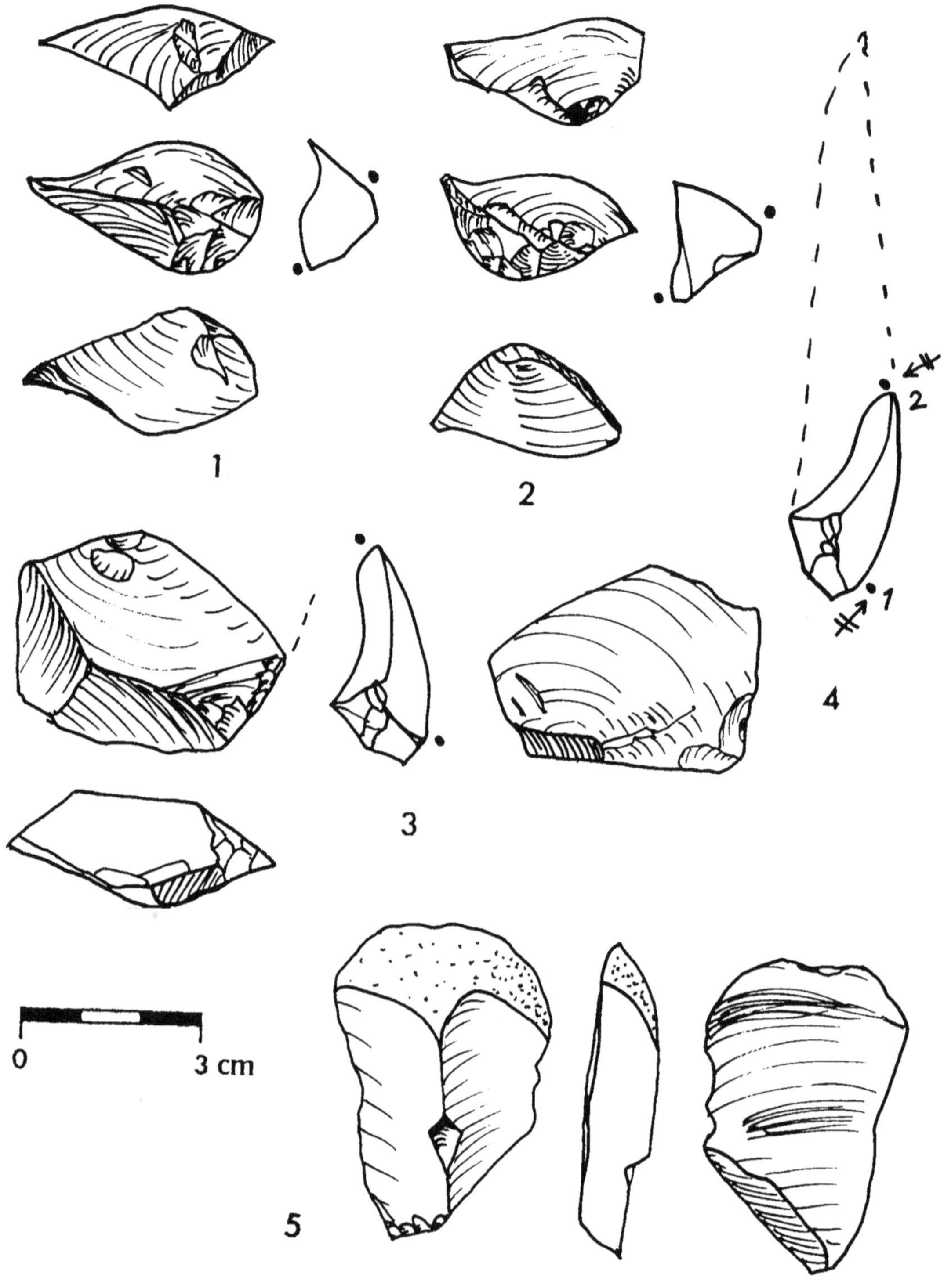

Fig. 6 1 and 2 - Tabun snap spalls from blanks snapped at similar length and thickness; 3 - Tabun snap on ventral face; 4 - possible reconstruction of the blank snapped by the spall in no. 3; 5 - a flake snapped with slight subsequent retouch.

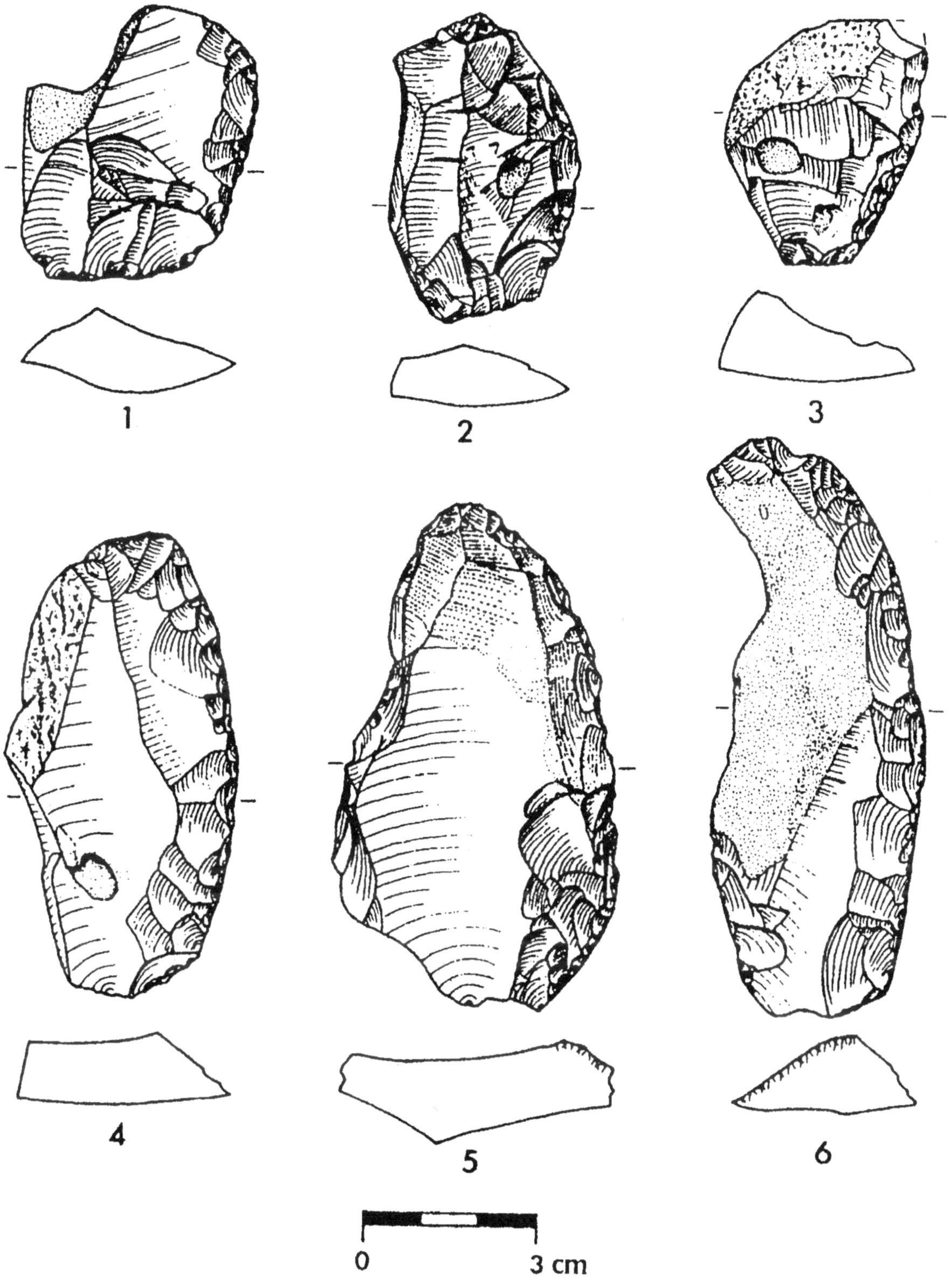

Fig. 7 Yabrudian racloirs

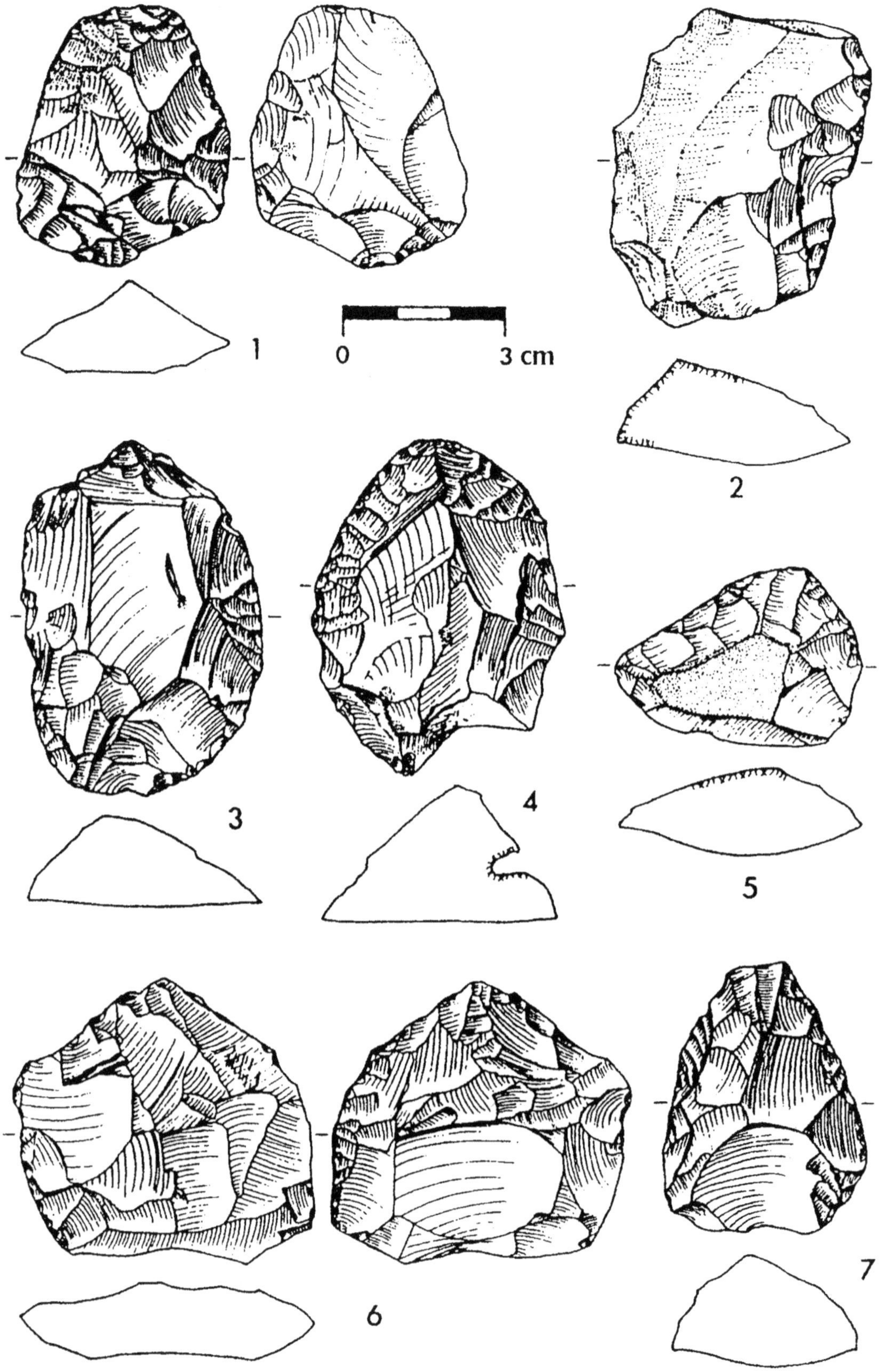

Fig. 8 Yabrudian racloirs. 1 - bifacial; 2 - simple straight; 3-5 and 7 - convergent, heavy retouch racloirs; 6 - bifacial piece.

Bibliography

Bar-Yosef, O., 1994 The Lower Paleolithic of the Near East. *J. World Prehistory* 8, 211-265.

Bordes, F., 1979 *Typologie du Paleolithique Ancien et Moyen.* 3d édition, CNRS, Paris.

Copeland, L., 1983 The Stone Industries, the Acheulo-Yabrudian of Bezez Cave, Level C. in: Roe, D. A. (ed), *Adlun in the Stone Age*, pp. 89-194. BAR IS 159, Oxford.

Garrod, D. A. E., 1956 Acheuleo-Yabrudien et pre-Aurignacien de la Grotte de Tabun, Mont Carmel: étude stratigraphique et chronologique. *Quaternaria* 3, 39-59.

Garrod, D. A. E. and Bate, D. M. A., 1937 *The Stone Age of Mount Carmel,* Vol. I. Clarendon Press, Oxford.

Goren-Inbar, N., 1994 The lower Paleolithic of Israel. in: Levy, T. E. (eds.), *The Archaeology of Society in the Holy Land*, pp. 93-109. Leicester University Press, London.

Grün, R., Stringer, C. B. and Schwarcz, H. P. 1992 ESR dating of teeth from Garrod's Tabun Cave collection. *J. Human Evolution* 20, 231-248.

Jelinek, A. J., 1977 A preliminary study of flakes from the Tabun Cave, Mount Carmel. *Eretz Israel* 13, 87-96.

Jelinek, A. J., 1982 The Middle Palaeolithic in the Southern Levant with comments on the appearance of modern Homo sapiens. in: Ronen, A. (ed.), *The Transition from Lower to Middle Palaeolithic and the Origin of Modern Man,* pp. 57-104. BAR IS 151, Oxford.

Jelinek, A. J., Farrand, W. R., Haas, G., Horowittz, A. and Goldberg, P., 1973 New excavations at the Tabun Cave, Mount Carmel, Israel, 1967-1972: a preliminary report. *Paléorient* 1, 151-183.

Kaufman, D., 1983 D101B: A Mushabian site in the Nahal Zin. in: Marks, A. E. (ed.), *Prehistory and Paleoenvironments in the Central Negev, Israel,* Vol. III, pp. 333-341. Southern Methodist University, Dallas.

McDermott, F., Grün, R. Stringer, C. B., and Hawkesworth, C. J. 1993 Mass-spectrometric U-series dates for Israeli Neanderthal – early modern Hominid sites, *Nature* 363, 252-254.

Mercier, N. and Valladas, H., 1994 Thermoluminescence dates for Paleolithic Levant. in: Bar-Yosef, O. and Kra, R. S. (eds.), *Late Quaternary Chronology and Paleoclimates of the Eastern Mediterranean,* pp. 13-20. Tuscon.

Mercier, N., Valladas, H., Valladas, G., Jelinek, A., Meignen, L., Joron, J. L., and Reyss, J. L., 1995 TL dates of burnt flints from Jelinek's excavations at Tabun and their implications. *J. Archaeological Science* 22, 495-509.

Mercier, N., Valladas, H., Froget, L., Joron, J. L. and Ronen, A., nd *Datation par la thermo-luminescence de la base du gisement de Taboun (Mont Carmel, Israël).* Compte Rendu Academie des Sciences, Paris (in press).

Ronen, A., 1992 The emergence of blade technology: Cultural affinities. In: Akazawa, T., Aoki, A. and Kimura, T. (eds.), *The Evolution and Dispersal of Modern Humans in Asia*, pp. 217-228. Hokusen-Sha, Tokyo.

Ronen, A., and Tsatskin, A., 1995 New interpretation of the oldest part of the Tabun Cave sequence, Mount Carmel, Israel. in: Ullrich, H. (ed.), *Man and Environment in the Palaeolithic.* ERAUL 62, pp. 265-281, Liège.

Rust, A., 1950 *Die Hohlenfunde von Jabrud (Syrien).* Karel Wacholtz, Neumunster.

Shifroni, A. and Ronen, A. nd. The 'Tabun Snap' from the Yabrudian/Acheulean interface at Tabun. *Prehistoria* 1 (in press).

DISCUSSION

Paul Callow: Can you give an indication of the incidents, the percentage of decortication or rather residual cortex, which remains on the Acheulean and the Yabrudian scrapers? Because, obviously this has implications for the degree of reduction of the pieces, both in respect of how far the cores have been worked down and how much secondary retouch was there.

Anat Shifroni: There are a lot of pieces with cortex, but at present I can not give exact numbers.

Joachim Schäfer: We have heard that there is no Levallois technique in the Acheulean and Yabrudian, but some of your pictures may have shown Levallois flakes struck from prepared cores.

Anat Shifroni: We do not have Levallois cores and we do not have Levallois flakes.

Avraham Ronen: There are a few accidental Levallois flakes. More plausibly, they are handaxe reduction flakes. While there is no Levallois technique in both Yabrudian and the underlying Acheulean at Tabun, the Upper Acheulean of Holon on the coastal plain and of Berekhat Ram on the Golan Heights, both older than 250,000 years, are clearly of Levallois technique.

ACHEULO-YABRUDIAN SEDIMENTS OF TABUN: A VIEW FROM THE MICROSCOPE

Alexander Tsatskin

Introduction

Being one of the richest Lower Palaeolithic sites in the southern Levant, Tabun Cave (Mount Carmel, Israel) is still a focus of intensive archaeological, sedimentological and geochronological research. Attempts to corroborate chronological findings with geological evidence were undertaken by Grün et al. (1991), who outlined the existence of numerous erosional unconformities within the cave fill suggesting a much older geological age than originally maintained (Jelinek et al., 1973). In the latest critical reviews Jelinek (1992) and Farrand (1994) admitted the complexities of the geological stratigraphy at Tabun, although still favoured earlier chronological estimations. A major breakthrough was recently achieved by thermoluminescence dating of Acheulo-Yabrudian layers at Tabun which rendered surprisingly old dates of about 350 ka ago (Mercier et al., 1995). These are supported by a $^{230}Th/^{234}U$ dating of a flowstone overlying the Acheulo-Yabrudian layer at the adjacent Jamal cave (Weinstein-Evron et al., 1999).

In the classical work by Garrod and Bate (1937), the Lower Palaeolithic layers F and E at Tabun were found to vary in thickness and nature in different parts of the cave (Fig. 1). In the centre of the middle chamber the sediments were drawn down into a karstic sinkhole. Near the walls the sediments were 'a hard red breccia, more extensive on the west side than on the east', but away from the rock they '...changed to a pale brown earth' (Garrod and Bate, 1937, 66). Jelinek et al. (1973) showed that layers E and F (Unit III) are basically composed of well-sorted aeolian sediments with bimodal distribution of particle size, sand originating from the adjacent coastal plain and silt from a distant desert source. After deposition, the wind-blown sediments of Unit III underwent strong weathering and phosphatization (Goldberg and Nathan, 1975), resulting from increased wetness in the cave. According to Goldberg (1973), this presumably took place when the Lower Mousterian Layer D formed.

Disturbances in the Acheulo-Yabrudian deposits at Tabun are indeed ubiquitous. Most striking is the subsidence of the layers into the karstic sinkhole which Farrand (1979; 1994) interpreted as the result of a lowered water table following sea level retreat. In the eastern side of the cave, secondary spring-fed deposits eventually fell down forming a cavity (Garrod and Bate, 1937). In the western side, Jelinek et al. (1973) described a major unconformity between Layer E and earlier deposits, designated first Bed 80, and later renamed Bed 90, or Unit XIV (Jelinek, 1982). This erosion episode was not detected in other parts of the cave fill. Earlier we have used refined methods of sedimentological analysis, primarily soil micromorphology (Courty et al., 1989), mainly in the eastern and central sectors of the middle chamber (Tsatskin et al., 1995; Ronen and Tsatskin, 1995). This paper attempts to trace that major unconformity across the whole of the middle chamber.

Materials and Methods

Undisturbed blocks taken from the Acheulo-Yabrudian layers E and F at 15 to 20 cm intervals were impregnated with polyester resin (Hillquist, Scotland) under vacuum. The blocks were prepared for petrographic thin sections by slicing to a thickness of 30 microns. The thin sections were examined by a polarizing transmitted light microscope (Olympus BH2). The micromorphological descriptions were made according to Bullock et al. (1985). Interpretations of the major processes were made according to Courty et al. (1989). The mineralogical composition was determined by Fourier Transform Infrared Spectrometer (Weiner et al., 1993).

Results

The western sector. The profile in the western sector (Squares 45a, 45b and 45c, Fig. 1) includes Acheulo-Yabrudian deposits above the unconformity from 5.50 m to 10.00 m depth, designated as Upper Unit, and Unit XIV of Jelinek (1982), designated as Lower Unit (Fig. 2). The Upper Unit is a dark-coloured, crumbly, poorly-bedded, silty loam with animal burrows and some rock fragments. The Lower Unit is an olive-grey sand, occasionally silty sand, well bedded, with no rock fragments. The contact between the units, defined by Farrand (1994) as a high-angle offset, shows in reality a wavy, subhorizontal appearance from the west wall of the cave eastwards for a distance of about 3 m, further dipping into the sinkhole and splitting into a series of lenticular layers of loose and cemented sand.

At 7.10 m depth thin sections of the Upper Unit in square 45A exhibit a loose, excremental, clayey microfabric, with only ca. 10% quartz sand. Charred materials and abundant chips of heated bones with reddish hue are characteristic. Some bones are yellowish and bleached, probably due to diagenesis under reduced conditions. Translucent phosphatic coatings in pores, as well as brown/reddish nodules of various sizes, composed primarily of montgomerite and crandalite, occur alongside sparitic calcite accumulations. In addition, we recognized brown silty-clay soil aggregates which increase in quantity close to the wall, where they acquire a reddish hue (Fig. 3A), characteristic of terra rossa soils from the Mount Carmel area. Some soil aggregates show enhanced birefringence or are broken apart by thin intersecting cracks, both of which may result from high temperature effects (Courtry et al., 1989). Thus, deposits of the Upper Unit contain large amounts of terra rossa materials and display abundant signs of anthropogenic and biological reworking.

At ca. 7.60 m below datum, close to the unconformity zone between the Upper and Lower Units (Fig. 2), deposits are strongly cemented, occasionally taking the form of a hardpan. In Squares 45A and 45B (Fig. 1) the hardpan splits into white calcitic, silty laminae and greyish or yellowish decalcified sandy laminae, dipping eastward at an angle of about 20°. In thin sections the sand is poorly rounded, similar to that in Layer E from the central part of the cave (Bull and Goldberg, 1985), and alternates with laminae of dark silty-clay and of brown-reddish materials, detected by FTIR as montgomerite. This pattern seems to result from runoff towards the centre of the cave.

At ca. 8.00 m depth, the Lower Unit (XIV) shows an entirely different microfabric. At the top it is sandy with more than 60% rounded fine sand; contains abundant brown pellets of montgomerite/crandallite composition and of hydroxyapatite lucid coatings and mosaics (determinations by FTIR). Coated sand grains and aggregates with ca. 100 µm thick coatings are abundant, in contrast to the Upper Unit (Fig. 3B). Adherence of clay and fine-grained phosphates to quartz sand in the form of coatings and their eventual roundness are possibly due to short-distance transportation under conditions of water saturation (inundation) alternating with drying episodes. This hypothesis will be discussed below.

The central sector. In the central sector, Layers E and F plunge sharply into the sinkhole cavity (Garrod and Bate, 1937; Goldberg, 1973). Between 7.50 and 9.00 m depth, the Acheulo-Yabrudian sediments are composed of dark-coloured beds, probably derived from reworking of hearths, and show unconformities due to recurrent karstic collapses (Jelinek, 1992; Farrand, 1979; 1994). Field observations show that these hearth-derived deposits laterally grade to the Upper Unit in the western sector. Below hearth-derived deposits appear yellowish-green bedded sand/silt sediments, strongly phosphatized in places, which are considered analogues of cemented sand dunes (kurkar) of the adjacent coastal plain (Goldberg, 1973). The contact between the hearth-derived and the kurkar-like deposits is tilted and contains cut-and-fill structures.

In petrographic thin sections, the kurkar-like sediments exhibit disturbed microlaminated fabrics, manifested by alternation of sandy, densely packed, laminae with those of sandy silt strongly impregnated with brown phosphatized matrix. This pattern, identical to that seen in the unconformity zone in the western sector (see above), may have originated from low-energy runoff. Interestingly, channels and pellets, i.e., biologically related features, were not found here, apparently because of severe diagenesis. At 10.70 m depth, discontinuous, faint, 0.1 mm thick carbonate coatings were found to overlie earlier phosphatic coatings in elongated pores. Taking into account that deposits in the centre of the chamber were mostly influenced by karstic processes and leaching, alternation of phosphatic and carbonate coatings indicates contrasting hydrological conditions. The correlation of this level with the western sector will be discussed below.

The eastern sector. This sector occupies squares 96, 97, 98 (Fig. 1). We have analyzed the profile from 6.80 m depth to bedrock at 11.80 m depth (Fig. 4). Only the profile between elevations 6.80 and 9.50 m is relevant to the present discussion. Around 7.50 m depth, deposits appear as a light brown, hard, crumply, clay loam, in which Garrod described 'pale streaky hearths pushed up and broken above the summit of a spring cavity' (Garrod and Bate 1937, 67). These 'hearths', which are abundant around 7.80 m below datum (Fig. 4), appear as elongated strips 3-5 cm thick, with yellowish speckles and greyish crusty rims. Below, in the interval between 7.80 m and 9.00 m depth, the deposits turn into a brownish loamy sand with minor truncations and subvertical rootlets 5-8 mm thick, as well as laminae of various colours and consistency, from loose to hard crusty appearance. In addition to the yellowish 'hearths' of Garrod, two dark-coloured hearths (Fig. 4) with an increase in charred materials and organic matter content have been reported (Ronen and Tsatskin, 1995).

In thin sections, deposits of the Upper Unit are composed mainly of calcareous clay mixed with ash, quartz sand (less than 10%) and abundant chips of bones. The bones show a yellow-orange hue, probably due to heating, and are sometimes secondarily calcified. Rounded terra rossa soil aggregates 0.5 mm thick are present. Secondary calcite deposits are found in the form of brown-stained micrite, occasionally mixed with 0.5 - 0.6 mm biological pellets, as well as pure sparite, micrite and microsparite infillings in larger channels. The yellow strips (Garrod's 'hearths'), in contrast, are completely decalcified and contain more than 20% sand, abundant soil aggregates, a few bleached bones, and phosphatic nodules with sharp boundaries. Coated sand grains enveloped by 0.1 - 0.15 mm thick phosphate-clay coatings, also occur.

Deposits in the lower part of the section turned out to be even more heterogeneous in thin sections than suspected from field observations. In contrast to the upper part, deposits below 7.80 m depth are overall sandy with more than 50% well-sorted sand, ubiquitous coated grains, and alternations of calcareous and calcite-free laminae. Interestingly, the abundance of pale-coloured chips of bones peaks here, often with thick hydroxyapatite coatings, as has been reported by Tsatskin et al. (1995) through a SEM/EDXA study. These are covered by sparitic calcite, indicating that accretion of carbonates post-dated transformation of bones into apatite coatings. Heat-fractured terra rossa aggregates occasionally occur, and hydroxyapatite fills in the fractures. In the area of 'black hearths' at 8.50 m depth (Fig. 4), terra rossa aggregates with phosphate-filled fissures have been also found, as well as coated grains. Here, diagenetic alteration of bones leads to fluidal type of phosphatization rather than the formation of brown nodules. Likewise, calcite is superimposed upon secondary phosphatic deposits in larger channels, testifying to recurrent environmental changes from wetter (phosphates) to drier (carbonates) conditions. Thus, micromorphological observations clearly indicate that a boundary between the upper, clayey deposits and the lower, more sandy deposits lies in Garrod's 'hearths' zone, at an altitude of ca. 7.80 m below datum.

Discussion

Key micromorphological features and their interpretation. Variation and relative abundance of individual micromorphologiocal features and suites of features in the Acheulo-Yabrudian sediments from the western and the eastern sectors of Tabun are summarized in Table 1. The central sector is not given here because of strong disturbances and maximal diagenesis of the deposits.

Signs of burning and biological activity. In the Acheulo-Yabrudian layers analyzed thus far, hearths are dubious as ash is not preserved, in contrast to the Mousterian layers at Tabun (Albert et al., 1999). Heating effects (charcoal, baked soil aggregates with enhanced birefringence, reddish chips of bones) can still be discerned, however, in thin sections. Anthropogenic and biological features show strong variation both in space and with depth. As seen in Table 1, rather similar deposits with higher ash-derived component occur above 7.80m depth in both sectors. In contrast, below this depth sediments from the two sectors substantially differ. While in the western sector the Lower Unit contains no hearth-derived materials, deposits in the eastern sector below 7.80m exhibit at least several laminae with indications of burning.

Aeolian sediments vs. terra-rossa soils. Goldberg (1973) has shown that Tabun Layers F and E are composed of aeolian sands and silts in variable proportions, whereas terra rossa soils have accumulated in the Mousterian layers C and B. Our examination of thin sections revealed that close to the walls, the Acheulo-Yabrudian deposits do contain quantities of terra rossa soil aggregates. On the other hand, while terra rossa aggregates are absent from the central sector in the middle chamber, concentrations of silty material are found embedded in the clayey matrix. It is plausible that these silt concentrations are partly derived from disintegration and diagenetic alteration of terra rossa. Terra rossa soils contain considerable amounts of aeolian quartz silt (Yaalon, 1997). In this way the silt in Layer E may be of both local origin (alteration of terra rossa) and a distant aeolian origin. Table 1 demonstrates that aggregates of intact terra rossa reach their maximum in deposits above 7.80 m depth on both sides of the cave. Below this depth, however, terra rossa is indistinguishable in the western sector but can be identified in the eastern sector. This may indicate a shift in depositional environment, with a stronger accumulation of local Mount Carmel soils than distant aeolian materials within Layer E at about 7.80 m depth, i.e., above the major unconformity in the western sector.

Coated grains and aggregates. Adherence of clay, silica and other materials to the surface of sand grains in Lower Layer E was first observed with scanning electron microscope by Bull and Goldberg (1985), who viewed this feature as post-depositional chemical process. Thin sections show that sand grains enveloped by brownish clay or phosphatic coatings 0.10-0.15 mm thick are common in the lower parts of both eastern and western profiles (Table 1). Coated grains and aggregates are found in the central sector as well. Similar ooid-like coated aggregates have been described in the neighbouring Jamal cave (Weinstein-Evron and Tsatskin, 1994) and in waterlogged palaeosols in the Carmel coastal plain (Tsatskin and Ronen, 1999). It is plausible that such features originated in layers which experienced pronounced wet/dry cycles and slight erosion. Hence, coated grains may indicate unstable regime of sedimentation and colluviation. As seen in Table 1, coated grains peak in the Lower Unit of the western sector and substantially increase in the lower part of the eastern sector. It is worth noting that in the eastern sector, at least

Features	Western Sector		Eastern Sector	
	Upper Unit 6.50-7.80 m	Lower Unit 7.80-8.80 m	Upper part 6.80-7.80 m	Lower part 7.80 - 9.00 m
Charred bones and organic materials	• • •	–	• • •	• • /•
Biological microstructure	• • •	–	• • •	• /–
Sand grains	•	• • •	•	• • •
Aggregates of terra rossa soils	• • •	–	• • •	• • /–
Coated grains and aggregates	• •	• • •	• /–	• • •
Carbonate secondary deposits	• •	–	• • •	• • /–
Phosphatic nodules and coatings	–	• • •	•	• • •

Codes used: – absent; • rare; • • common; • • • abundant

Table 1. Micromorphological Features in the Western and Eastern Sectors of the Middle Chamber at Tabun.

two major episodes of erosion may be distinguished, manifested by an increase in coated grains in elevations 7.80-8.00 m and 8.50-8.70 m (Fig. 4; not shown in Table 1)).

Carbonate deposits. Carbonate neoformations are abundant near the walls of the cave and in close vicinity to the spring cavity. We were able to identify two major forms of carbonates: 1) pure infillings in pores and channels of both sparitic and micritic calcite; and 2) brown-stained concentrations in the matrix, incorporating faecal pellets and soil microaggregates. The latter clearly post-date biologically related features. Localized precipitation of carbonates obviously results from the circulation of cave water saturated with CO_2. We would expect this process to keep on going uninterruptedly in a karstic limestone cave such as Tabun. However, this is not the case. Even close to the cave walls, carbonate deposits alternate with phosphatized laminae.

Phosphatic features. Secondary phosphate minerals, derived from transformation of bones and guano, constitute an inseparable component of cave deposits (Schiegl et al., 1996). Goldberg and Nathan (1975) showed the existence of at least two forms of phosphate minerals at Tabun: hydroxyapatite ($Ca_5(PO_4,CO_3)_3$) and montgomerite ($Ca_4MgAl_4(PO_4)_6{\cdot}OH_4{\cdot}12H_2O$). In a current project in collaboration with S. Weiner we have identified by FTIR additional mineralogical phases, details of which are beyond the scope of this paper. In thin sections diverse phosphatic features are recognized: chips of bones in various states of alteration, complex opaque nodules, fine-grained phosphate in the matrix, and translucent coatings (Tsatskin et al., 1995). It was also shown that lucid coatings are composed of hydroxyapatite, while concentrations in nodules or in the matrix are composed of Al(Mg)-phosphates like montgomerite or crandalite. Although the geochemistry of phosphate minerals in the cave is complex, it seems clear that Al-phosphates can replace hydroxyapatite after complete removal of calcite by groundwater (Flicoteaux and Lucas, 1984). It is reasonable then to attribute phosphate neoformation to episodes of inundation which favored reduced conditions as compared to episodes of increased water saturation with eventual precipitation of carbonates. As shown in Table 1, the co-existence of phosphatic and carbonate features is characteristic of the sediments below 7.80 m depth in the eastern sector. The western sector, instead, displays complete phosphatization of the deposits in the Lower Unit. However, around the major unconformity in the western sector, phosphate/carbonate alternation does occur. This also appears, although less prounced, at 10.50 m depth in the slumped block of the central section described above. Such contrasting geochemical events testify to intermittent inundation episodes controlled by changes from wetter to drier conditions. Assuming that pronounced secondary phosphatization is linked to inundation, we will now turn to the question of stratigraphical correlations.

Correlation of the Different Sectors at Tabun

Strong variations in the nature of sediments within the cave (Garrod and Bate, 1937), subsidence in the centre and high-angle offset in the western side, all render spatial relationships within Layers E and F complex and difficult to interpret. Farrand (1994, 37) holds that 'a continuous section from Layer G up through F into E remains intact on the east side of Garrod's excavated area', while in the western side Layer E slumped down against sediments which Jelinek (1982) equated with Layer G.

Micromorphological studies of the Acheulo-Yabrudian layers allow us to tentatively propose a different scenario. In the western sector, the deposits display contrasting microfabric both above and below the major unconformity at 7.8 - 8.0 m depth. The same is true for the deposits, albeit more variegated, in the eastern sector. Above the 7.80 - 8.00 m horizon on both sides of the cave the sediments are similar in nature - rich in terra rossa soil materials, with abundant carbonate accumulation and no phosphatic coatings or infillings. Below the 7.80 - 8.00 m horizon the sediments in the two sectors, though not identical, are basically sand-rich, decalcified, with signs of strong inundation. Compared to the western sector with its major unconformity zone at ca. 7.80 m depth, the eastern sector displays at least two less pronounced erosional surfaces, at ca. 7.80 and 8.70 m depth. In both sectors of the middle chamber a roughly similar trend is recorded - from an earlier episode of inundation, through wet/dry cyclicity to a relatively massive input of terra rossa.

The present-day configuration of the middle chamber may help clarify the spatial differentiation of the sedimentological processes in the eastern and western sectors. The western side has a lower ceiling and recedes deeper into the cliff, offering a more sheltered, and hence a more attractive, habitat for cave animals. During periods of inundation, large amounts of guano which would have accumulated here resulted in particularly strong phosphatization and an overall post-depositional alteration of earlier sediments, as is the case in other cave sites (Courty et al., 1989; Bar-Yosef et al., 1992). The eastern side of the middle chamber, on the other hand, is less protected and more aerated, hence less humid than the western side. Consequently, the differences in secondary phosphatization between these sectors may reflect varying geochemical conditions linked to varying inundation effects inside the cave.

Interestingly, later episodes of wetter climate within Layer E (Jelinek, 1992) have not reached the extent of inundation marked by the unconformity above Unit XIV which, as suggested here, extends across the middle chamber of Tabun. It is thus tempting to link this major erosion episode to the subsidence in the central part of the cave.

Conclusions

The complexity of the spatial relations between the Acheulo-Yabrudian deposits in different parts of Tabun Cave, established by former geological investigations, is confirmed by micromorphological analysis. Micromorphology allowed us to study the cave's history of depositional and post-depositional processes in detail, and to re-examine the correlation of Jelinek's Unit XIV with other sedimentary facies. We were able to demonstrate the clear differences in microfabric of deposits below and above the major truncation surface in the west sector (Farrand, 1994). Similar differences, albeit less striking, have been also found in the eastern side, approximately at the same depth as in the west. Hence it is suggested that a) both sectors went through the same sequence of events and are roughly contemporaneous; and b) ca. 7.80 m below datum, within Layer E, lies a highly significant chronostratigraphic boundary in the geological history of Tabun.

Wind-blown sand was the major material to have accumulated in the cave below the 7.80 m line. Above this line the mode of deposition drastically changed in both sectors of the middle chamber, due to a relatively massive input of terra rossa. This could have resulted from a reshaping of the cave's roof. The presence of terra rossa soils at Tabun was previously known only from the Mousterian layers C and B, believed to have accumulated through the chimney in the inner chamber (Goldberg, 1973; Jelinek et al., 1973). Our findings prove that terra rossa may have accumulated in the cave much earlier, at least since Layer E. It has to be stressed that in the slumped block of the central sector, terra rossa soils no longer exist as such. Severe diagenesis may have destroyed the clay, leaving only the quartz silt contained in the terra rossa (Yaalon, 1997).

Before the accumulation of terra rossa, the middle chamber was subjected to a major phase of inundation, followed by desiccation and erosion. During this phase bones were partially altered or completely disintegrated under reduced conditions, phosphatic nodules were formed and the matrix was impregnated by fine-grained phosphates in sloping situations. We tentatively suggest that at the time of desiccation, the central part began to slowly subside, thus forming a depression in the ancient surface topography and enhancing accumulation of low-energy runoff from the sides of the chamber. At some point in time this subsidence event ended in a drastic collapse of a stronger magnitude than subsequent episodes of karstic collapse recognized by Jelinek (1992). The evidence that the erosion coincided with desiccation is provided by distortions of the earlier substrate, the presence of desiccation cracks and the alternation of phosphatic and carbonate laminae seen in thin sections.

Micromorphological analysis of different sedimentary facies at Tabun shows that the major turnover in the geological history of the cave during the Acheulo-Yabrudian occupation is recorded at a depth of 7.80 m on both sides of the middle chamber. We anticipate that future research, coupled by absolute dating, will enable us to recognize both earlier and later erosional events during the long occupational history of the cave.

Acknowledgments

I thank Prof. M. Weinstein-Evron and Dr. D. Kaufman for fruitful discussions and help in preparing the manuscript. My special thanks are due to Prof. A. Ronen who has encouraged micromorphological studies in the Tabun Cave project, to Anat Shifroni and the students of the University of Haifa who helped in the field work. Prof. S. Weiner kindly provided the FTIR facilities in the Weizmann Institute of Science.

Author's address: Zinman Institute of Archaeology, University of Haifa, Haifa 31905, Israel. tsatskin@research.haifa.ac.il

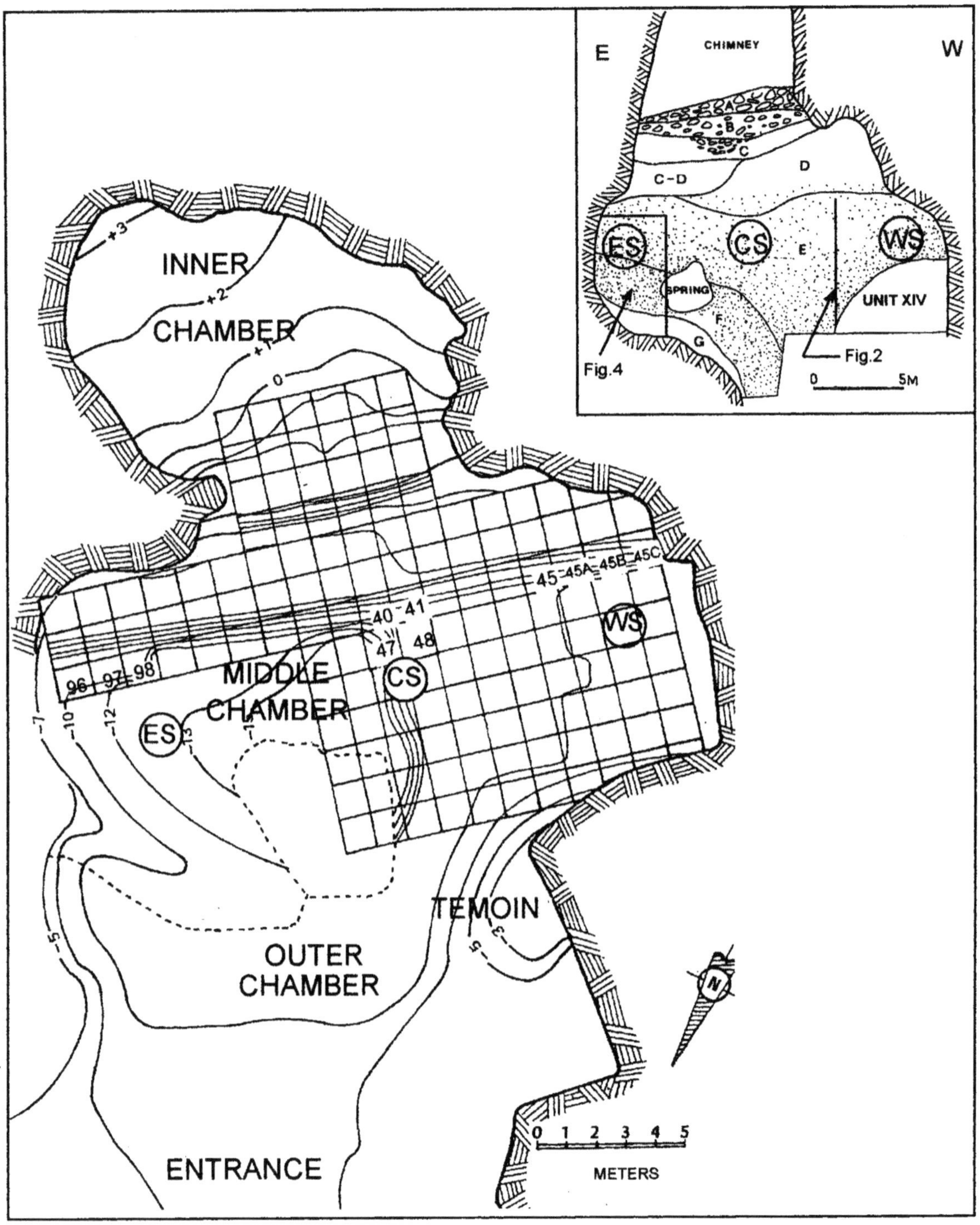

Fig. 1. Tabun cave, plan and schematic section. ES - eastern sector; CS - central sector; WS - western sector.

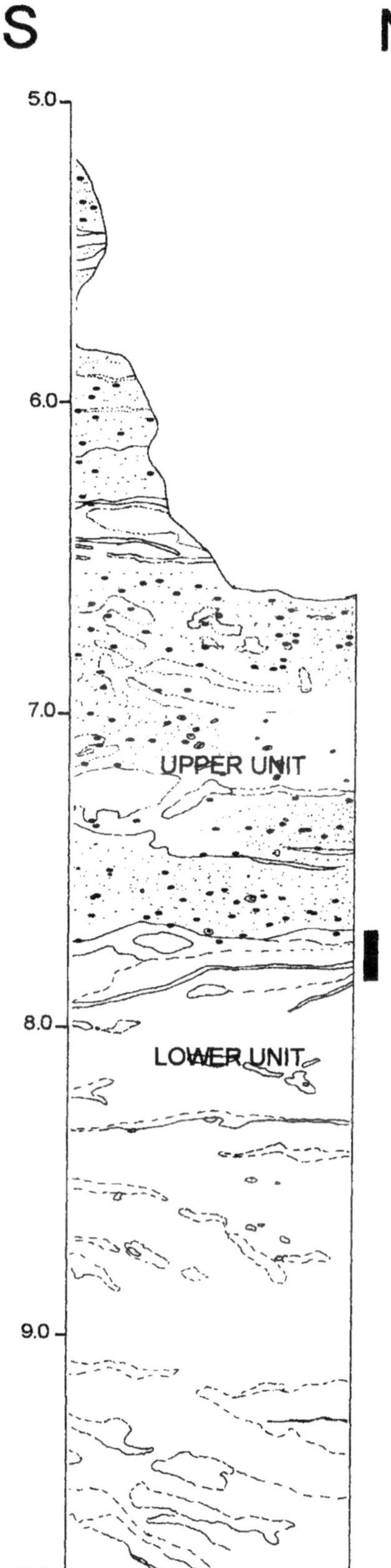

Fig. 2. Stratigraphical section in the western sector (west side of square 45). The black bar indicates the zone of unconformity.

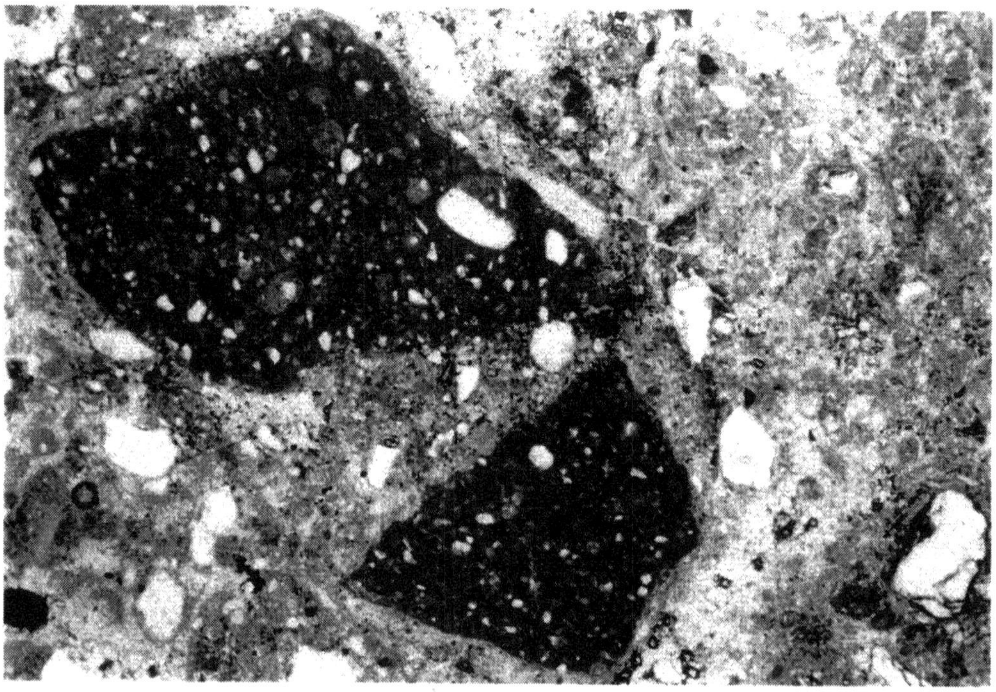

A

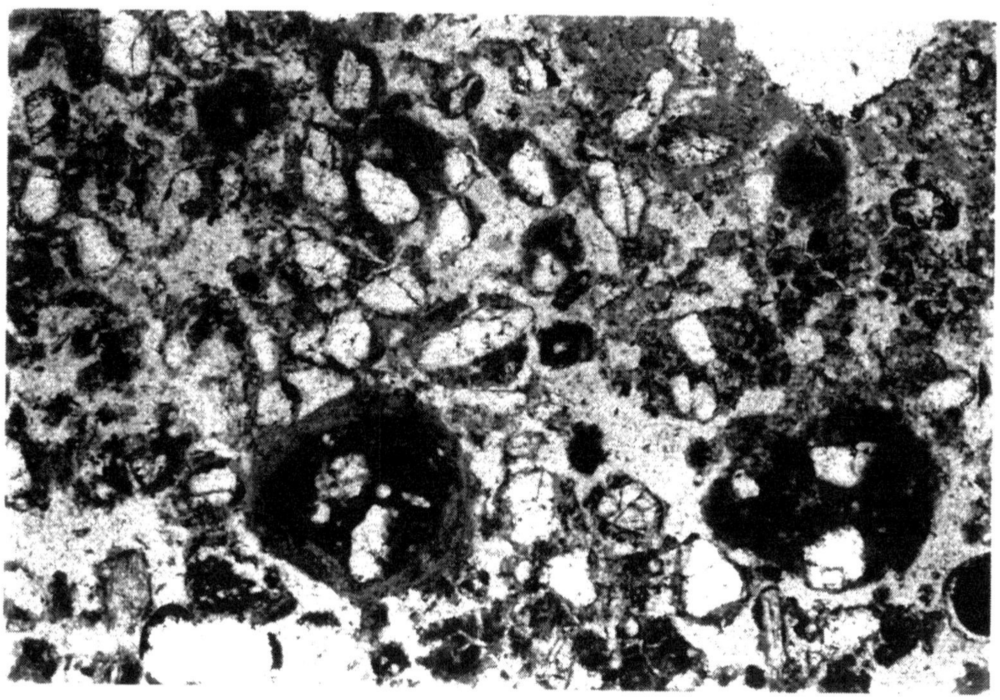

B

Fig. 3. Photomicrographs of Yabrudian deposits in the western sector, plane polarized light, frame length 2.6 mm: A - disrupted terra rossa aggregate in a clayey, biologically reworked matrix of the upper unit, Square 45c, ca. 7 m depth; B - coated sand grains and aggregates in a phosphatized matrix of Unit XIV, ca. 8 m depth.

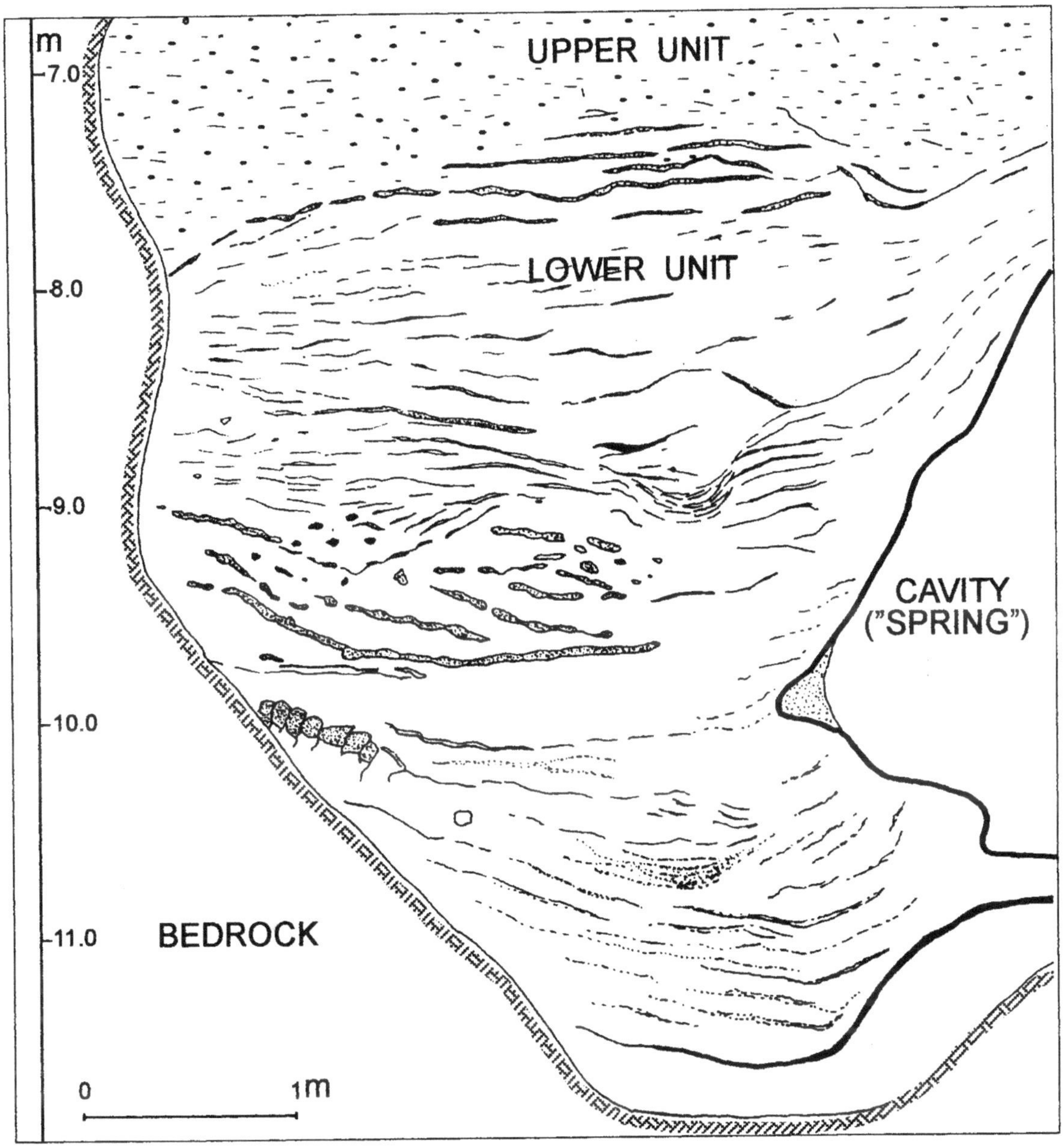

Fig. 4. Stratigraphical section in the eastern sector (squares 96-98).

Bibliography

Albert, R. A., Lavi, O., Estroff, L., Weiner, S., Tsatskin, A., Ronen, A. and Lev-Yadun, S., 1999. Mode of occupation of Tabun Cave, Mount Carmel, Israel during the Mousterian period: A study of sediments and phytoliths. *Journal of Archaeological Science* 26, 1249-1260.

Bar-Yosef, O., Vandermeersch, B., Arensburg, B., Belfer-Cohen, A., Goldberg, P., Laville, H., Meignen, L., Rak, Y., Speth, J.D., Tchernov, E., Tillier, A.-M. and Weiner, S., 1992. The excavations in Kebara Cave. *Current Anthropology* 33, 497-550.

Bull, P. A. and Goldberg, P., 1985. Scanning electron microscope analysis of sediments from Tabun Cave, Mount Carmel, Israel. *Journal of Archaeological Science* 12, 177-185.

Bullock, P., Fedoroff, N., Jongerius, A., Stoops, G. and Tursina, T., 1985. *Handbook for Soil Thin Section Description.* Waine Research, Wolverhampton.

Courty, M.-A., Goldberg, P., and Macphail, R., 1989. *Soils and Micromorphology in Archaeology.* Cambridge University Press, Cambridge.

Farrand, W. R., 1979. Chronology and palaeoenvironment of levantine prehistoric sites as seen from sediment studies. *Journal of Archaeological Science* 6, 369-392.

Farrand, W. R., 1994. Confrontation of geological stratigraphy and radiometric dates from Upper Pleistocene sites in the Levant. In: O. Bar-Yosef and R. S. Kra (eds.), *Late Quaternary Chronology and Paleoclimates of the Eastern Mediterranean.* Michigan: Braum-Brunfield Inc., pp. 33-54.

Flicoteaux, R. and Lucas, J., 1984. Weathering of phosphate minerals. In: Nriagu, J. O. and Moore, P. B. (eds.), *Phosphate Minerals*, pp. 292-317.

Garrod, D. A. E. and Bate, D. M. A., 1937. *The Stone Age of Mount Carmel. Excavations at the Wady El-Mughara, Volume I.* Clarendon Press, Oxford.

Goldberg, P., 1973. *Sedimentology, Stratigraphy and Paleoclimatology of et-Tabun, Mt Carmel, Israel.* Ph.D. Dissertation, Dept. Geological Sciences, University of Michigan, Ann Arbor.

Goldberg, P. and Nathan, Y., 1975. The phosphate mineralogy at et-Tabun cave, Mt. Carmel, Israel. *Mineralogical Magazine* 40, 253-258.

Grün, R., Stringer, C. B. and Schwarcz, H. P., 1991. ESR dating of teeth from Garrod's Tabun cave collection. *Journal of Human Evolution* 20, 231-248.

Jelinek, A. J., 1982. The Tabun Cave and Paleolithic Man in the Levant. *Science* 216, 1369-1375.

Jelinek, A. J., 1992. Problems in chronology of the Middle Paleolithic and the first appearance of early modern *Homo sapiens* in Southwest Asia. In: Akazawa, T., Aoki, K., Kimura, T. (eds.), *The Evolution and Dispersal of Modern Humans in Asia.* Hokusen-Sha, Tokyo, pp. 253-275.

Jelinek, A. J., Farrand, W. R., Haas, G., Horowitz, A. and Goldberg, P., 1973. New excavations at the Tabun Cave, Mount Carmel, Israel, 1967-1972: a preliminary report. *Paléorient* 1, 151-183.

Mercier, N., Valladas, H., Valladas, G., Reyss, J. L., Jelinek, A., Meignen, L. and Joron, J. L., 1995. TL dates of burnt flints from Jelinek's excavations at Tabun and their implications. *Journal of Archaeological Science* 22, 495-509.

Ronen, A. and Tsatskin, A., 1995. New interpretation of the oldest part of the Tabun cave sequence, Mount Carmel, Israel. In: H. Ullrich (ed.), *Man and Environment in the Paleolithic.* ERAUL (Liège) 62, 265-281.

Schiegl, S., Goldberg, P., Bar-Yosef, O. and Weiner, S., 1996. Ash deposits in Hayonim and Kebara Caves, Israel: macroscopic, microscopic and mineralogical observations, and their archaeological implications. *Journal of Archaeological Science* 23, 763-781.

Tsatskin, A. and Ronen, A., 1999. Micromorphology of a Mousterian paleosol in aeolianites at the site Habonim, Israel. *Catena* 34, 365-384.

Tsatskin, A., Weinstein-Evron, M. and Ronen, A., 1995. Weathering and pedogenesis of wind-blown sediments in the Mount Carmel Caves, Israel. *Quaternary Proceedings* 4, 83-93.

Weiner, S., Goldberg, P. and Bar-Yosef, O., 1993. Bone preservation in Kebara Cave, Israel, using on-site Fourier Transform Infrared Spectrometry. *Journal of Archaeolohgical Science* 20, 613-627

Weinstein-Evron, M. and Tsatskin, A., 1994. The Jamal Cave is not empty: Recent discoveries in the Mount Carmel caves, Israel. *Paléorient*, 20/2, 119-128.

Weinstein-Evron, M., Tsatskin, A., Porat, N. and Kronfeld, J., 1999. A $^{230}Th/^{234}U$ date for the Acheulo-Yabrudian layer in the Jamal Cave, Mount Carmel, Israel. *South African Journal of Science* 95, 186-188.

Yaalon, D. H., 1997. Soils in the Mediterranean region: what makes them different? In: Mermut, A. R. Yaalon, D. H., Kapur, S. (eds.), Red Mediterranean Soils. *Catena* 28, 157-169.

DISCUSSION

Daniel Kaufman: There is considerable variability in terms of the environmental facies in the different portions of Tabun. Is that variability one of the reasons why there are so many difficulties dating Tabun?

Alexander Tsatskin: The dating should take into consideration the geological setting in the site. The site is very complex indeed because of different kinds of disturbances.

Nicolas Rolland: If I heard you correctly, you mentioned that the anthropogenic remains contain ashes. Is that correct? Could these be attributed to fire making?

Alexander Tsatskin: Yes.

Nicholas Conard: My question is also related to the statement that calcite forms in some cases are related to ash that could be of anthropogenic origin. Based on the work of Steve Weiner and Solvieg Schiegl at Kebara, siliceous aggregates were a key for identifying the use of fire. They allowed at least crude approximations to be made for the amount of wood (and grass and other materials) burned. Is it possible to just take such a small indicator like a thin section showing these calcite forms you mentioned, and claim they are anthropogenic?

Alexander Tsatskin: I suggest to relate this microscopic feature to the results of diagenesis of ashes. I should mention that we have two different types of calcite and they are important even though they are found in microscopic quantities. The work of Weiner and Solvieg Schiegl is a different case. They dealt actually with much younger sediments and the environmental conditions after the deposition of ashes were different within the same cave, at the same site. Steve Weiner speaks about siliceous aggregates as the most stable component of ash. In the Tabun sediments we have less stable components, being much older than Kebara.

THE PRE-AURIGNACIAN AND AMUDIAN AS AN INTRA-YABURDIAN EPISODE

Leonid B.Vishnyatsky

Introduction

All the known Yabrudian assemblages are confined to the northern and southern Levant, where they are wedged between the Acheulean and Early Levantine Mousterian. At nearly all the sites where Yabrudian assemblages have been found, they interstratify with industries deviating markedly from Lower/Middle Palaeolithic standards. These industries are widely known under the labels 'Pre-Aurignacian' (Yabrud 1, Layers 13 and 15) and 'Amudian' (Tabun, bed 75; Abri Zumoffen, lower levels; El Masloukh, levels A and B; and possibly Zuttiyeh). They look rather strange against the background of finds from both underlying and overlying layers. In these industries, blades and blade-flakes become dominant among blanks, the Lower/Middle Palaeolithic component of tool-kits either decreases or disappears, and tools characteristic of the Upper Palaeolithic (hereafter UP), particularly the early UP, appear in relatively great numbers. Since their identification, Pre-Aurignacian and Amudian have been regarded as either: directly connected (Bordes 1955; 1958; Howell, 1959; Wright, 1966) or unconnected with the UP (Jelinek, 1990), related and 'closely ... resembling each other' (Garrod, 1970, 226; see also Garrod, 1962) or different (Bordes, 1977; Copeland, 1975; 1983), intrusive (Bordes, 1955; Rust, 1950, 1958) or autochthonous and constituting a division of the Yabrudian (Copeland, 1975; 1983; Jelinek, 1982; 1990; Shmookler, 1983). In this paper I will attempt to consider each of these positons. The present analysis will be based first of all on the results of first-hand examinations of the Amudian of Tabun (Jelinek's collection) and the Pre-Aurignacian of Yabrud (Rust's collection). I will begin with a description of the Pre-Aurignacian of Yabrud, and use these data as the basis for subsequent comparisons and analysis.

Pre-Aurignacian Assemblages of Yabrud I

Layer 15

In October of 1996, when I worked with the collection of artefacts obtained by A. Rust from Layer 15 of Yabrud 1 (Rust, 1950), it consisted of 927 items. Of these, 117 can be defined as cores and core-like pieces, 192 are unretouched blades and fragments, 484 are unretouched flakes (including blanks, waste flakes, chips, resharpening spalls, and core rejuvenation flakes), and 134 are tools. I was unable to locate 13 of the artefacts depicted in Rust's illustrations (Rust, 1950, Fig. 32, 1, 4; Fig. 33, 6; Fig. 36, 4-5, 8-10; Fig. 37, 2, 5, 7, 10, 13).

Most of the cores are small, and either exhausted or secondary (on flakes). Very few of them could serve further for the production of blanks longer than 3 - 4 cm. In contrast, the overwhelming majority of blades is much longer (see Table 3). It appears as if the people who left this industry tried to exploit the raw material as much as possible. It is due to this fact that the cores are not very informative in terms of the methods of flaking and core preparation employed. However, in many cases it can be seen that the manufacturers either selected or created convex flaking surfaces to obtain elongated flakes.

The unretouched blades include 18 objects with high, triangular cross sections, two of which can be defined as true (i.e., intentionally prepared) crested blades. The rest can be described as 'natural' crested blades (at least some of these served to initiate the process of core reduction), and 174 are proper, or blank, blades. The latter have, as a rule, a rather regular shape and dorsal scar patterns. These characteristics undoubtedly represent the most desirable (for the given industry) end-product of core splitting. With the exception of 34 small (less than 3 cm long) fragments, the blank blades retain some significance for analysis (the combined metrical data for unretouched and retouched blades are presented in Tables 2 and 3). Many of these blades display what can be considered macroscopic use wear traces on one or both edges (see also Narr and Lass, 1995).

Judging by the size and shape of unretouched flakes, only one fifth can be defined (formally) as blanks. Four fifths are characterized by elongated proportions (though their length rarely exceeds 5 cm). There are no massive flakes suitable for producing Yabrudian scrapers or any other bifacial tool. The data presented here is in general agreement with that described by Bordes (1977), although there are some differences (these I consider inevitable, as evaluation is invariably somewhat subjective). The composition of the tool set is shown in Table 1.

Tool	N
Distal parts of handaxes	8
Sidescrapers	4
Burins	43
Endscrapers	17
Burins-endscrapers	3
Retouch truncated blade	3
Retouched blades	19
Backed blade	1
Notches	6
Chisel-like tools	4
Retouched flakes and indeterminate tool fragments	26

Table 1. Yabrud I, Layer 15. Tools.

The distal parts of handaxes have been described in detail by Rust (1950). Rust considered these remnants of artefacts which, he thought, had been extracted from Layer 18 by Layer 15 people and used as cores. Though not without its problems (the most serious of which results from our poor understanding of the site stratigraphy) this explanation is confirmed by the fact that only tips (2.5 – 4 cm long) of handaxes are present in the collection of Layer 15. This finding is in line with the observation that most cores are heavily reduced, suggesting some shortage of raw material.

There is only one Yabrudian sidescraper in the collection (Rust, 1950, Fig. 35,1). The other tools of this category are represented by a fragment of a single convex sidescraper, a double straight-convex sidescraper, and a subquadrangular sidescraper worked along the whole perimeter by alternate retouch so that two parallel working edges are dorsal and the other two are ventral.

Of the 43 tools defined here as burins, only 26 are more or less typical. Most of these (21) are angle burins on broken blades and flakes. In three cases the burin blow was struck from the retouched (endscraper-like) platform, and in one case from a plain flake butt. One of the typical burins can be defined as dihedral. In addition, there are at least 17 atypical (flat) burins. This number may be increased at the expense of some artefacts classified as cores and secondary cores.

Endscrapers are well expressed as a rule, except six so-called 'high endscrapers' (Rust, 1950, Fig. 35, 2-7) which are morphologically ambiguous. Some can equally be defined as cores. In other cases, what are considered working ends may actually represent either faceted butts or retouched platforms designed for burin blows (three of the burins described above have just such platforms). The remaining endscrapers are on blades (6) and flakes (5), all with convex working ends.

Though not numerous, combined tools (burins-endscrapers) and retouched truncated blades are very clearly formed and leave no doubt as to their typological definition. The same applies to notches. Conversely, chisel-like tools are ambiguous. There is similar unclarity regarding classification of the only backed blade.

One of the most important and conspicuous features of the Layer 15 collection is the presence of a relatively large number of retouched blades. There are 11 blades with intermittent retouch along one edge, 6 blades with continuous retouch along one edge, and 2 blades with continuous retouch along both edges. Formally speaking, the tools classified as belonging to the latter two groups may also be defined as single and double sidescrapers on blades. However as the general typological context here is not a Mousterian one, I prefer to describe these artefacts as retouched blades. It should be noted that both the average length and width of the retouched blades exceed those of unretouched blades (66.0 mm versus 59.4 mm, and 26.3 mm versus 23.2 mm, respectively). This may signify that larger blades were used and resharpened most intensively.

Tables 2 and 3 present metrical data derived from 161 blades, both retouched and unretouched. Other data indicate that 114 (71%) of the blades are intact and 47 (29%) are broken. Furthermore, 37 (23%) retain parts of primary cortex on the dorsal surface, and 132 have distinct dorsal scar patterns. Of the latter, about 80% displayed unipolar parallel flaking, indicating the major role of this techique in blade production. The second most common technique was bipolar flaking, as represented by 15% of items (all calculations relate to the measured blades only, and do not take into account the 34 smallest fragments and several tools on blades). Characteristics of the blade butts are given in Table 4.

Layer 13

At the time I worked with the collection, it included 95 objects: 5 secondary cores and core fragments, 31 unretouched flakes (mostly waste), 51 unretouched blades (including one which may be defined as a crested blade) and 8 tools (1 ambiguous and 2 well-expressed angle burins on break, 3 atypical endscrapers, and 2 retouched flakes). I was unable to locate one of the artefacts depicted in Rust's drawings (Rust, 1950, Taf. 40:2). Of 50 blades, 27 (54%) are intact and 23 (46%) are broken. Seven (14%) retain parts of cortex on the dorsal face. Forty-two have clear dorsal scar patterns suggesting that, as in Layer 15, unipolar parallel flaking and bipolar flaking predominated (75% and 20%, respectively) (Tables 2, 3 and 4).

Layer	15	13	10	15+13
Max. thickness	9.1 (161)	7.1 (49)	9.1 (25)	8.6 (210)
Mid. thickness	8.0 (141)	6.2 (22)	7.9 (25)	7.8 (163)
Max. width	24.2 (161)	19.4 (49)	30.0 (25)	23.1 (210)
Mid. width	22.5 (141)	16.7 (22)	27.5 (25)	21.7 (163)
Base thickness**	5.0 (106)	4.1 (23)	5.5 (20)	4.9 (129)
base width***	12.7 (106)	9.9 (23)	17.6 (20)	12.2 (129)
Max. w/th	2.6	2.7	3.3	2.6
Mid. w/th	2.8	2.6	3.4	2.7
Bas. w/th	2.5	2.4	3.2	2.4
Bas.w/mid.w	0.5	0.5	0.6	0.5

Table 2. All Measurable Blades.*

* All measurements are in mm. The number of measured specimens is given in brackets.
Mid. = Measured at the middle part of the blade.
Max. = Measured at the widest or thickest part of the blade.
** When taking punctiform butts into account, the mean values decrease: 4.7 (115) for Layer 15; 3.7 (27) for Layer 13; 4.6 (142) for a combined sample.
*** When taking punctiform butts into account, the mean values decrease: 11.8 (115) for Layer 15; 8.6 (27) for Layer 13; 11.2 (142) for a combined sample.

Layer	15 (107)	13 (26)	10 (23)	15+13 (133)
Length	60.0	49.1	79.8	58.0
Max. width	23.4	18.5	29.8	22.5
Mid. width	22.0	16.6	27.3	21.1
Max. thickness	9.2	6.9	9.1	8.8
Mid. thickness	7.9	6.2	8.0	7.6
L/max. w	2.5	2.6	2.6	2.5
L/mid. w	2.7	3.1	2.9	2.8
Max. w th	2.5	2.6	3.2	2.5
Mid. w/th	2.7	2.6	3.4	2.7

Table 3. Intact Blades*.

* All measurements are in mm. The number of measured specimens is given in brackets.
Mid. = Measured at the middle part of the blade.
Max. = Measured at the widest or thickest part of the blade.

Butt Type:	Smooth	Faceted +Dihedr.	Punctiform +Linear	Natural
Layer				
15	72 (65%)	15+5 (18%)	9+8 (15.2%)	2 (1.8%)
13	17 (53%)	7 (22%)	8 (25%)	---
15+13	89 (62%)	27 (19%)	25 (17.5%)	2 (1.5%)

Table 4. Blade Butt Types.

Pre-Aurignacian, Amudian and Upper Palaeolithic

How much do the Amudian and Pre-Aurignacian have in common with the true UP? This question can be approached from both typological and technological points of view.

Typology

Typologically, the Amudian of Tabun is clearly remote from the UP norm. This is because this industry: (1) retains numerous bifaces and sidescrapers (including many of the Yabrudian type); and (2) lacks, or has only very scarce and atypical endscrapers and burins. Other UP tool types are also rare, except backed knives, which are not only relatively numerous but are sometimes of quite Chatelperronian appearance.

The Amudian of Masloukh also is dominated by sidescrapers, with backed knives following them in number (Shmookler, 1983). A somewhat different picture is present in Abri Zumoffen (Copeland, 1983; Garrod and Kirkbride, 1961). There, bifaces and characteristic Yabrudian sidescraper forms (Quina types) are virtually absent, whereas backed knives (including some specimens indistinguishable from Chatelperron points), supplemented with a number of burins, comprise the bulk of the tool set. Most UP-looking of all these industries is, of course, the Pre-Aurignacian of Yabrud I, with its famous burins and endscrapers. However, even here some reservations should be noted regarding the UP appearance of some of these tools. As mentioned above, true burins are not as numerous in Layer 15 as we used to think. Most of the so-called 'high endscrapers' are also represented by atypical and apparently accidental forms. Nevertheless, the collection contains no less than 25 morphologically pronounced burins (these are mainly angle burins on break), a dozen indisputable endscrapers, 3 tools combining these two types (burins - endscrapers), and around 20 continuously or intermittently retouched blades.

Technology

Despite the fact that both Pre-Aurignacian and Amudian reveal the combination of a high percentage of blades with non-Levallois methods of flaking (Meignen, 1994; Nishiaki, 1989), their technology can by no means be defined as UP. True prismatic cores are absent. I found them neither in Tabun nor in Yabrud. Even more significant is the fact that though platform faceting was not characteristic of these industries (Table 5), the non-marginal flaking absolutely dominated and neither exterior platform trimming nor abrasion were used to secure direct hit to the wanted point of percussion. It appears that the latter methods of core platform preparation came into wide use only in the UP (at the beginning of this period or some time after). Hopefully, this difference can serve as one of the most revealing criteria for distinguishing between pre-UP and UP cores and blanks (Nekhoroshev, 1993).

The evidence described above leads to the conclusion that the Amudian and Pre-Aurignacian industries are both typologically and technologically rather distant from UP standards. Furthermore, neither can be called 'fully Upper Palaeolithic'. At the same time, there is another, more intriguing side to the question. Though not as advanced (derived) as they are sometimes thought to be, both the Amudian and particularly the Pre-Aurignacian provide a picture of what one might expect to observe in an ideal 'transitory' Middle/Upper Palaeolithic industry. This is true in both typological and technological respects.

On one hand, both Pre-Aurignacian (Yabrud I) and some Amudian (Abri Zumoffen) assemblages are characterized by a decline or nearly total disappearance of Lower-Middle Palaeolithic tool types. On the other hand, tools of UP types become either pronounced (Tabun, Bed 75, Units I2 and I1; Masloukh, Levels A and B) or predominant (Abri Zumoffen, Levels 21-11 in Trench A and 8-5 in Trench B; Yabrud I, Layers 13 and 15). Sometimes the latter phenomenon is thought to be a direct consequence of the emphasis laid on blade production (e.g., Jelinek, 1990, 85). However, this explanation contradicts what we know about many other industries. For example, in Mousterian assemblages such as Warwasi (Iran) and Obi-Rakhmat (Uzbekistan), blades are extremely numerous but the typology remains pure Middle Palaeolithic. In the inventories of these two sites, UP tool types (such as burins, endscrapers or backed and truncated blades) are represented by rare, atypical, and most probably, accidental items.

As to technology, it would not be an exaggeration to say that in Layers 15 and 13 of Yabrud blades actually represent almost the only type of blanks. The

	Yabrudian Tabun, Lower E	MP	UP	Masloukh A	Masloukh B	Yabrud I 15	Yabrud I 13	Yabrud I 13+15	Yabrud I 10
Faceted butts (%)		62	21	21.4**			18	22	19
W/th		3.5	2.7			2.8	2.7	2.78	3.48
Mid. w	24.47	23.4	18.2	20	19.3	22.5	16.7	21.7	27.5
Base th	5.97	8.0	3.8	4.4	4.4	5.0	4.1	4.9	5.5
Base w	16.0	16.7	9.5	11.4	12.0	12.7	9.9	12.2	17.6
Base w / Mid w	0.66	0.71	0.51	0.58	0.63	0.56	0.59	0.56	0.64
Base w / Base th	3.31	2.88	2.5	2.83	2.97	2.54	2.42	2.49	3.2

Table 5. Blade Characteristics of Several Yabrudian, MP, UP, Amudian and Pre-Aurignacian Assemblages (Data from Shmookler, 1983; Lamdan and Ronen, 1989; Ronen, 1992; data for Yabrud I, Layers 15, 13 and 10, added by the author)*.

* All measurements are in mm.

** IF of Masloukh combined for A and B, and relates to blades and flakes together.

overwhelming majority of flakes are, judging by their size and shape, by-products and waste. Though the technology of blade production in the Pre-Aurignacian and the Amudian is not yet UP in character, in most respects the blades themselves are closer to UP than to Mousterian blanks, or lie at an intermediate position. For the Amudian of Masloukh this conclusion is confirmed by metrical and other data published by Lamdan and Ronen (1989) and Ronen (1992). The same seems to be true for the Pre-Aurignacian of Yabrud. A comparison with data published for several Near Eastern Middle and Upper Palaeolithic assemblages (*ibid.*) reveals that in all metrical characteristics, Pre-Aurignacian blades are closer to the Masloukh and/or UP samples than to the Middle Palaeolithic ones (Table 5). Blades from Layer 13 are metrically even closer to the UP norms than those of Layer 15. Blades from Layer 10, which is the first (earliest) Levallois-Mousterian assemblage of Yabrud 1, are in all their dimensions very close to blades from the other Near Eastern Mousterian sites.

The fact that the w/th and base w/th ratios calculated for the Pre-Aurignacian blades are considerably smaller than those calculated for the Mousterian and Yabrudian blades probably signifies (given that the thickness is approximately constant, as is evidenced by almost equal figures for Layers 15 and 10) that the former were obtained from considerably more convex flaking surfaces. Such surfaces closely resemble UP cores. This supposition is confirmed by the presence of two prepared crested blades and a number of 'natural' crested blades and blade-flakes in Layer 15, at least some of which served to initiate the reduction sequences aimed at blade production.

Thus, both the Amudian and the Pre-Aurignacian provide evidence of changes going in one and the same direction: towards blade-oriented non-Levallois (and probably volumetric) blank production, light-weight tool-kits and more economic use of raw material (i.e., from MP to UP technology and tool composition). It is no wonder then, that some scholars saw (and continue to see) in these industries the roots from which the later and true UP cultures, such as the Aurignacian and Chatelperronian, may have arose (Anikovich, 1991; Garrod, 1937; Wright, 1966). However, neither the stratigraphic positions of the Amudian and Pre-Aurignacian nor their supposedly very old age favour such an hypothesis. The absence of any intermediate links which would connect these industries with the earliest UP renders this hypothesis highly improbable.

Invasion or Local Transformation?

The only possibility to preserve the Amudian and Pre-Aurignacian as a probable source (or sources) of the later UP would be to consider them non-local, external cultures that invaded the region and, after a short stay, left to an unknown destination. However, this scenario is strongly contradicted by the following evidence: (1) in Tabun we have quite compelling proof of continuous local evolution throughout the entire Yabrudian sequence; (2) neither starting nor terminal points (areas) of the supposed migration(s) are discernible in archaeological records; and (3) in the absence of their Yabrudian context, the Amudian and the Pre-Aurignacian have little in common (they may seem similar only against the background of the pure Yabrudian) and appear to be culturally unrelated phenomena. Hence, to assert that these cultures were foreign one has to assume that two alien groups, each withan advanced stone inventory, arrived at the same, very small area, at the same period. This contention is hardly plausible. Therefore, it is possible to conclude that both direct and indirect data support the idea of local, Yabrudian roots for the Amudian and Pre-Aurignacian.

Was the Transformation Gradual or Abrupt?

According to Jelinek, there is evidence in Tabun of 'a single gradual increase and decrease in the frequency of characteristic Amudian elements relative to other tool and flake types' (Jelinek, 1990, 87). According to Rust, nothing in the lower layers of Yabrud foreshadowed the arrival of the Pre-Aurignacian (Rust, 1950). The excavations carried out by Solecki in the 1960s showed that while Rust's cultural stratigraphy could not be duplicated in every detail, his description of the general succession of industrial development was correct (Solecki, 1970, 209). Nonetheless, it seems very strange that while the collections of Layers 15 and 13 are dominated by blades and Upper Palaeolithic tool types, the material derived from Layers 16 (92 objects, i.e., 7 more than indicated by Rust) and 14 (about 500 objects) contains neither the former nor the latter. Instead, the material is entirely indicative of flake industries, in which typically Yabrudian sidescrapers are the most conspicuous tools.

What does this finding mean? Were the corresponding cultural transformations really so abrupt? Or is this just a result of the sorting of the excavated materials? We will probably never be able to answer these questions with certainty. Nevertheless, it should be emphasized that even if one mixes the collections of the four layers in one box and then analyses the combined collection as a homogenous unit, the conclusion regarding its transitory character will remain unshaken. Moreover, under these circumstances it would become obvious that the Pre-Aurignacian of Yabrud, like the Amudian of Tabun, has its roots in the Yabrudian culture, and therefore represents an autochtonous phenomenon.

Dating Problems and the Notion of an Intra-Yabrudian Episode

The chronological position of the Pre-Aurignacian and the Amudian always was, and continues to be, a matter of controversy (Bar-Yosef, 1994; Bordes; 1955, 1960; Farrand, 1970; 1994; Garrod, 1956; Heinzelin, 1966; Jelinek, 1992; Mercier, et al., 1995). It seems that the more Th/U, ESR, TL and other dates we have for the Yabrudian and succeeding Mousterian levels of Tabun, Yabrud I, Zuttiyeh and El Kowm, the less we are able to say about their absolute and even relative ages. Dates obtained for the same layers and industries are often quite incompatible, and choices made by different authors depend mainly on subjective factors and preferences. Nevertheless, the long cultural-stratigraphic sequences of Yabrud I and Tabun leave no doubt that the Amudian and Pre-Aurignacian levels are wedged in between the pure Yabrudian levels. At sites with incomplete sequences, the Amudian and Pre-Aurignacian either underlie (Abri Zumoffen) or overlie (Masloukh) Yabrudian levels. Therefore, taking into account both the stratigraphic position of these industries and their basic similarity in terms of direction of cultural transformation, it is reasonable to consider the Amudian and Pre-Aurignacian as contemporaneous and representative of a specific episode of a rather substantial (albeit gradual) change within the Yabrudian. This change most likely reflects some critical ecological and/or social situation experienced by the local population.

The notion of an 'Intra-Yabrudian Episode' and the designation of the Pre-Aurignacian and Amudian as the industries of this episode would seem appropriate on the basis of current evidence. Although some absolute dates contradict this view and suggest different ages for the Intra-Yabrudian industries, priority in the present case should be given to archaeological and stratigraphic data particularly as TL and other methods are still in the making.

The Intra-Yabrudian assemblages, together with Howiesons Poort in southern Africa and the Seclinien of north-west Europe, provide good examples of what has been called 'running ahead of time' in the development of Palaeolithic industries (Vishnyatsky, 1994). In each of these examples we have to deal with cultural transformations resembling those characteristic of the Middle/Upper Palaeolithic transition (at least in the realm of stone working). It can not be ruled out that the causes of change in all these cases (including here the Middle/Upper Palaeolithic transition) also were very similar. In order to reveal these causes, what we need most is to understand the relationship between stone technologies and modes of subsistence. This end remains, however, distant.

Acknowledgements. This work was supported by the Research Support Scheme of the OSI\HESP, Grant No. 569/1996. I am very grateful to Prof. G. Bosinski for allowing me access to the Yabrud 1 material. In addition I owe Prof. G. Bosinski, as well as Dr. O. Joris, my gratitude for the help received during my visit to Monrepos.

Author's address: Institute for the History of Material Culture, St. Petersburg, Russia. paleo@mail.dux.ru

Bibliography

Anikovich M. V. 1991. *The Early Stage of the Upper Palaeolithic in East Europe.* St. Petersburg (in Russian).

Bar-Yosef O. 1994. The contributions of Southwest Asia to the study of the origin of modern humans. In: M. H. Nitecki and D. V. Nitecki (eds.), *Origins of Anatomically Modern Humans.* New York, Plenum Press: 23-66.

Bordes F. 1955. Le Paléolithique inferieur et moyen de Jabrud (Syrie) et la question du Pré-Aurignacien. *L'Anthropologie* 59: 486-507.

Bordes F. 1958. Le Passage du Paléolithique moyen au Paléolithique superieur. In *Hundert Jahre Neanderthaler.* Köln-Graz: 175-181.

Bordes F. 1960. Le Pré-Aurignacien de Yabrud (Syrie) et son incidence sur la chronologie du Quaternaire au Moyen Orient. *Bulletin of the Research Council of Israel. Section G. Geosciences* 9G (2-3): 91-103.

Bordes F. 1977. Que sont le Pré-Aurignacien et l'Iabroudien? In: Arensburg, B. and O. Bar-Yosef (eds.), *Eretz-Israel* 13, Moshe Stekelis Memorial Volume: 49*-55*.

Copeland L. 1975. The Middle and Upper Paleolithic of Lebanon and Syria in the light of recent research In: Wendorf, F. and A. E. Marks, (eds.), *Problems in Prehistory: North Africa and the Levant.* Dallas, Southern Methodist Univ. Press: 317-350.

Copeland L. 1983. The Palaeolithic industries at Adlun. In: Roe, D. (ed.), *Adlun in the Stone Age: The Excavations of D. A. E. Garrod in the Lebanon,* 1958-1963. Oxford, BAR IS 159: 89-260.

Farrand W. R. 1970. Geology, climate and chronology of Jabroud rockshelter I. *Fundamenta.* Reihe A. Band 2: 212-223.

Farrand W. R. 1994. Confrontation of geological stratigraphy and radiocarbon dates from Upper Pleistocene sites in the Levant. In: Bar Yosef, O. and R. S. Kra (eds.), *Late Quaternary Chronology and Paleoclimates of the Eastern Mediterranean.* Radiocarbon 1994, 33-53. University of Arizona, Tuscon.

Garrod D. A. E. 1937. The Near East as a gateway of prehistoric migration. *Early Man.* London, J.B. Lappincott: 33-40.

Garrod D. A. E. 1956. Acheuleo-Yabroudien et 'Pre-Aurignacien' de la Grotte de Taboun (Mont Carmel). Etude stratigraphique et chronologique. *Quaternaria* 3: 39-59.

Garrod D. A. E. 1962. The Middle Palaeolithic of the Near East and the problem of Mount Carmel man. *Journal of the Royal Anthropological Institute* 92: 232-251.

Garrod D. A. E. 1970. Pre-Aurignacian and Amudian: A comparative study of the earliest blade industries of the Near East. *Fundamenta.* Reihe A. Band 2: 224-229.

Garrod D. A. E. and D. Kirkbride 1961. Excavation of the Abri Zumoffen, a Paleolithic rock-shelter near Adlun, South Lebanon, 1958. *Bulletin du Musee de Beyrouth* 16: 7-48.

Heinzelin J. de. 1966. Révision du site de Yabroud. *Ann. Arch. de Syrie.* Vol 16: 157-165.

Howell F. C. 1959. Upper Pleistocene stratigraphy and early man in the Levant. *Proc. Amer. Philos. Soc.* 103: 1-65.

Jelinek, A. J., 1982. The Middle Palaeolithic in the Southern Levant with comments on the appearance of modern *Homo sapiens.* In: Ronen, A. (ed.), *The Transition from Lower to Middle Palaeolithic and the Origin of Modern Man,* BAR IS 151, Oxford, 57-104.

Jelinek A. J. 1990. The Amudian in the context of the Mugharan Tradition at the Tabun Cave (Mount Carmel), Israel. In: Mellars, P. (ed.), *The Emergence of Modern Humans.* Edinburgh, Edinburgh Univ. Press: 81-90.

Jelinek A. J. 1992. Problems in the chronology of the Middle Paleolithic and the first appearance of early modern *Homo sapiens* in Southwest Asia. In: Akazawa T., K. Aoki and T. Kimura (eds.), *The Evolution and Dispersal of Modern Humans in Asia.* Tokyo, Hokusen-sha: 253-275.

Lamdan M. and Ronen A. 1989. Middle and Upper Palaeolithic blades in the Levant. In: Hershkovitz, I. (ed.), *People and Culture in Change.* Oxford: BAR IS 508(i): 29-36.

Meignen L. 1994. Paléolithique moyen au Proche-Orient: le phénomène laminaire. In: Révillion, S. and A. Tuffreau (eds.), *Les Industries Laminaire au Paléolithique Moyen.* Paris, CNRS: 125-159.

Mercier N., H. Valladas, G. Valladas, J.-L. Reyss, A. Jelinek, L. Meignen and J.-L. Joron. 1995. TL dates of burnt flints from Jelinek's excavations at Tabun and their implications. *Journal of Archaeological Science* 22: 495-509.

Narr K. J. and Lass G. 1995. Jabrud: Mikroskopische Gebrauchsspuren. *Quartär.* B.45/46: 37-74.

Nekhoroshev P. E. 1993. On the methodology of studying the Lower Palaeolithic stoneknapping technique and technology. *Rossijskaya Arkheologia* 3: 100-119 (in Russian).

Nishiaki Y. 1989. Early blade industries of the Levant: the placement of Douara IV industry in the context of the Levantine early Middle Palaeolithic. *Paléorient* 15: 215-229.

Ronen A. 1992. The emergence of blade technology: Cultural affinities. In: Akazawa T., K. Aoki and T. Kimura (eds.), *The Evolution and Dispersal of Modern Humans in Asia.* Tokyo, Hokusen-sha: 217-228.

Rust A. 1950. *Die Hohlenfunde von Jabrud (Syrien).* Neumunster, Karl Wachholtz.

Rust A. 1958. *Über Kulturen der Neanderthalers in Nähen Osten. Hundert Jahre Neanderthaler.* Köln-Graz.

Shmookler L. 1983. *Masloukh revisited: The Amudian layers of the coastal site in Lebanon.* Unpublished manuscript.

Solecki R. S. 1970. A sketch of the Columbia University archaeological investigations at Yabroud (Syria). *Fundamenta.* Reihe A. Band 2: 199-211. Cologne.

Vishnyatsky L. B. 1994. 'Running ahead of time' in the development of Palaeolithic industries. *Antiquity* 68: 134-140.

Wright G. A. 1966. The University of Michigan archaeological collections from et-Tabun, Palestine: Levels F and E. *Papers of the Michigan Academy of Science, Arts, and Letters* 51: 407-423.

DISCUSSION

Vadim Ranov: The report of Leonid Vishnyatsky is very interesting. I would like to ask your ideas about the appearance of the Amudian and other blade industries inside the Yabrudian. Is it due to immigration, or the evolution of technology?

Leonid Vishnyatsky: I do not think that it was immigration. I think that it is not a coincidence that very similar phenomena took place in the Near East, in South Africa and in north-west Europe. It is not just a coincidence that the industries classified as Seclinien in Europe and Howieson's Poort in Africa are roughly coeval. And I think that this coincidence in time can be used as one of the proofs that the assemblages classified by me as Intra-Yabrudian are also coeval. Moreover, in all three cases we have very similar phenomena, with innovations very similar to the Upper Palaeolithic. I think that in all these cases we must look for the same explanation, for the same cause of cultural change. I do not know for certain what were these causes, but in my opinion the local populations have experienced some ecological and demographic changes.

Vadim Ranov: We do not have a clear ecological change in your Intra-Yabrudian occurrences. Perhaps it is economical causes.

Leonid Vishnyatsky: To say that it is an economic cause is to say nothing, because economical changes must be caused by another change. They can not just happen. As to your remark that we have no proofs for an ecological change, we do not know the exact date of the Intra-Yabrudian Episode, so how can we know why it happened? We can only guess.

Joachim Schäfer: I think that we do not have a single Middle Palaeolithic blade industry, we have several Middle Palaeolithic blade industries. We can not put together the 20 to 50 cm blades of Obi-Mazar in one context with the 5 cm blades of Toenchesberg. If we look closer at blade industries, we might find different techniques. For example, Toenchesberg may be somewhat closer to prismatic core technique than Obi-Mazar. We must divide all these phenomena.

Leonid Vishnyatsky: In the Obi-Rahmat collection, which is partly kept in our institute, we have in fact lots of prepared cores. But when I was talking about the Middle Palaeolithic blade industries, I did not mean the Obi Rahmat or Toenchesberg assemblages. I meant Seclin and Riencourt les Bapaume and some other French industries, may be Rocourt too. As to Seclin and Riencourt les Bapaume, I have no doubt that these are real blade industries and what is even more important, and more interesting, is that both have some significant numbers of Upper Palaeolithic-like tools.

Gerhard Bosinski: It is not only Seclin and Recourt, but Toenchesberg, St. Germain de Vaux and Kabazi as well.

Leonid Vishnyatsky: No. I think that in the Crimean peninsula we have just another thing which has nothing to do with the material that I was talking about. Firstly, the Crimean materials which you mentioned are very late and they are not overlain by any Middle or Lower Palaeolithic levels. Secondly, the percentage of blades and blade tools in Kabazi is not so high.

Gerhard Bosinski: I should correct you. Kabazi tools are overlain by 'Staroseljan' type industry.

Leonid Vishnyatsky: Yes, but among Kabazi tools there are about 20% blades. This is not a blade industry for me. It does not affect my Near Eastern materials.

Avraham Ronen: I want to comment about Leonid Vishnyatsky's notion of the Intra-Yabrudian blade events. These blades are really nothing like the Mousterian blades. They have more in common with the Upper Palaeolithic than with anything we know from the Lower or Middle Palaeolithic. It is an anomaly. Even a keen observer like Bordes was sure that the Pre-Aurignacian of Rust is an early Aurignacian, a Levantine precursor of the French Aurignacian. He was upset by the large time gap, and consequently he strongly squeezed the Levantine Middle Palaeolithic.

I would like to stress an additional fact. There are only two cultural occurrences in the Levant that are restricted to caves: the Yabrudian and the Levantine Aurignacian. In the Mediterranean climatic zone, these two never occur in the open. Did those people never leave the cave, or did they make a different industry outside? Could they change tradition and technology? Nicolas Rolland suggested that the Yabrudian scrapers may have had some special function. Was that function reserved only for cavities or caves?

The Lower Palaeolithic blade assemblages are intriguing. Skinner found blades of the Pre-Aurignacian/Amudian type in the cave of Masloukh, but he could not believe that they belonged to his Yabrudian site. He came up with an unfounded scenario involving the chimney of the cave and an hypothetical open-air, Upper Palaeolithic site above the chimney, to account for those blades. The Intra-Yabrudian blades are far closer to the Upper Palaeolithic ones than any Middle Palaeolithic blade products in the Levant.

Leonid Vishnyatsky: I would like to make a reservation about the Upper Palaeolithic nature of the Amudian blades. The preparation of striking platforms is not the same as in the Upper Palaeolithic. Amudian and Pre-Aurignacian blades have thick butts.

PART III

THE MICOQUIAN AND QUINA

TOWARD MODERN HUMANS: YABRUDIAN AND MICOQUIAN, 400 – 50 kyears ago.
Edited by A. Ronen and M. Weinstein-Evron. BAR S850, 2000.

THE BACKED BIFACE ASSEMBLAGES OF EAST CENTRAL EUROPE

Jan Michał Burdukiewicz

One of the most difficult problems of archaeology is stone tool terminology and typology. How we can define and classify very old stone artefacts? How we can compare artefact assemblages? Which criteria are more significant, stylistic or technological ones? Middle Palaeolithc tools, called 'side-scraper-knife', 'prondnik' or 'Keilmesser', 'backed biface' are a good example of such controversies from methodological as well as taxonomic points of view.

Krukowski proposed the earliest classification of backed biface assemblages from central Europe. Based on their similarity, Krukowski (1939-1948) identified a number of artefact assemblages found in southern Poland as a unique local taxonomic unit. He termed these assemblages, which contained handaxes and other bifacial tools, the 'Prądnik Cycle' (after the Prądnik river - an affluent of the Vistula River). The most characteristic items were the bifacial tools, which were therefore termed *prądniks* (pronounced 'prondnik' in Polish).

Chmielewski (1969), who used the name 'Micoquo-Prondnikian', later developed Krukowski's concept of a particular taxonomic unit to describe the assemblages. Unlike Polish archaeologists, Czech researcher Valoch preferred the term 'Micoquian' to describe Moravian and Polish sites, and stressed their central European character (Valoch, 1968; 1988). Günther (1964) and Bosinski (1967) also focused on the central European traits, although others (e.g., Schild, et al., 1988, 154) used more general terms such as 'Final Acheulean' and 'Post-Acheulean'.

East-central European sites containing backed bifaces (i.e., Micoquian assemblages, see Fig. 1) have typically been caves in the Cracow-Częstochowa Upland and Moravia regions (Figs. 2 and 3, respectively). These caves are characterized by a stratigraphic arrangement of the assemblages and well-preserved faunal and floral remains. Recently, previously unknown Micoquian open-air sites were discovered in central Poland (Schild, et al., 1988; Cyrek, 1996; 1997) and Moravia (Oliva, 1991). Unfortunately, numerous sites have been excavated without a detailed stratigraphic analysis, or have been published very briefly.

Bifacial tools

In his original description of the Prondnikian, Krukowski (1939-1948, 55-56) paid special attention to bifacial tools (Fig. 4). According to this account, prondniks are *pièces arquées* of which some are *pointes asymétriques*. Prondniks are divided into two main groups: typical prondniks, and knife-like prondniks (or *prondnitshaks*). The latter group contains several varieties, including side-scrapers. Krukowski viewed prondniks dynamically, and described them as unfinished, finished, reworked or as remnants. He maintained that prondniks are diversified, and are present in several Middle Palaeolithic industries. Sobczyk (1992) recently conducted a detailed analysis of Krukowski's approach.

In the 1960s, Kowalski (excavator of the Ciemna Cave) proposed another classification of prondniks which focused on the so-called para-burin scar, an attribute which had been previously identified by Krukowski (Kowalski, 1967). An additional description was suggested during this period by Chmielewski (1969: 372-375). In contrast to other investigators, Chmielewski defined prondnik knives more broadly than 'typical prondniks'. He regarded these as a particular type of biface, which could be ordered chronologically. According to Chmielewski, younger forms could be distinguished by presence of more extensive retouched surfaces, thinner butts and para-burin scars.

Unfortunately, in the various typological approaches, backed bifacial tools were usually classified as 'type fossils' of the central European Micoquian Complex. Classification of these tools was simplified to a few selected types according to several sites in central and eastern Europe (Bosinski, 1967). The prondnik variety of tools have been reduced to one type alone (Fig. 5). However, Kozłowski (1972) and Sobczyk (1975) conducted numerical analyses of the qualitative and quantitative attributes of prondniks, which demonstrated that these tools are much diversified, thereby disproving the narrow definition of prondniks. The view that prondniks are diverse is an adjustment of Krukowski's earlier hypothesis. It would appear that prondniks represent a wide variety of Middle Palaeolithic bifacial tools. Consequently, it has been rather difficult to establish spatio/temporal diagnostic types.

Schild and Wendorf (1977, 35-43) conducted a formal complex classification of backed bifaces from Europe and northern Africa and compared it with the above-mentioned 'type fossils' with rather discouraging results. Like Krukowski, these investigators included several side-scrapers and small handaxes in the backed bifaces group (*Fäustel* Type X - see Figure 6). According to their results, classification of the so-called 'typical prondniks' (in the narrow definition of Bosinski (1967) and Kowalski (1967) should be limited to specimens with either a broken back or convex half-back (Schild and Wendorf, 1977, 41).

Somewhat later Kozłowski and Kozłowski (1977, 291), seeing a contradiction in the various definitions of Prondnik 'type fossils', introduced the new term, 'Ciemna knives', to replace 'prondniks' in Chmielewski's broader definition (beside parallel types of bifacial knives from

Germany - Fig. 5). They argued that the term Ciemna knife better fits Kozłowski (1972) model definition of prondnik, as well as prondnik groups 3 - 5 of Sobczyk's (1975) numerical classification. Unfortunately, Kozłowski and Kozłowski (1977) used the term Wylotne or 'asymmetric knives' for backed knives in order to stress their difference from 'Ciemna knives'. Numerous bifacial knives of various types are asymmetric (Figs. 5-7). In later years asymmetry was used as the main attribute by which it was possible to distinguish all bifacial knives from handaxes (Kulakovskaya et al., 1993).

This extended discussion of bifacial tool classification on the basis of specific diagnostic features leads to the conclusion that there are two main approaches in the literature to date:

♦ In the first approach, bifacial tools are seen as a wide and diversified group related to handaxes by the presence of bifacial preparation, but separated from them by lack of a continuous, sharp, worked edge toward the base, which is a natural or prepared back. The shape of bifacials overlaps partly with that of handaxes and partly with that of knives (Krukowski, 1939-1948 - See Fig. 4). Recently, such bifacials have been more precisely called 'backed bifaces', with an asymmetrical or symmetrical shape (Schild and Wendorf, 1977, 35).

♦ In the second approach, knives with various local names (Fig. 5) are defined as asymmetrical bifacial knives. This holds whether the biface is symmetrical or asymmetrical. All bifacial tools are called knives. The degree of their 'bifaciality' is held to allow bifacial knives to be distinguished from various side scrapers (Kulakovskaya et al., 1993, 10).

The first approach, which appears to be a continuation of the broad concept of prondnik tools introduced by Krukowski, better fits the empirical data and follows a general classification of bifaces (Bordes, 1961). Numerous bifacial knives from several Micoquian assemblages are symmetrical, however adherents of the second approach do not distinguish them from Micoquian knives (Fig. 7). The presence of a back, as opposed to the symmetry of a specimen, is a less controversial feature by which to distinguish backed bifaces from other biface groups. Furthermore, there are several sites outside Europe containing such tools which have never been associated with the Micoquian complex (Schild and Wendorf, 1977, 35-43).

Ad hoc classifications shown on Fig. 5 need justification by 'proper evaluation of attributes' used for analysis (Kozłowski, 1972; Schild and Wendorf, 1977; Sobczyk, 1975; 1992). The term 'backed biface' is applied in this paper to describe a diagnostic feature common to several assemblages in central Europe, traditionally associated with the 'Micoquian', 'Late Acheulean' or 'Post-Acheulean'. In general, linking backed biface assemblages from central Europe with the western European Micoquian requires a more advanced analysis than simple typological similarity.

Typical handaxes are a much less controversial group of artefacts than bifacial knives. Frequently found cordiform, oval, triangular and bottle shaped handaxes are all seen as diagnostic tools for 'Micoquian' and 'Late Acheulean' assemblages. As early as 1961, Bordes preferred the more exact term 'backed biface' for several specimens of handaxes of lanceoloate or Micoquian shape (called *Faustkeilshaber* in German). There are several types of very small handaxes, a number of which (*Fäustel* X) can also be classified as backed bifaces (Fig. 5).

The taxonomic position of several leaf points, very similar to handaxes made of tabular flint (*Faustkeilblatt*) and bifacial scrapers (*Blattschaber*) is uncertain. The similarity of these artefacts has been held by some researchers to indicate that Micoquian may be linked with Middle Palaeolithic assemblages containing leaf points.

Locality and Time Period

The backed biface assemblages in central Europe derive from the northern and southern sides of the central European mountain chains: the Sudetes and the Carpathians. Such assemblages were primarily found at locations in northeast Moravia, northwest Slovakia, and the Cracow-Częstochowa Upland. The two main site aggregations of backed biface assemblages were found in chalky uplands with good raw material for tool production, as well as numerous caves and shelters for camping. The depression between the Carpathian and Sudeten mountains, called the Moravian Gate, served as an easy pathway between northern and southern regions of sites of backed biface assemblages (Fig. 1).

Deep valleys with variable ecological conditions enabled survival of hunter-gatherer groups in the changeable climatic conditions during the Early Weichselian period (Fig. 9). Even during cold periods, when upland changed into steppe or tundra, patches of forest survived in valleys.

The chronology of the backed biface sites has been established according to the stratigraphy of cave sediments, typological ordering and, on occasion, absolute dating (Fig. 8). Chmielewski (1969, 385; 1975, 86) proposed that the evolution of Micoquo-Prondnikian assemblages during the Early Weichselian period in central Europe could be divided in three phases:

1. Late Acheulean assemblages with Micoquian elements (Clause, Cracow-Zwierzyniec, Cracow-Wawel, Koziarnia Cave layer 20 - beginning of Weichselian, before Brörup Interstadial).
2. Lower Micoquo-Prondnikian assemblages (Wylotne Shelter layer 7/8 and Ciemna Cave layer 6 - Brörup Interstadial, Wylotne layer 6, Okiennik Cave Piekary II, Piekary I and Wylotne layer 5 - between Brörup and first cold maximum of the Weichselian).
3. Upper Micoquo-Prondnikian assemblages (Ciemna cave layer 5, Buhlen - first cold maximum of the Weichselian).

This evolutionary interpretation was rejected by Kozłowski and Kozłowski (1977). These investigators linked the Micoquian and Mousterian of the Acheulean Tradition

(MTA) within one taxonomic unit they called, 'Middle Palaeolithic Assemblages with Handaxes' (MPAH). According to this view, assemblages with bifacial tools from central Europe were divided into:

1. Micoquian assemblages characterized by the Levallois technique (absent in central Europe).
2. Micoquian assemblages characterized by a non-Levallois technique (classic group from La Micoque and the Bockstein group).
3. Assemblages with asymmetric knives and *Faustkeilblätter* (Königsaue assemblages, Ciemna assemblages).
4. Mousterian of the Acheulean Tradition.
5. Assemblages of Starosele type.
6. Assemblages of Sukhaya Metchetka type.
7. Assemblages of Wylotne type.

The chronology of MPAH was established as ranging from the Eemian Interglacial to the middle Weichselian. During excavation of an open-air site at Zwoleń, Schild obtained several TL datings ranging from 95,000 to 60,000 years B.P. (Schild et al., 1988, 152). According to Kozłowski and Kozłowski (1977, 59) each taxonomic unit evolved in its own way, although all could be described as decreasing in bifacial elements and increasing in flake and blade tools over time.

Kozłowski and Kozłowski (1977) also distinguished Bockstein, Ciemna and Wylotne assemblages in the eastern part of central Europe (Fig. 2). The characteristic features of these assemblages were as follows:

- *Bockstein group:* (Okiennik Cave, Piekary I, Piekary III and Cracow-Wawel). The central features are: discoid cores, less numerous single platform cores for flakes, Micoquian handaxes and numerous handaxes made on tabular flint (*Faustkeilblätter*), bifacial knives of Bockstein type, less numerous Klausenische and Königsaue types, rare Ciemna types, side and transversal scrapers with Quina retouch, bifacial side scrapers, round scrapers (or *groshaks* – a name derived from a small Polish coin called the *grosz*), rare leaf points, and also rare denticulated tools;
- *Ciemna assemblages:* (Ciemna Cave layer 6, Okiennik, Wierzchowska Cave, Pietrowice Wielkie 23, and influence into the Ků;na Cave, Buhlen layer IIIa). Central features are single platform cores for flakes, less numerous discoid or Levallois cores, side scrapers some of which are with bifacial retouch, and bifacial knives of Ciemna type. A possible evolution from Bockstein to Ciemna layer 6 assemblages has been suggested by Kozłowski and Kozłowski (1977);
- *Wylotne assemblages:* (Ciemna Cave layer 8, Cracow–Kopernik Street and Wylotne Shelter). These assemblages are characterized by single platform cores for flakes, rare discoid and globular cores, numerous side scrapers which frequently display bifacial retouch, rounded scrapers (*groshaks*) original backed knives (described as asymmetrical), big triangular and cordiform handaxes, raclettes and denticulated tools. These assemblages may be eastern European in origin (Kozłowski and Kozłowski. 1977).

Valoch suggested the association of some assemblages containing leaf points, i.e., the Cracow-Zwierzyniec and Nietoperzowa Cave layer 13d (described by Chmielewski as Mousterian of Acheulean Tradition, or 'Pre-Szeletian'), with the Micoquian tradition. Such an association would be similar to that which is held to occur between the Rörshein and the Kösten sites, represented of the youngest sequence of Micoquian in Germany (Valoch, 1988, 62). Valoch further proposed that the central European Micoquian originated in Eemian industries like Tata, which contain a number of bifacial tools (Valoch, 1988, 64).

Unfortunately, evidence of central European sites with backed bifaces is rather poor. Sites have been either partly excavated or there are only preliminary reports published. Therefore, discussion regarding the classification of these assemblages is far from being concluded. What can be stressed is that there is general agreement that the assemblages must include bifacially worked tools classified as handaxes or knives. The traditional classification of bifacial knives (Bockstein, Klausenische, Königsaue, Prondnik, Schambach, Starosiele, or recently, Ciemna) is rather controversial, as shown above. Schild and Wendorf's suggestion regarding the development of a formal classification could be a satisfactory solution, but it needs a new morpho-metrical analysis.

The concept proposed here of temporal ordering of central European backed biface assemblages is also controversial. Stratigraphical sequences from several cave sites are very important (Fig. 8), but without taxonomy, it is difficult to establish the general evolution of the assemblages. The most complex sites are Kůlna Cave, which contains four layers with backed bifaces, and the Wylotne Shelter, which contains three such layers. Unfortunately, neither shows any clear stylistic or technological evolution. Possibly, the triangular backed bifaces are older than the trapezoid bifaces (Fig. 9).

In general, central European sites are similar to Micoquian sites in western Europe, however some assemblages (of Wylotne type) probably originated in eastern Europe. The Kůlna Cave and other Moravian sites are taxonomically between southern German and Polish backed biface assemblages.

Palaeogeographic and Palaeoecological dTata

Valoch (1988) and Musil (1988) carried out a reconstruction of the palaeogeographic and palaeo-ecological settings of Moravian sites with backed bifaces and Madeyska (1981) did the same for Polish sites. The environment of southern Poland and Moravia since the end of the Eemian Interglacial to the middle Weichselian changed from deciduous and coniferous forests to steppe and tundra. In general, the environment of central Europe was less humid than in the Netherlands. *Pinus*, *Salix* and *Betula* were much more frequent than other trees (Fig. 9, compare Starkel 1977).

The warmer interstadials of the early Weichselian were quite mild and humid. Plant remains indicate the presence of pine-birch forests, which were later replaced by tundra or steppe plants. The variability of plant communities was strongly connected to the landscape of the highland area, which was characterized by deep valleys and high plains. Different plant communities corresponded well with limited forest areas, as well as steppe and shrub tundra. In cold stadials, periglacial processes and eolian sedimentation limited plant cover. In cave sediments, two or three weathering horizons indicate at least two short phases of climatic amelioration (Madeyska, 1979, 24). In Moravia, at least three interstadials have been distinguished in the Kůlna Cave (Musil, 1988; Valoch, 1988; 1989).

The faunal assemblages from numerous sites include species of forest, tundra, steppe or steppe-tundra and eurytropic. The occurrence of different species varied with locality and time period. Tundra and steppe species dominated in the colder periods of the early glacial, while forest species are more numerous in the warmer periods of the last glacial (Fig. 9).

In cave sediments, evidence of a decrease in temperature is revealed in the appearance of animals that live in a mixed environment (e.g., *Bos*, *Bison* and *Canis lupus*), as well as cool steppe species (*Equus caballus*, *Ochotona pusilla*). Both of these are associated with several types of forest animals, a finding which indicates the presence of trees in some of the deep valleys.

A detailed study of several caves in southern Poland (Madeyska 1969; 1979; 1981) described palaeoecological conditions according to the following sedimentological data: 1 - quantity of the limestone debris (1a - debris of over 4 cm Ø in Nietoperzowa Cave 15 m from entrance, 1b - debris of over 2 cm Ø in Nietoperzowa Cave 10 m from entrance, 1c - debris of over 2 cm Ø in Koziarnia Cave); 2 - roundness coefficient of the limestone particles (2a - Koziarnia, 2b - Nietoperzowa); 3 - humus content in the sediments of Nietoperzowa Cave; 4 - porosity of the limestone rubble in the sediments of Koziarnia Cave; 5 - particles creaked after their deposition in Nietoperzowa Cave; 6 - diagnostic rodent fauna; 7 - percentage of arctic (7a) and forest (7b) species of the Nietoperzowa Cave (Fig. 8). The backed bifaces had been found in various layers of cave sediments from the early Weichselian (Fig. 8). These layers initiated in the transitional period between the climatic optimum of the Eemian Interglacial and the first arctic period of the Weichselian. According to Madeyska (1979, 21) two or three warmer phases, represented by soil formation or weathering horizons, can be distinguished (Figs. 8 and 9) during this period.

In the stratigraphy of the Kůlna Cave, Musil included Middle Palaeolithic layers in the purported Ecological Cycle II (layers L-I; or 7d-6a of Valoch's stratigraphy), which was interpreted as early Weichselian by both researchers. According to Musil (1988: 249-250), Valoch's layers 9b and 8a, which contained backed bifaces, should be included in into his Ecological Cycle I of the Eemian interglacial age. Valoch argued that these layers are not older than early Weichselian (Valoch, 1989). A transition from the interglacial to the glacial period was established between layers 11 and 10, which revealed a domination of steppe species (Valoch, 1989, 9). Layers 9b, 8b and 7c were interpreted as interstadials with warmer steppe or even forest species. The backed biface assemblage from layers 9b and 8a were associated with a warmer forest-steppe environment. The youngest layers with backed bifaces (7a and 6a) were described as cold steppe environments occurring at the end of the Lower Pleniglacial (Musil, 1988, 238).

A comparison of the ecological conditions of the backed biface sites from southern Poland and Moravia indicates that better conditions prevailed on the southern side of the Carpathians. There, Middle Palaeolithic sites are relatively numerous, but backed biface assemblages are rare. This would seem to indicate ecological preferences of cultures characterized by the backed biface assemblages. Possibly, these groups went to the south only during cooler periods.

The palaeoecological conditions for the backed biface assemblages from the western part of central Europe, as well as eastern Europe, are rather similar. Unlike the eastern part of central Europe, the western part was characterized by a scarcity of arctic species. This may be the result of the more humid climate of the west (Madeyska, 1981).

Conclusions

The concept of temporal ordering of the backed biface assemblages in the eastern part of central Europe proposed here is still conjectural. Stratigraphical sequences or, rather, a single hypothesized sequence for which evidence from different areas was gathered, show general changes of the backed biface assemblages (Figs. 8 and 9). The most complex sites are Kůlna Cave, which contains four backed biface assemblage layers, and Wylotne Shelter, which contains three such layers. The sequence is supported by simple stylistic evolution of backed bifaces. In the older period, triangular (Bockstein) backed bifaces were more frequent. In the younger period, they were replaced by trapezoid (Ciemna, Wylotne) backed bifaces (Fig. 9). Further morpho-metrical analyses should solve the problem of more precisely ordering of the assemblages. Information concerning the spatial distribution of the backed biface assemblages of central Europe is very limited. Main site aggregations are located in the area divided by the central European mountain ridge (Carpathians and Sudetes). The sites of biface assemblages are located primarily in northeast Moravia, northwest Slovakia and the Cracow-Częstochowa Upland. Both main site agglomerations of backed biface assemblages have been found in chalky uplands with good raw material for tool production, and numerous caves and shelters for temporary encampments.

A comparison of the ecological conditions of the backed biface sites from southern Poland and Moravia indicates that somewhat better conditions were available on the southern side of the mountains. However during warmer periods ecological conditions of the northern area were similar to the southern one. There, Middle Paleolithic sites are relatively numerous, but backed biface assemblages are rare. This would seem to indicate ecological preferences of hunters-gatherers characterized by the backed biface assemblages. Possibly, these groups went to the south during cooler periods, when resources were less available in the north.

Numerous sites with backed bifaces in central Europe are similar to those in Germany and France. The assemblages similar to Wylotne Shelter probably originated in eastern Europe. The assemblages from Kůlna Cave and other Moravian sites are taxonomically between the south German and Polish backed biface assemblages. Further research requires additional sites with well-preserved artefact concentrations and good stratigraphical evidence.

Author's address: Jan Michał Burdukiewicz, Department of Archaeology, University of Wroclaw, ul. Szewska 48, PL-50-139 Wroclaw, Poland. jmbur@poczta.onet.pl

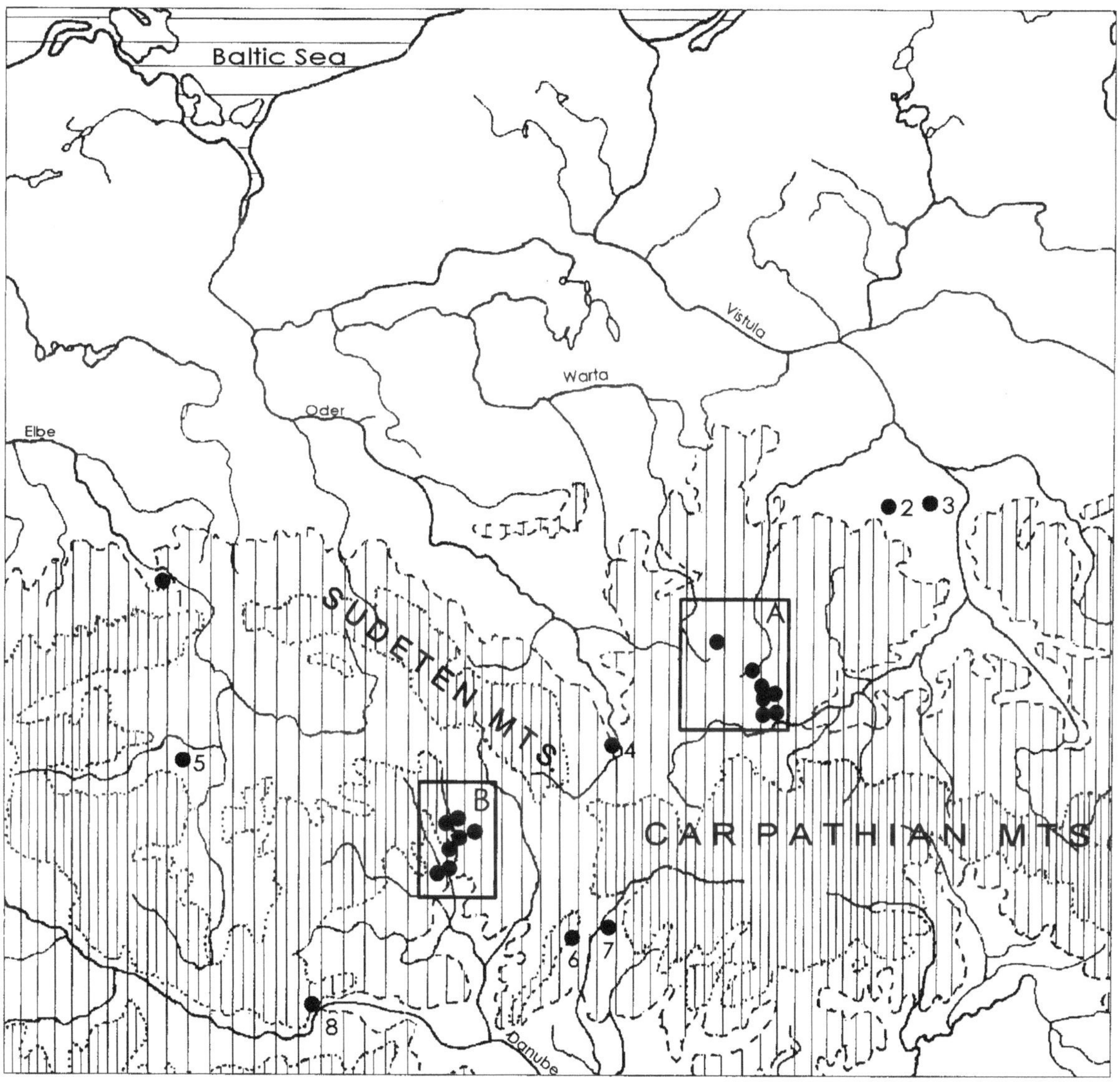

Fig. 1. Sites with backed bifaces (Micoquian) in the eastern part of Central Europe: A - sites near Cracow ; B - sites near Brno. 1- Dresden-Plauen; 2 – Kobylany; 3 – Zwoleń; 4 - Pietrowice Wielkie; 5 - Tmáň (Ve vratech); 6 – Myjava; 7 – Zamarovce; 8 - Gudenus Cave.

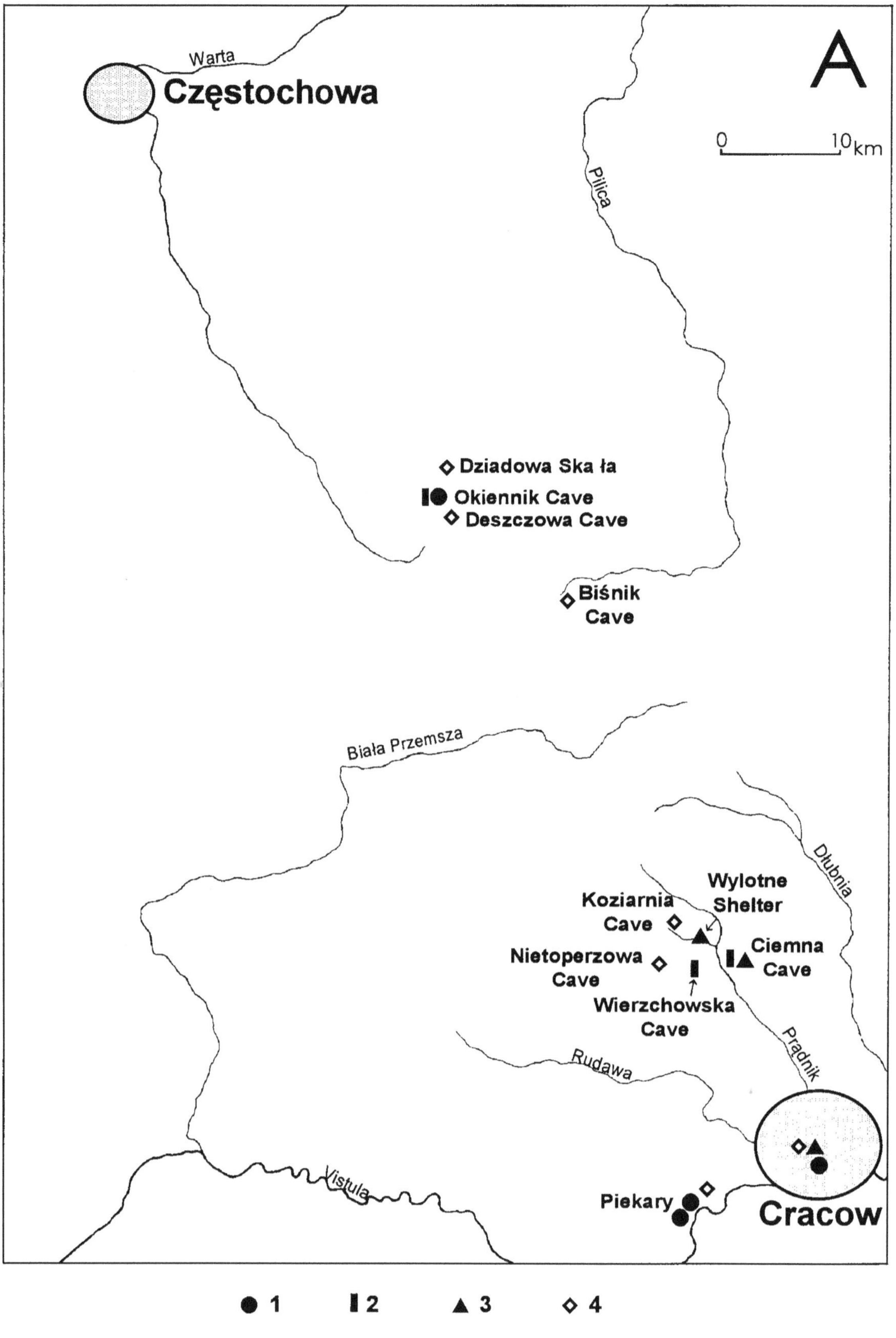

Fig. 2. Assemblages with backed bifaces in the Cracow- Częstochowa Upland (after Kozłowski and Kozłowski (1977) and Cyrek (1997)): 1 – Bockstein; 2 – Ciemna; 3 – Wylotne; 4 - other collections.

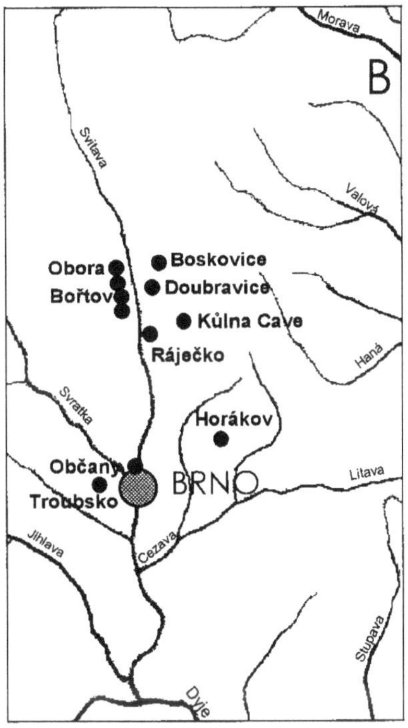

Fig. 3. Sites with backed bifaces in Moravia (after Oliva, 1991).

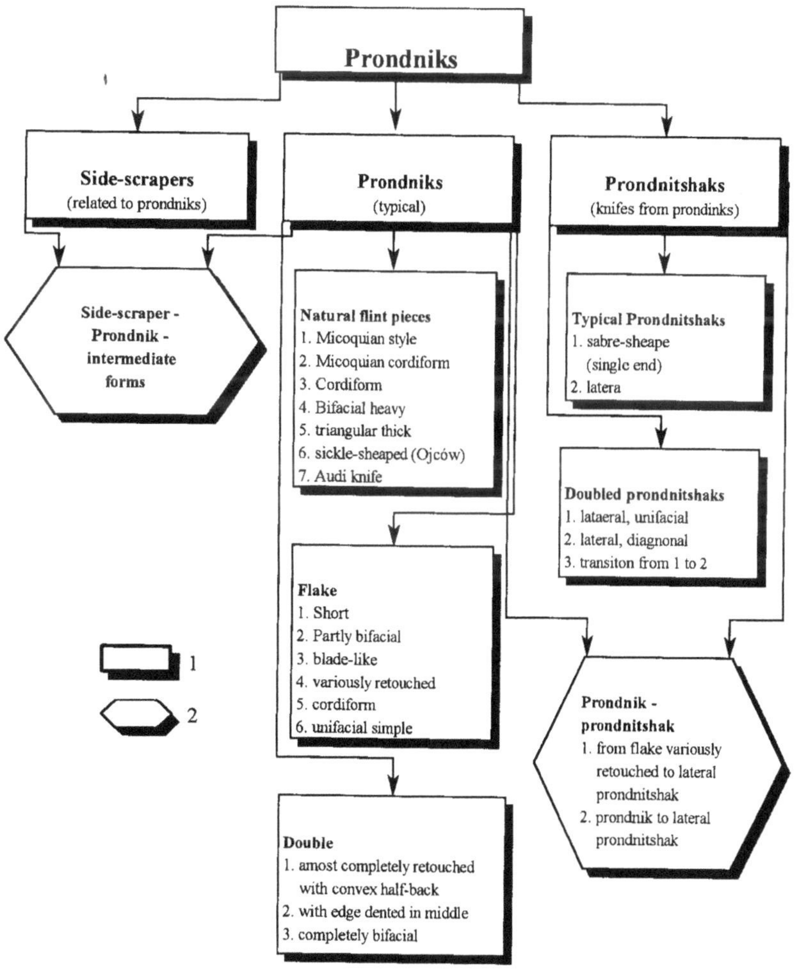

Fig. 4. Classification of prondniks (backed bifaces) by Krukowski (1939): 1 - typical forms; 2 - transitional forms.

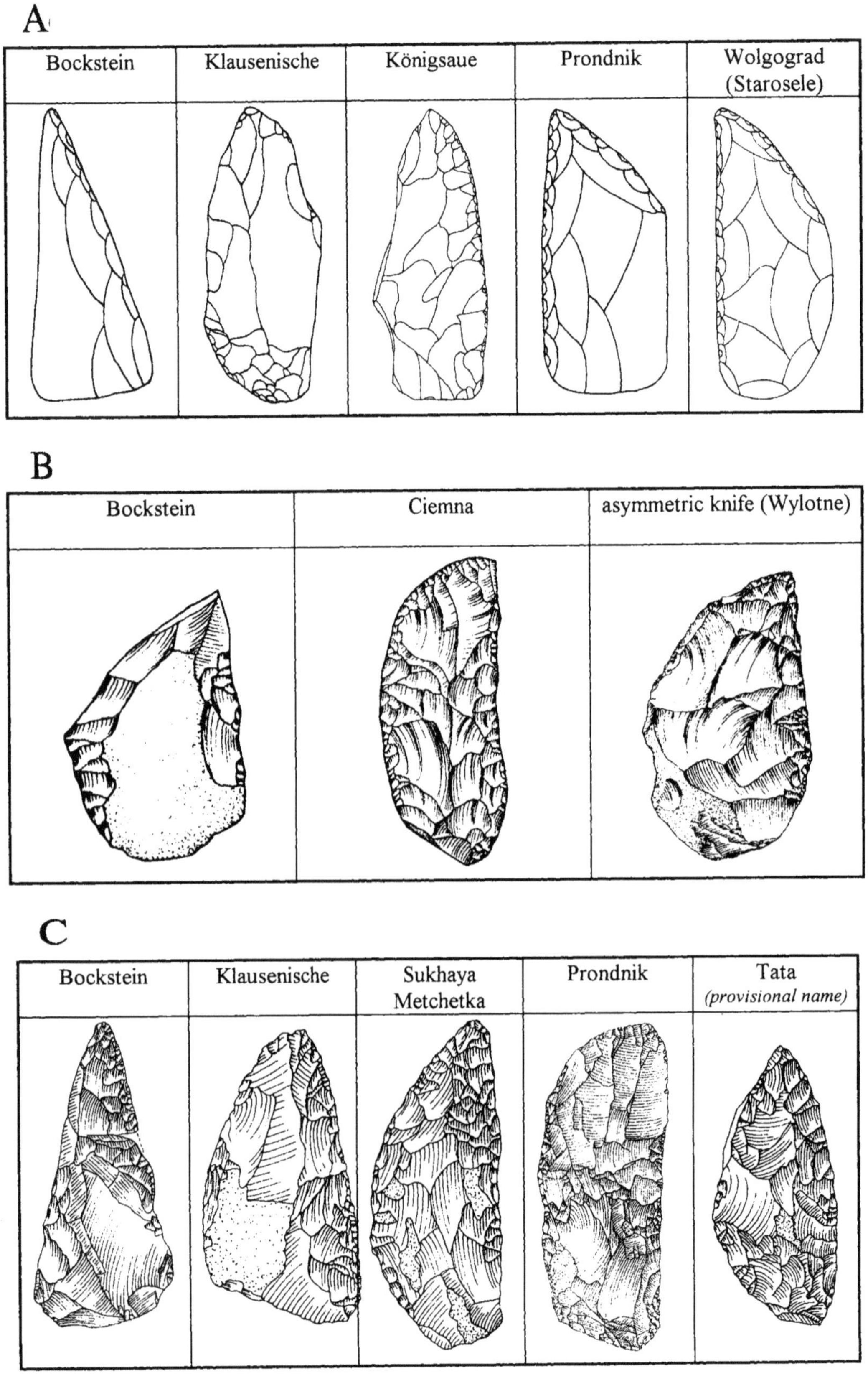

Fig. 5. Classification of backed bifaces (Micoquian knives): A - according to Bosinski (1967) and Mania and Toepfer (1973); B - according to Kozłowski and Kozłowski (1977) ; C -according to Kulakovskaya et al. (1993).

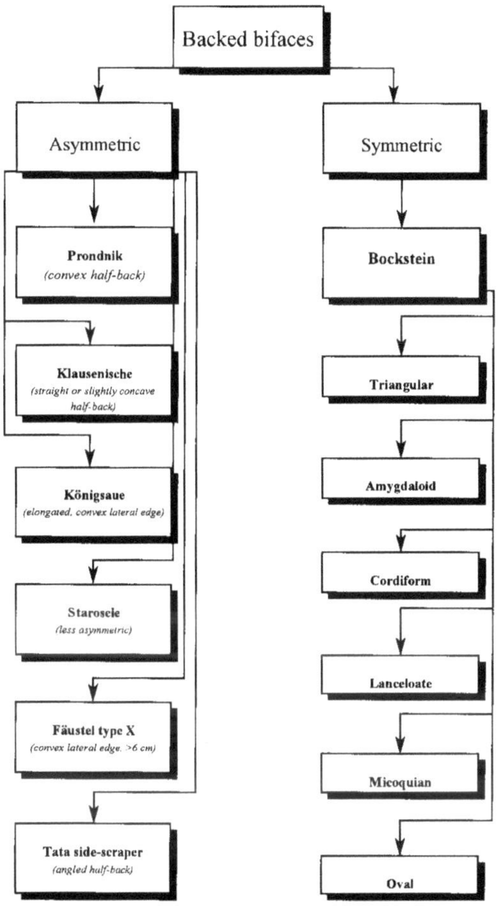

Fig. 6. Position of selected central European backed bifaces in the classification of Schild and Wendorf (1977).

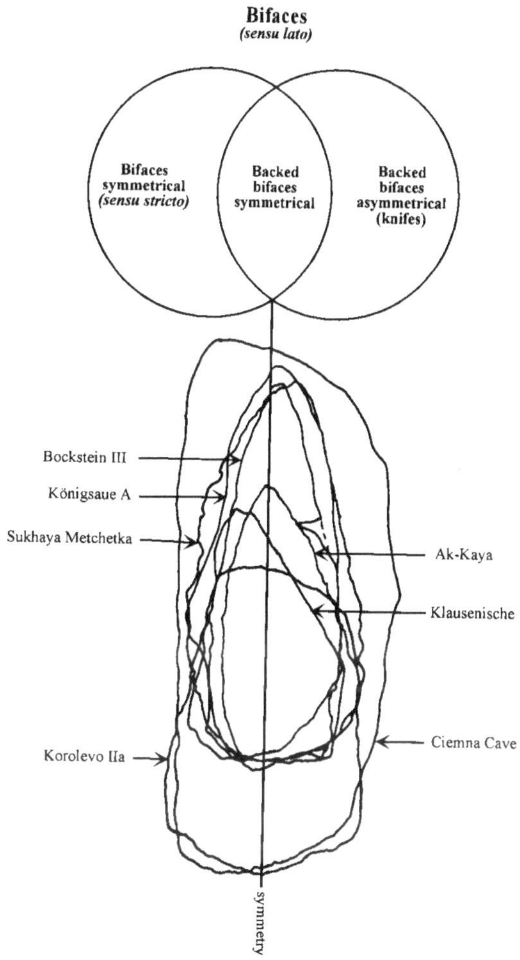

Fig. 7. Diagram of bifaces. Backed bifaces - bifacial knives and their symmetry (compare with Fig. 5).

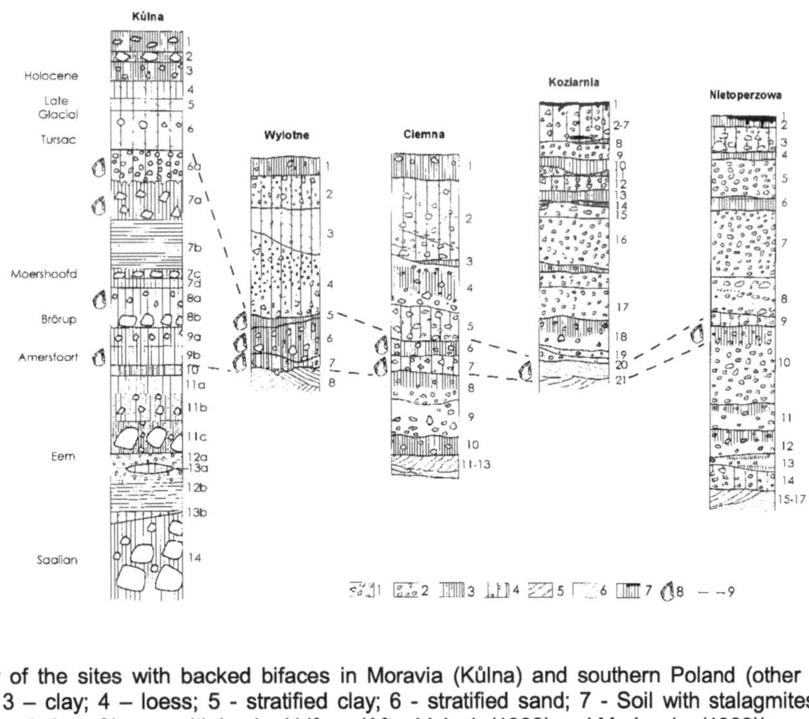

Fig. 8. Stratigraphy of the sites with backed bifaces in Moravia (Kůlna) and southern Poland (other sites): 1 - limestone debris; 2 – gravel; 3 – clay; 4 – loess; 5 - stratified clay; 6 - stratified sand; 7 - Soil with stalagmites; 8 - Backed biface assemblages; 9 - correlation of layers with backed biface (After Valoch (1988) and Madeyska (1969)).

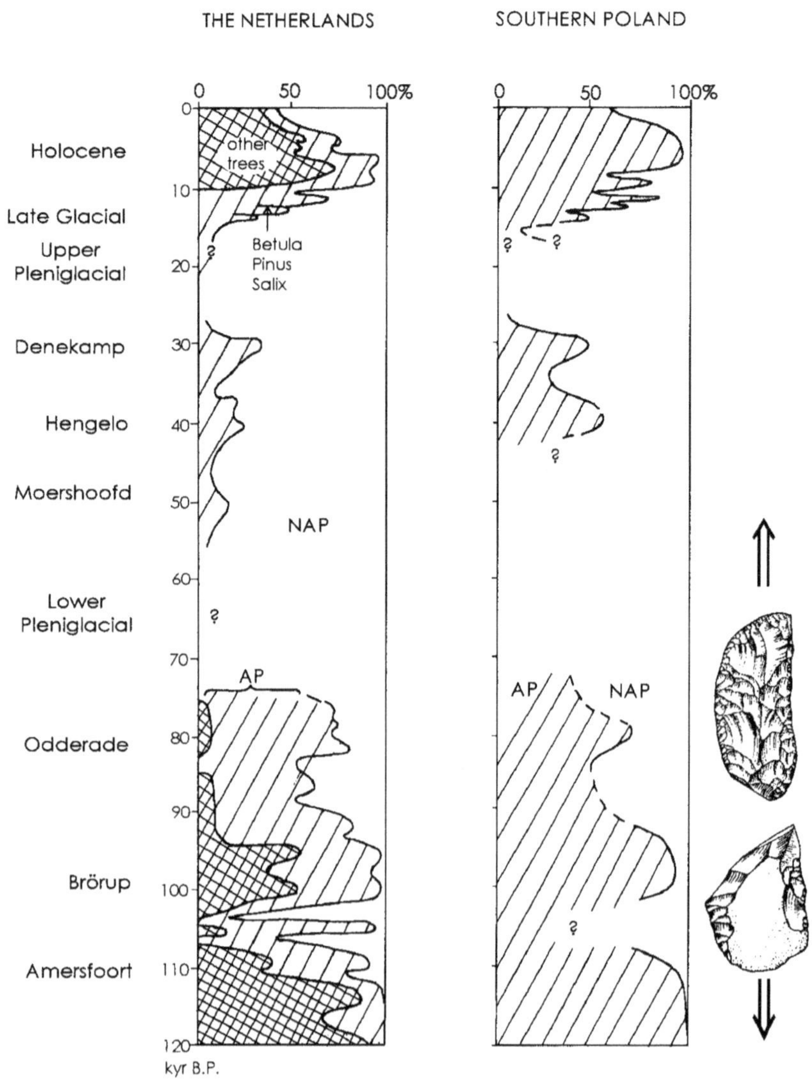

Fig. 9. Possible chronology of the backed biface assemblages in the eastern part of central Europe (simplified pollen diagrams from the Netherlands and Poland according to Starkel (1977)).

Bibliography

Bordes F. 1961. Typologie du Paléolithique ancien et moyen, *Publications de l'Institut de Préhistoire de l'Université de Bordeaux*, Mémoire No. 1.

Bosinski G. 1967. Die mittelpaläolithischen Funde im westlichem Mitteleuropa, *Fundamenta*, Series A, vol. 4, Böhlau Verlag, Köln-Graz, pp. 206.

Chmielewski W. 1969. Ensembles micoquo-prondnikiens en Europe centrale, *Geographia Polonica* 17, 371-386.

Chmielewski W. 1975. Paleolit środkowy i górny. In: W. Chmielewski and W. Hensel (eds.), *Prahistoria ziem polskich*, vol. I, *Paleolit i mezolit*, Ossolineum, Wrocław, pp. 9-158.

Cyrek K. 1996. Wstępne wyniki badań wykopaliskowych w Jaskini Deszczowej w Kostkowicach (st. 4), woj. częstochowskie w 1994 roku, *Łódzkie Sprawozdania Archeologiczne* 2, 5-21.

Cyrek K. 1997. Jaskinia Biśnik, Strzegowa, gm. Wolbrom, woj. katowickie, *Infromator Archeologiczny* 1997 (in press).

Fridrich J. 1982. *Středopaleolitické osídleni Čech*, AÚ ČSAV, Praha, pp. 151.

Gábori M. 1976. *Les civilisations du Paléolithique moyen entre les Alpes et l'Oural*, Akadémiai Kiadó, Budapest, pp. 247.

Günther K. 1964. Die altsteinzeitlichen Funde der Balver Höhle, *Bodenaltertümer Westfaliens*, 8.

Kulakovskaya L., J. K. Kozłowski and K. Sobczyk 1993. Les couteaux micoquiens du Würm ancien, *Préhistoire Européenne*, vol. 4, p. 9-32.

Kowalski S. 1967. *Zagadnienie przejścia od paleolitu środkowego do górnego w Polsce południowej w aspekcie elementów postępu technicznego*, III Sympozjum paleolityczne, Kraków, vol. 1, p. 3-7.

Kozłowski J.K. 1972. On the typological classification of Stone Age artifacts, *Sprawozdania Archeologiczne*, vol. 24, p. 455-466.

Kozłowski J. K. and S. K. Kozłowski 1977. *Epoka kamienia na ziemiach polskich*, PWN, Warszawa, pp. 388.

Krukowski S. 1939-1948. Paleolit (in:) *Prehistoria ziem polskich. Encyklopedia Polska*, vol. 4, p. 1-117.

Madeyska T. 1969. Situation stratigraphique des ensembles micoquo-prondnikiens, *Geographia Polonica*, vol. 17, p. 387-393.

Madeyska T. 1979. The environment of Middle and Upper Palaeolithic cultures in Poland, *Quaternary Studies in Poland*, vol. 1, p. 15-28.

Madeyska T. 1981. Środowisko człowieka w środkowym i górnym paleolicie na ziemiach polskich w świetle badań geologicznych, *Studia Geologica Polonica*, vol. 69, pp. 125.

Mania D. and V. Toepfer 1973. Königsaue. Gliederung, Ökologie und mittelpaläolithische Funde der letzter Eiszeit, *Veröffentlichungen des Landesumusem für Vorgeschichte Halle*, vol. 26, Berlin, pp. 164.

Musil R. 1988. Ökostratigraphie der Sedimente in der Kůlna-Höhle, (in:) Valoch K. 1988. Die Erforschung der Kůlna-Höhle 1961-1976, *Anthropos*, vol. 24, Brno, p. 215-255.

Oliva M. 1991. The Micoquian open-air site of Rájecko 1. The land use in the Moravian Middle Palaeolithic, *Anthropologie* (Brno), vol. 29, p. 45-61.

Schild R. Z. Sulgostowska, A. Gautier, A. Bluszcz, H. J. Jensen, H. Królik and J. Tomaszewski 1988. The Middle Palaeolithic of the North European Plain at Zwoleń, (in:) *L'Homme de Neandertal. Actes du colloque international de Liège 4-7 décembre 1986*, vol. 8, La mutation, ERAUL, No. 35, p. 149-167.

Schild R. and F. Wendorf 1977. *The Prehistory of Dakhla Oasis and adjacent desert*, Ossolineum, Wrocław, pp. 259.

Sobczyk K. 1975. Problem prądnika w świetle taksonomii numerycznej, *Sprawozdania Archeologiczne*, vol. 27, p. 255-269.

Sobczyk K. 1992. Stefana Krukowskiego koncepcja prądnika, (in:) J. Lech and J. Partyka (eds.), *Prof. Stefan Krukowski 1890-1982. Działalność archeologiczna i jej znaczenie dla nauki polskiej*, Ojcowski Park Narodowy, Ojców, p. 103-117.

Starkel L. 1997. The paleogeography of mid- and east Europe during last cold stage, with west European comparisons. *Philosophical Transactions of the Royal Society of London*, B. 290, 351-372.

Valoch K. 1968. Evolution of the Palaeolithic in Central end Eastern Europe, *Current Anthropology*, vol. 9, p. 351-368.

Valoch K. 1988. Die Erforschung der Kůlna-Höhle 1961-1976, *Anthropos*, vol. 24, Brno, p. 5-199.

Valoch K. 1989. Osídleni a klimatické zmény v posledni dobé ledové na Moravé, *Časopis Moravského Muzea*, vol. 74, p. 7-34.

DISCUSSION

Nicholas Conard: Jan Burdukiewicz gave us an excellent historical review of the problems of the bifacial industries from the Weichselian period, the diversity within them, and the problems of classifying them. I see a lot of similarities with the problems we were discussing in West-Central Europe. Many of the excavations are caves where we do not have the stratigraphy we would like, where assemblages may be mixed because of inherent taphonomic problems as well as excavation techniques, particularly in Germany. Most of the excavations are older excavations, and it sounds like you have a lot of the same problems. The general question is what we should call these industries that are so rich in Keilmesser and similar forms. And how can we subdivide them if we care to, and what can we do to clarify what you described as the pessimistic situation. What can we do to make this situation better?

Jan Burdukiewicz: Concerning the classification of these Backed bifaces, as they are called by Romuald Schild, probably the best name is Keilmesser or backed knife, though we do not know whether they were really knives. They belong to the backed biface assemblages, or backed biface tradition, using this term not as a specific term but as a diagnostic one. Some assemblages are very poor in these tools. This is usually the first approach. *Ad-hoc* hypotheses were offered by several researchers, specially the genius work by Krukowski already before the war. But his term Prodnik was misunderstood in western Europe. This is a special type, not including the whole Keilmessergroup, and it created a lot of troubles. As a result, Kozlowski decided to use another term, Ciemna knife, but in fact this is no solution either. The same is true for the sharpening scar. Is it a special way of thinning? I suggest to form an international group to discuss these problems and make some detailed attribute analysis of some well-excavated assemblages, to try and solve these problems.

Nicholas Conard: You mentioned one possible chronological marker, the so-called para-burin technique. Is that typically late? Is that just a working hypothesis?

Jan Burdukiewicz: It is more frequent in later cases.

Gerhard Bosinski: Concerning the confusion in nomenclature: In my thesis I defined the Bockstein and Prodnik-knives. When Kozlowski read it, he told me that a Prodnik-knife has to be bowed in the upper part, and carry a sharpening blow. Consequently, I changed the nomenclature, and introduced the Klausenische-knife. I reserved the term Prodnik for the type which is now also called Ciemna.

Avraham Ronen: This discussion shows difficulties in bridging our varying levels of knowledge on different parts of the world, which is a bit of a handicap.

TOWARD MODERN HUMANS: YABRUDIAN AND MICOQUIAN, 400 – 50 kyears ago.
Edited by A. Ronen and M. Weinstein-Evron. BAR S850, 2000.

PREMIÈRE ÉVIDENCE DE L'ÂGE PRÉ-ÉEMIEN DES INDUSTRIES À PIÈCES BIFACIALES ASYMÉTRIQUES EN EUROPE CENTRALE

Edelgarda M. Foltyn, Eugeniusz Foltyn, Janusz K. Kozłowski

Introduction

Les industries à pièces bifaciales asymétriques ont été définies en Europe centrale et orientale. En Allemagne, G. Bosinski a distingué les industries à outils bifaciaux, comme celles de Bockstein, de Buhlen et de Klausennische, les rapprochant du Micoquien classique occidental (Bosinski, 1967;1970). Plusieurs industries semblables ont été caractérisées en Pologne (celles de Pradnik, de Wylotne et 'skalski', cf. Chmielewski, 1969; Kozłowski et Kozłowski, 1977; 1996), en Moravie (Valoch, 1995; Boëda, 1995), en Hongrie, oú ces industries sont connue sous la dénomination de Babonyen (Ringer, 1983), en Ukraine transcarpatique (Koulakovskaya, 1989; 1995) et en Crimée (Kolosov et al., 1993).

Toutes ces industries datent du Würmien (Vistulien) ancien. Ce n'est que le Babonyen de Hongrie qui pourrait éventuellement évoluer pendant tout le stade isotopique 5, y compris I'Eemien (5e). Cette datation serait justifiée surtout grâce aux observations stratigraphiques effectuées à Malyi, près de Miskolc, où cette industrie a été trouvé dans un paléosol éemien (Ringer et al., 1995). Par contre, les fouilles récentes de la Micoque montrent la permanence d'une tradition technique entre les niveaux 3, 4, 5 et 5', datant d'environ 300,000 ans BP et la couche 6 - l'archétype Micoquien - qui serait probablement contemporaine du début de la dernière glaciation (Delpech et al., 1995). L'ancienneté de la tradition des pièces bifaciales asymétriques, évoluant à partir d'une conception trifaciale du débitage et de façonnage en Occident (Boëda, 1991), incitait certains auteurs à suggérer une diffusion entre l'Europe occidentale et l'Europe centrale.

Dans ce contexte, les découvertes centre-européennes qui pourraient attester d'un âge pré-éemien pour les industries à pièces bifaciales asymétriques sont d'une importance particulière. Dans cette communication, nous allons présenter une découverte de quelques formes bifaciales dans un contexte stratigraphique suggérant un âge pré-éemien, probablement de la glaciation de Warthe ou de la fin de l'interstade de Ohe. Cette découverte provient du site Dzierzyslaw I en Haute Silésie, jusqu'à présent connu comme site du Paléolitique supérieur ancien avec des niveaux szélétiens (Kozłowski, 1964; 1996) et bohuniciens (Bluszcz et al., 1994).

Le site

Le site Dzierzyslaw I est situé dans la zone limitrophe entre la Silésie et la Moravie, au Nord-Ouest de la Porte de Moravie et des Carpates occidentales. Du point de vue morphologique, il s'agit d'une élévation dite Czarna Gora (288 m au-dessus du niveau de la mer), formée par l'accumulation fluvio-glaciaire et glaciaire du Rissien (glaciation de la Pologne Centrale ou de l'Oder). Cette élévation fait partie d'une zone de paysage d'origine fluvio-glaciaire sur le Plateau de Glubczyce, entre Pietraszyn, Kietrz, Rozumice en Pologne et Chuchelna, Trebom, Hnevosice et Kobzice dans la République Tchèque. Cette zone est délimitée par les vallées de Rozumicki Potok et de Morawka, affluents de Troja. L'élévation en question est formée par des sables et graviers d'origine fluvio-glaciaire, reposant sur les graviers du Pliocène et les marnes du Badenien et recouverts par plusieurs nappes de loess, surtout sur les versants. Le relief fossile est caractérisé par plusieurs cuvettes dues au dégel de glace morte. Par contre, dans le relief des nappes de couverture loessique, nous observons la présence de mouillères.

Stratigraphie du site

La séquence stratigraphique du site (Fig. 2):

1. Sol Holocène,
2. Loess bipartite homogène avec lentilles sableuses, concrétions de carbonates et cailloutis. La base de ce loess pourrait être stratifiée ou remplacée par la formation d'un sol initial ou de gley de toundra. Ce loess a rempli la dernière génération des coins de glace. Un os d'*Equus caballus* provenant de la base de cette nappe loessique a été daté du ^{14}C de 19,650±2200 ans (Gd-10233). Il s'agit donc de la période du Pléniglaciaire superieur.
3. Une série de solifluxion composée de loess, sable, limon et de matière organique, stratifée au sommet, et des langues et des buttes fluées à la base. Cette solifluxion contenait également des charbons de *Pinus silvestris* et a donné plusieurs datations TL: 22,000±3000 BP (GdTL-347), 36,500±5500 BP (GdTL-349), 12,5000±55000 BP (GdTL-346) et 240,000±50,000 ans BP (GdTL-341).
4. Le sol à gley de toundra développé sur un sédiment limonoargileux riche en carbonates, recoupé par les coins de glace de génération plus ancienne. Ce gley de toundra passe à un horizon hydromorphe lié à la stagnation d'une congère ou à la présence d'une mouillère. Tout est flué à cause du développement d'une structure du type palsa. Le sédiment sur lequel ce sol est développé a été daté par TL de ≥75,000 ans BP (GdTL-351).
5. Le loess stratifié avec traces de pédogénèse au sommet et avec des structures de liquéfaction; à l'intérieur d'une mouillère, ce loess passe au limon jaunâtre, daté par TL de ≥100,000 ans BP (GdTL-350).
6. Loess finement lamellaire, sableux, daté par TL de 180,000±35,000 ans BP (GdTL-348). Cette

nappe de loess est partiellement recoupée par l'érosion formant un ravin, qui pourrait être daté du dernier interglaciaire (Eemien).

7. Sables à grains fins et moyens, partiellement avec cailloutis (Kozłowski, 1964; Bluszcz et al., 1994; Foltyn, 1994).

Entre le loess finement lamellaire de la couche 6 et le loess stratifié de la couche 5, nous observons un hiatus sédimentaire important dû à l'érosion, marqué non seulement par une discontinuité de structure laminaire, mais également par des ravins fossiles. À l'extérieur du site, à l'intérieur des cuvettes dues au dégel de glace morte, le loess stratifié de la couche 5 repose sur le paléosol éemien. Il s'agit d'un paléosol lessivé sous-jacent à un horizon gleyifié, avec des horizons A1-A3-B1 bien développés. Ce paléosol repose sur les trois horizons de petits sols humoferriques datant de la fin de la glaciation de Warthe (cf. Fajer et al., 1993).

Les pièces bifaciales de la couche 6

Le premier artefact lithique a été trouvé dans la couche 6 pendant les fouilles de 1962 (Kozłowski, 1964). Il s'agit d'un éclat débité à partir d'un nucléus discoide avec une retouche fine denticulée. A l'époque, le loess laminé a été considéré comme provenant du Würm ancien; donc cette trouvaille a été rapportée au Moustérien. Les fouilles suivantes de 1989-1992 (Bluszcz et al., 1994; Foltyn, 1994), étendues sur $27m^2$ (sondages I/89 et II/92), ont fourni quelques autres objets lithiques provenant de la couche 6, notamment:

1. Le genre de 'Blattschaber' (Fig. 3a) avec un dos légèrement convexe abattu par une retouche abrupte et semi-abrupte. Les deux faces portent une retouche envahissante plate à partir du bord tranchant. Dans la partie distale, on observe la présence de coups de burin transversaux. Dimensions: longueur 7.7 cm, largeur 5.4 cm et épaisseur 2.1 cm.
2. Fragment thermique d'une pièce bifaciale asymétrique à retouche plano-convexe. Sur la face entièrement conservée, les négatifs centripètes assez larges et profonds sont bien visibles. Par contre, sur la face convexe abîmée par fracture thermique, il reste seulement un fragment de bord tranchant. Ce bord porte, dans la partie distale, le négatif d'un coup de tranchet latéral, typique des couteaux-racloirs de Pradnik (Kowalski, 1969; Bosinski, 1969; Desbrosse et al., 1976). Dimensions: longueur 8.3 cm, largeur 4.5 cm, épaisseur 1.4 cm (Fig. 3e).

3, 4. Deux esquilles de retouche dont les dimensions sont: 0.7 x 0.6 x 0.05 cm et 0.5 x 0.7 .x 0.1 cm.

A ces objets il faut ajouter l'éclat provenant d'un nucléus discoide, déjà mentionné, dont la face dorsale est formée de quatre négatifs convergents, au talon facetté, portant une retouche fine denticulée sur les deux bords.

L'ensemble de ces artefacts est très restreint, mais ils proviennent d'une couche située à une profondeur de plus de 2 m, donc difficilement accessible dans les sondages. Il est possible que le centre de concentration du matériel lithique dans la couche 6 se situe entre les sondages I/62, I/89 et II/92 (Fig. 1).

Conclusions

A Dzierzyslaw I les trouvailles de la couche 6 ont prouvé, pour la première fois en Europe centrale, la présence de pièces bifaciales asymétriques du type Micoquien dans des dépôts pré-éemiens. La datation TL de cette couche, de 180,000±35,000 ans BP (GdTL-348), indique qu'il s'agit de loess ancien supérieur correspondant au stade Isotopique 6 (stade de Warthe) ou du stade isotopique 7 (fin de l'interstade de Lublin-0he). La limite entre les stades 6 et 7 correspond à environ 195,000 ans BP.

Malgre le nombre trs restreint des artefacts, il s' agit de pièces assez typiques. L'éclat ci-haut mentioné confirme l'application du débitage doiscoide a direction cordale, dont proviennent les éclats du type des pointes pseudo-levallois. Cette technique de débitage est bien attestée dans plusieurs ensembles du Micoquien de l'Est (Boëda, 1995). Les deux outils bifaciaux façonnes sur des fragments de silex local erratique montrent l'application de la methode mixte de façonnage. La pièce du type Blattschaber est façonnée du côté du bord tranchant par une methode biconvexe. Par contre, le dos est plano-convexe, donc ressemble aux pièces à section rhomboidale; mais, dans ce cas, la section est triangulaire puisque le bord tranchant est biconvexe. Cette pièce ressemble donc à plusieurs formes typiques du Micoquien de l'Est, organisées autour d'un dos opposé au bord tranchant (Delpech et al., 1995). Il s' agit des formes plus anvancées que les bifaces asymetriques du type de Bockstein, car le dos intentionellement retouché ressemble plutôt aux couteaux bifaciaux de type Klausennnische (Koulakovskaya et al., 1993).

L'autre couteau-racloir bifacial est façonné par la technique plano-convexe, mais son caractère particulier est donné surtout par la présence de coup de tranchet lateral (Desbrosse et al, 1976). Cet attribut morphologique, lié au réavivage du bord tranchant, est particulièrement caractéristique des industries micoquiennes aussi bien en Europe centrale (Kowalski, 1969) qu'en Europe occidentale (Bosinski, 1969; Farizy, 1995; Koulakovskaya et al., 1993). En plus, ces outils caractérisent généralement les ensembles plus récentes datés du Würm ancien (Vistulanien), plutôt même d'une phase avancée de cette période.

La continuation des fouilles à Dzierzyslaw pourrait sans doute apporter des précisions concernant l'industrie de la couche 6. Rappellons que ce site a déjà fourni des trouvailles importantes du Paléolithique supérieur ancien: Szélétien dans la couche 3 et Bohunicien dans la couche 4. Ces occupations datent de l' Interpleniglaciaire, la couche 4 etant l'equivalent de la base du sol de Komorniki.

Authors' address: Institute of Archaeology, Jagellonian University, 31007 Krakow, Poland. KOZLOWSK@argo.hist.uj.edu.pl

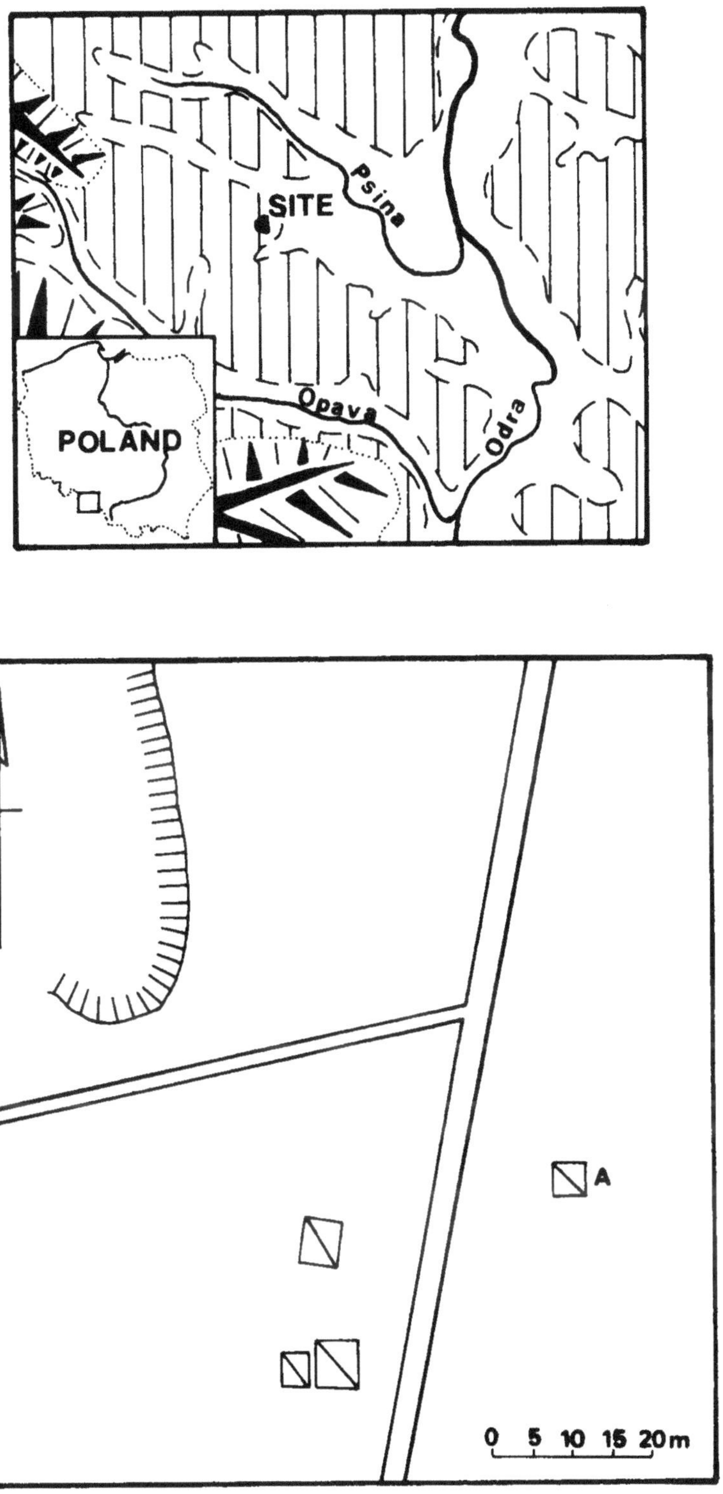

Fig. 1. Dzierzyslaw I. Localisation du site et situation des sondages (A)

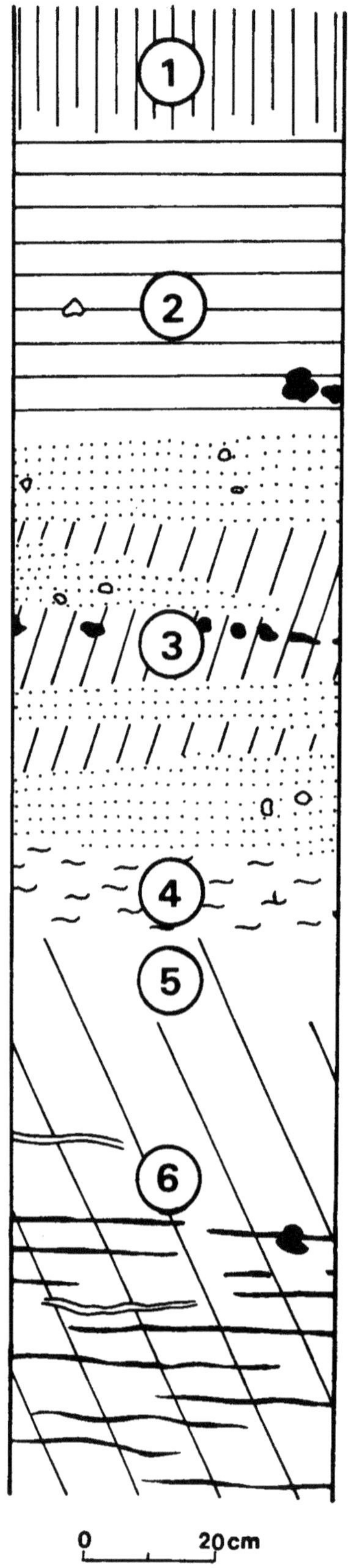

Fig. 2. Dzierzyslaw I. Séquence stratigraphique (couches 1-6).

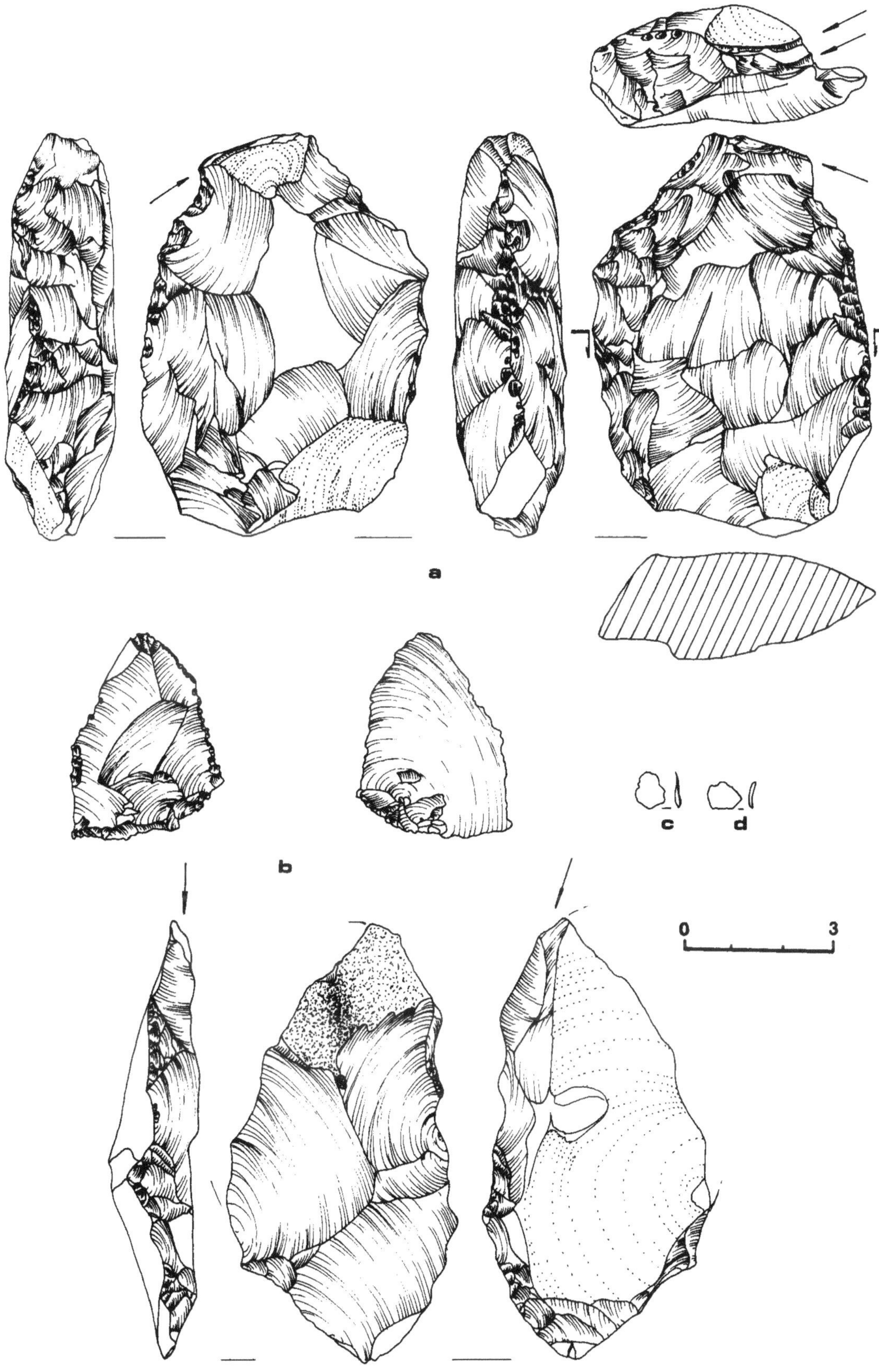

Fig. 3. Dzierzyslaw I. Trouvailles lithiques de la couche 6:
a - racloir de type 'Blattschaber'; b – éclat; c,d - esquilles; e - couteau asymétrique bifacial.

Bibliographie

Boëda, E., 1991. La conception trifaciale d'un nouveau mode de taille paléolithique. *Les premiers européens*. Paris, 251-263.

Boëda, E., 1995. Caractéristiques techniques des chaînes opératoires lithiques des niveaux micoquiens de Kulna (Tchécoslovaquie). *Les industries à pointes foliaciées d'Europe Centrale. Supplément de Paléo* 1, 57-72.

Bluszcz, A., Kozłowski, J. K. and Foltyn, E., 1994. New sequence of EUP leaf point industries in Southern Poland. *Préhistoire Européenne*, 6, 197-222.

Bosinski, G., 1967. Die mittelpalaolithische Funde im westlichen Mitteleuropa. *Fundamenta*, A/4. Köln. Böhlau Verlag.

Bosinski, G., 1969. Eine Variante der Micoque-Technik am Fundplatz Buhlen. *Jahresschrift für Mitteldeutsche Vorgeschichte,* 53, 59-74.

Bosinski, G., 1970. Bemerkungen zu der Grabung D. Peyrony's in La Micoque. *Frühe Mensch und Umwelt Festschrift Alfred Rust.* Edited by K. Gripp, R. Schütrumpf and H. Schwabedissen, pp. 52-56. Köln, Böhlau Verlag.

Chmielewski, W., 1969. Ensembles micoquo-prondnikiens en Europe Centrale. *Geographia Polonica*, 17, 371-386.

Delpech, F., Geneste, J.-M., Rigaud, J. Ph. et Texier, J. P., 1995. Les industries antérieures à la dernière glaciation en Aquitaine septentrionale: chronologie, paléoenvironnements, technologie, typologie et économie de subsistance. *Les industries à pointes foliacées d'Europe Centrale.* Supplément de Paléo, 1, 133-163.

Desbrosse, R., Kozłowski, J.K. et Zuate y Zuber, J., 1976. Prondniks de France et d'Europe centrale. *L'Anthropologie*, 80, 156-167.

Farizy, C., 1995. Industries charentiennes à influences micoquiennes, l'exemple de l'Est de la France. *Les industries à pointes foliacées d'Europe Centrale.* Supplément de Paléo, 1, 173-178.

Fajer, M., Pawelczyk, W. and Waga, J. M., 1993. Osady i struktury kopalnych zaglebien z Dzierzyslawia kolo Kietrza na Plaskowyzu Glabczyckim. *II Seminarium Badan Czwartorzedowych.* Poznan.

Foltyn, E., 1994. Tymczasowe sprawozdanie z badan interdyscyplinarnych na stanowisku Dzierzyslaw I, wojewodztwo opolskie, w 1992 roku. *Badania Archeologiczne na Gornym Slasku i w Zaglebiu Dabrowskim w latach 1991-1992.* Katowice, 212-221.

Kolosov, Y. G., Stepanchuk, V. N. and Chabai, V. P., 1993. *Ranniy paleolit Kryma.* Kiev, Naukova Dumka.

Koulakovskaya, L., 1989. *Mustierskiye kultury karpatskogo baseyna.* Kiev, Naukova Dumka.

Koulakovskaya, L., 1995. Aspects typologiques des industries micoquiennes: le site Korolevo en Ukraine. *Les industries à pointes foliaciées d'Europe centrale.* Supplément de Paléo, 1, 207-212.

Koulakovskaya, L., Kozłowski, J. K. et Sobczyk, K., 1993. Les couteaux micoquiens du Würm ancien. *Préhistoire Européenne*, 4, 9-32.

Kowalski, S., 1969. Zagadnienie przejscia od paleolitu srodkowego do gornego w aspekcie elementow postepu technicznego. *Swiatowit*, 30, 39-46.

Kozłowski, J. K., 1964. Stanowisko gornopaleolityczne Dzierzyslaw I pow. Glubczyce na Gornym Slasku w swietle badan przeprowadzonych w 1962 roku. *Wiadomosci Archeologiczne*, 30, 461-477.

Kozłowski, J. K. and Kozłowski, S. K., 1977. *Epoka kamienia na ziemiach polskich.* Warszawa, PWN.

Kozłowski, J. K. et Kozłowski, S. K., 1996. *Le Paléolithique en Pologne. Grenoble.* Millon.

Ringer, A., 1983. *Babonyen - eine mittelpalaolithische Blattwerkzeugindustrie in Nordostungarn.* Dissertationes Archaeologicae, II.11. Budapest.

Ringer, A. , Kordos, L. and Krolopp, E. , 1995. Le complexe Babonyen-Szélétien en Hongrie du Nord-Est dans son cadre chronologique et environnemental. *Les industries à pointes foliaciées d'Europe Centrale.* Supplément de Paléo, 1, 27-30.

Valoch, K., 1995. La variabilité typologique du Paléolithique moyen de la grotte Kůlna en Moravie *Les industries à pointe foliaciées d'Europe Centrale.* Supplément de Paléo, 1, 73-78.

TOWARD MODERN HUMANS: YABRUDIAN AND MICOQUIAN, 400 – 50 kyears ago.
Edited by A. Ronen and M. Weinstein-Evron. BAR S850, 2000.

LE MICOQUIEN ET LES PREMIÈRES INDUSTRIES LAMINAIRES EN MORAVIE

Karel Valoch

Le Micoquien centre-européen, défini par G. Bosinski (1967), représente un phénomène important du Paléolithique moyen. Sur le territoire de la République Tchèque, il est représenté par la grotte de Kůlna dans le Karst morave dont le remplissage a livré des sédiments bien stratifiés du Würm ancien (Valoch, 1988). Les industries paléolithiques se trouvaient dans les couches 9b, 8a, 7d, 7c, 7-alpha, 7a et 6a. Seules trois couches (9a, 8b et 7b), ne présentaient aucun contenu archéologique. Du point de vue de l'appartenance culturelle, le complexe tout entier tombe dans le Micoquien même si les couches 8a et 7d n'ont livré que des collections restreintes et donc difficiles à classifier. Le cadre chronologique du complexe en question est délimité approximativement par les dates 70 ka pour la couche 9b (Rink et al., 1996) et 45 ka pour la couche 7a (Mook, 1988). La couche la plus récente - 6a - ayant livré une industrie assez riche, n'est pas datée. Son âge, cependant, doit osciller entre 45 et 40 ka, sa faune étant nettement boréale et témoignant de l'apogée du premier pléniglaciaire avant l'oscillation climatique du Würm moyen.

Une analyse détaillée a été rendue possible grâce aux ensembles 7c, 7-alpha, 7a et 6a: la petite collection 9b contient deux artefacts tout à fait typiques permettant de dater le début du Micoquien dans l'interstade Odderade à la limite des stades isotopiques 5 et 4. Les traits technologiques de ce Micoquien sont tout à fait typiques. La méthode Levallois de la réduction des nucléi n'a pratiquement pas été employée. L'indexe IL égale zéro dans les couches 7c et 6a, alors qu'il atteint 2.76 dans la couche 7-alpha. Il est intéressant de constater que dans la couche 7a, les secteurs particuliers de la surface fouillée présentent des différences dans l'apparition des élements Levallois. Dans le secteur D1, près de l'entrée de la grotte, l'indexe IL atteint une valeur supérieure à 4.49, tandis que dans le secteur voisin D2, il se rapproche de zéro et dans le secteur suivant D3, il est de 1.47. Dans les secteurs de la partie médiane de la grotte, l'indexe IL n'est pas représenté dans les secteurs F et H, tandis que dans les autres secteurs (E, G1 , G2) il oscille entre 0.61 et 0.25. Ainsi, la méthode Levallois n'a pas été systématiquement utilisée et il semble plutôt que les produits Levallois soient accidentels.

Le schéma opératoire du débitage de type discoïde est caractéristique du Micoquien local (Fig. 1, 12-13) (Boëda, 1995). Le nombre des nucléi classifiables n'est pas très élevé (6a - 23 pièces, 7a - 122 pièces, 7-alpha - 10 pièces, et 7c - 16 pièces) et leur morphologie ne confirme pas la prédominance supposée des formes discoïdes. En principe, les nucléi peuvent être divisés en formes discoïdes (uni- et bifaciales) et 'protoprismatiques' uni- et bipolaires; un petit groupe est constitué par les nucléi irréguliers, et ceux vraiment Levallois font complètement défaut. L'absence des éléments Levallois n'est donc pas conditionnée par la prédominance absolue des nucléi discoïdes; à côté de la chaîne opératoire discoïde, celle 'protoprismatique' non-Levallois fut utilisée. Les pointes pseudo-Levallois de même que les lames-éclats à crête sont représentées dans le débitage. Pour le façonnage des pièces bifaciales, un schéma opératoire spécifique a été utilisé (Boëda 1995). En principe, le débitage est en éclats, les lames véritables étant faiblement représentées : 6a, 5.7%; 7a, 5.36%; 7-alpha, 2.53%; et 7c, 3.39% du nombre total des artefacts retouchés et non retouchés.

La composition typologique des industries est fortement influencée par la présence d'outils bifaciaux typiques du Micoquien centreeuropéen même si leur nombre n'est pas très élevé (Fig. 2). Il s'agit de bifaces de type acheuléen, de bifaces micoquiens, de bifaces foliacés, de petits bifaces et de couteaux de type Prondnik et Bockstein. L'indexe IBif est, pour 6a, 6.46; pour 7a, 7.96; pour 7c, 6.74 et pour 7-alpha, 15.49. Dans les secteurs particuliers de la couche 7a, l'indexe IBif oscille considérablement entre 3.55 en G2 et 12.16 en D1. L'outillage sur éclats est dominé par les racloirs (Fig. 1, 1-9) et il y a aussi des racloirs Quina typiques (Fig. 1, 10). Les pointes moustériennes apparaissent seulement dans les secteurs D3 et E de la couche 7a. Les types du Paléolithique supérieur sont représentés partout par 5% - 17%. Dans la couche la plus récente, 6a, ils représentent 13%. Les encoches et denticulés constituent 20% - 30% de la totalité des outils.

Dans toutes les couches micoquiennes de Kůlna on a découvert une quantité considérable d'os d'animaux fracturés, parmi lesquels de vrais outils ont pu être distingués. Il s'agit de quelques 150 compresseurs-retoucheurs et d'une cinquantaine d'éclats d'os nettement usés dans le sens longitudinal ou sur une des extrémités. Le secteur D3 de la couche 7a a livré un fragment de côte fort usé et arrondi sur les deux extrémités (Fig. 1, 11). Les côtes de mammouth usées de la même façon constituent un élément courant de l'industrie osseuse du Pavlovien (par example, á Predmosti ; Valoch, 1982 ; 1988). Une étude archéo-zoologique détaillée des os d'animaux provenant de Kůlna est actuellement en cours (Patou-Mathis et Auguste).

Le Micoquien morave occupe une position géographique importante parmi les sites concentrés dans les environs de Cracovie dans la Pologne du Sud et les grottes dans l'Allemagne du Sud. La Porte de Moravie (Moravie du Nord), le seule corridor entre le Massif de Bohème et les Carpates, servant probablement même à cette époque-là à établir un système de communication entre les plaines du nord et de l'est et les régions du Danube.

Les premières industries laminaires en Moravie sont en rapport avec le Micoquien du point de vue chronologique, mais ni culturel ni génétique. Il y a 40 000 ans apparaissait, dans les environs de Brno, un nouveau technocomplexe dont la production lithique était liée aux sources locales de silicite des calcaires jurassiques de Stránská skála. Le trait le plus frappant de cette industrie est l'utilisation de la chaîne opératoire laminaire Levallois récurrente. Ce technocomplexe a été appelé Bohunicien d'après le site éponyme Brno-Bohunice (Valoch et al., 1976; Oliva, 1984). Bohunice excepté, les ateliers ont été fouillés sur le plateau de Stránská skála, sites III1,2, IIIa et IIa (Svoboda et al., 1991). Les dates radiocarbone oscillent entre 38 et 43 ka. A proximité de Stránská skála se trouve une vaste station de surface, Lisen-Ctvrtê, ayant livré près de 20 000 artefacts.

La technologie du Bohunicien est donc caractérisée par le schéma laminaire Levallois le plus souvent bipolaire. Il semble cependant que le schéma laminaire non-Levallois fut également utilisé (Fig. 5). Une étude technologique détaillée sur Stránská skála III-1 sera effectuée par Z. Nerudová et P. Neruda (Valoch et al., sous press). L'ensemble des types du Bohunicien consiste en deux parties: en premier lieu, des lames, éclats et pointes Levallois non retouchés (certains portant une retouche partielle) (Fig. 3, 1-21) dont le pourcentage pouvant atteindre jusqu'à 50% de la totalité des outils. Les outils retouchés ont l'apparence du Paléolithique supérieur. Les grattoirs sur bout de lames et éclats sont le plus souvent plats et leurs bords ne sont pas retouchés (Fig. 4, 1-5), les formes carénoïdes sont rares (Fig. 4, 6). Les burins de tous types sont peu fréquents, ceux sur lame cassée et sur méplat naturel prédominent (Fig. 4, 7-11). Plus souvent que des perçoirs apparaissent différents becs, les lames tronquées et lames retouchées sont omniprésentes, le couteau typique de Châtelperron apparaît sporadiquement. Les racloirs sont intéressants: appliqués sur les éclats Levallois relativement minces, ils ne peuvent pas être traités par la retouche épaisse ou écailleuse; leur nombre est analogue à celui des grattoirs (Fig. 4, 14). De Bohunice proviennent aussi quelques pointes foliacées (Fig. 4, 15-16). Les encoches sont toujours bien fréquentes, les denticulés sont moins abondants mais ensemble ils constituent la partie essentielle, parfois même la majorité des outils retouchés. On ne peut cependant pas leur attribuer une grande importance pour la classification de l'industrie car iles ont probablement été créés plus ou moins accidentellement lors de l'utilisation des artefacts. Il s'agit souvent de différents éclats et fragments de lames aux encoches isolées ou avec une lignée continue d'encoches.

Le Bohunicien est considéré comme une industrie de transition du Paléolithique moyen au Paléolithique supérieur où la technologie représente la composante archaïque, du Paléolithique moyen, et la typologie correspond en principe au Paléolithique supérieur. L'existence du Bohunicien n'est prouvé, pour le moment, que dans les environs de Stránská skála où se trouvait la source de la matière première utilisée. La forte influence du Bohunicien sous forme du schéma opératoire Levallois peut être observée dans plusieurs stations de surface du Szélétien dans un périmètre de 10 - 15 km au sud et dans les industries des environs d'Ondratice á 40 km au nord-est de Stránská skála. Dans toutes ces industries le schéma Levallois est lié à la matière première de Stránská skála, il est rarement appliqué à d'autres roches (par example, radiolarite). Les produits Levallois dans le cadre de l'Aurignacien de Hradsko en Bohême centrale (Vencl, 1977) peuvent être considérés comme une reminiscence problable du Bohunicien. Récemment, l'industrie du Bohunicien a été constatée à Dzierzyslaw (Pologne du Sud) non loin de la frontière avec la Silésie (Kozłowski et Kozłowski, 1996).

Dans les pays voisins de l'Europe centrale, les industries de ce genre ne sont pas connues; en Ukraine, le Bohunicien apparaît dans la couche inférieure du site Kulycivka (Demidenko and Usik, 1993). La couche supérieure 1 de la succession du Moustérien dans l'Abri du Maras en Ardèche (sud-ouest de la France) présente une composition typologique analogue et est peut-être du même âge (Combier, 1967). Une analogie parfaite de notre Bohunicien est constituée par l'industrie de Boker Tachtit dans le désert du Negev, tant du point de vue technologique et typologique que de celui de la position chronologique (Marks, 1983).

Les origines du Bohunicien n'ont pas encore été prouvées. A la différence du Levant où les industries de type Boker Tachtit ou Ksar Akil phase A représentent une étape naturelle de l'évolution du Levallois-Moustérien (par example, Rosh Ein Mor) et constitue le courant principal du Paléolithique moyen local, en Europe centrale de telles industries ne sont pas connues. Au cours du Paléolithique moyen, la technique Levallois apparaît à Korolevo en Ukraine Transcarpatique, mais le Micoquien, prévalant dans la région centreeuropéen, ne s'en sert pas. Dans la plaine de l'Europe septentrionale cependant, les industries laminaires apparaissent dès le début du dernier glaciaire (Rheindahlen B1, Rocourt, Séclin et autres sites de la France du nord), utilisant tant le schéma laminaire Levallois que celui non-Levallois (Révillion, 1994).

Parmi ces industries laminaires et le Bohunicien il y a une grande distance chronologique, de sorte qu'une relation génétique ne peut être envisagée. D'autre part, il existe du Levallois-Moustérien tardif en Europe du sud-est (la grotte Temnata en Bulgarie) et en Ukraine (Molodova V), présentant les traits technologiques identiques même si les types du Paléolithique supérieur y font défaut. Le Levallois-Moustérien de type Molodova constitue probablement une base pour la naissance des industries similaires de la région (Kulyçivka).

Ce court aperçu permet de déduire que les industries du type Bohunicien, Kulycivka, Boker Tachtit, Ksar Akil A ou Maras couche 1 se constituaient polycentriquement, de façon convergente, indépendamment les unes des autres, comme le dernier stade d'évolution des industries laminaires du Paléolithique moyen de schéma Levallois et non-Levallois, et qu'elles représentent une transition vers le Paléolithique supérieur. Leur application dans ses cultures évoluant assez vite n'était probablement pas grande (par example, la tradition Levallois au Szélétien ou dans l'Aurignacien du Levant). Les restes osseux des hommes-créateurs de ces industries pourraient sans doute éclaircir maints problèmes, si l'on avait la chance de les découvrir.

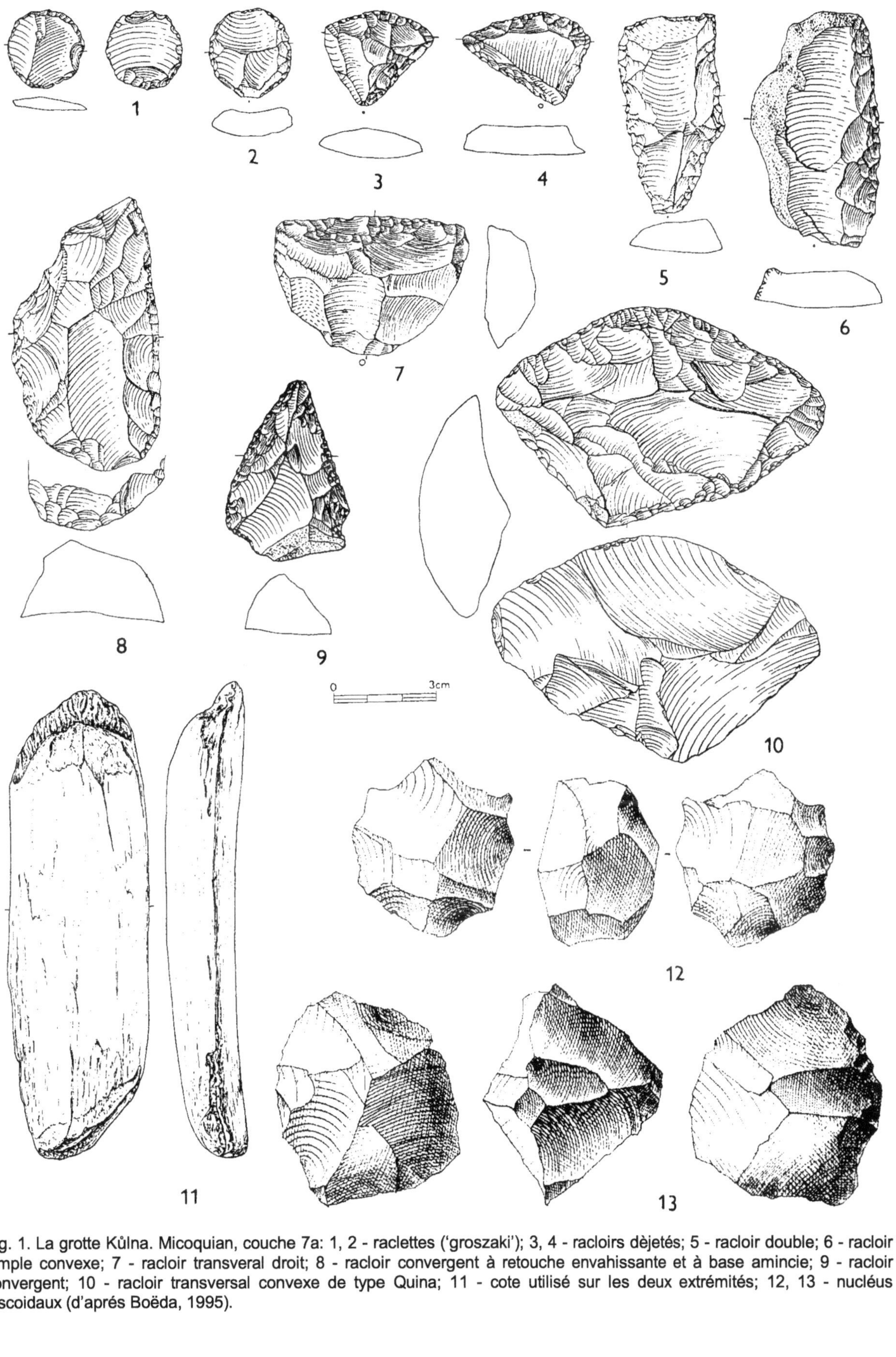

Fig. 1. La grotte Kůlna. Micoquian, couche 7a: 1, 2 - raclettes ('groszaki'); 3, 4 - racloirs dèjetés; 5 - racloir double; 6 - racloir simple convexe; 7 - racloir transveral droit; 8 - racloir convergent à retouche envahissante et à base amincie; 9 - racloir convergent; 10 - racloir transversal convexe de type Quina; 11 - cote utilisé sur les deux extrémités; 12, 13 - nucléus discoidaux (d'aprés Boëda, 1995).

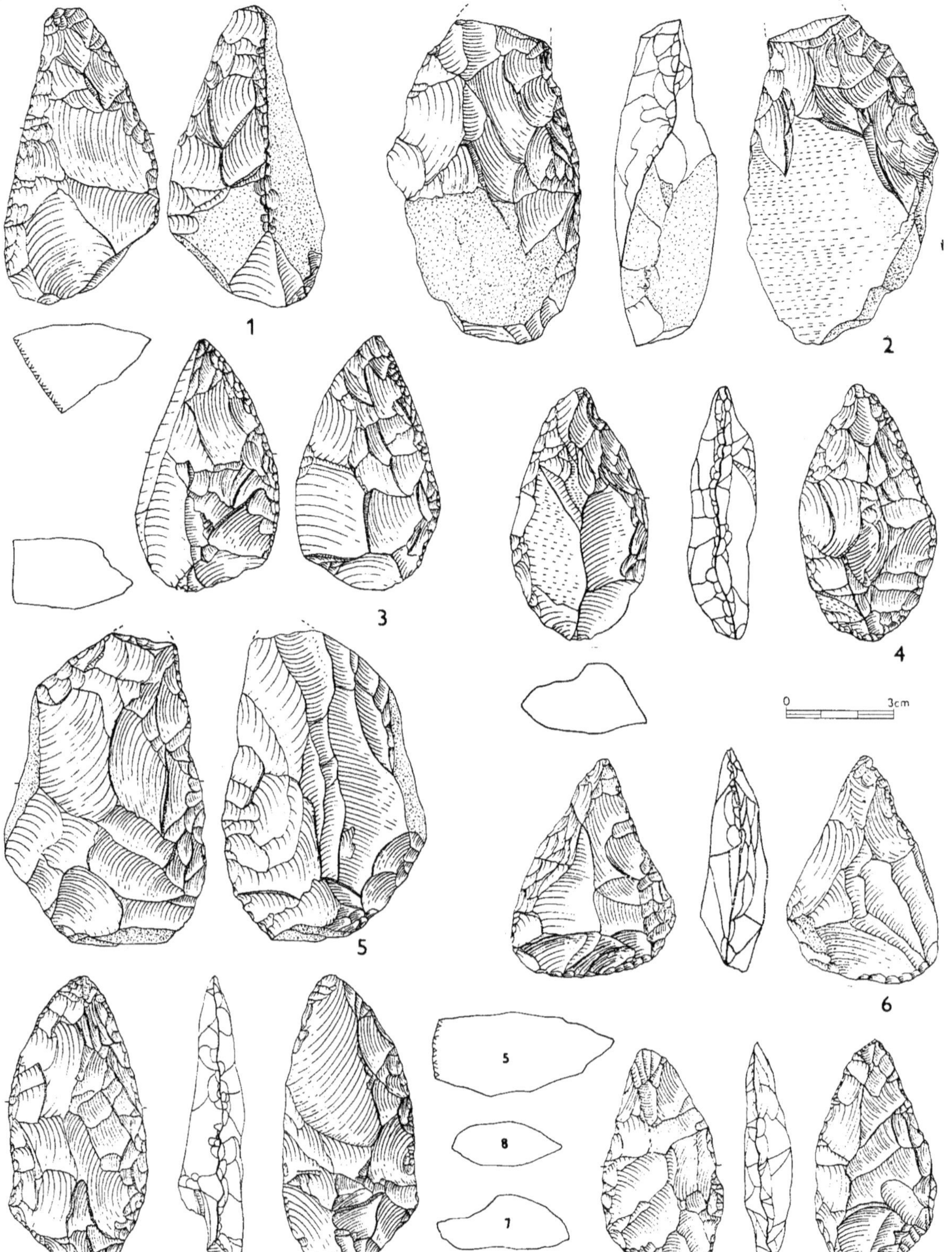

Fig. 2. La grotte Kůlna. Micoquian, couche 7a: 1, 3, 5 - bifaces-couteaux ('Faustkeilmesser'); 2 - biface á pointe cassée; 4 - biface foliacée. Couche 7 alpha: 6, petit biface ('Faustel'). Couche 7c: 7, biface foliacée. Couche 9b: 8, pointe foliacée.

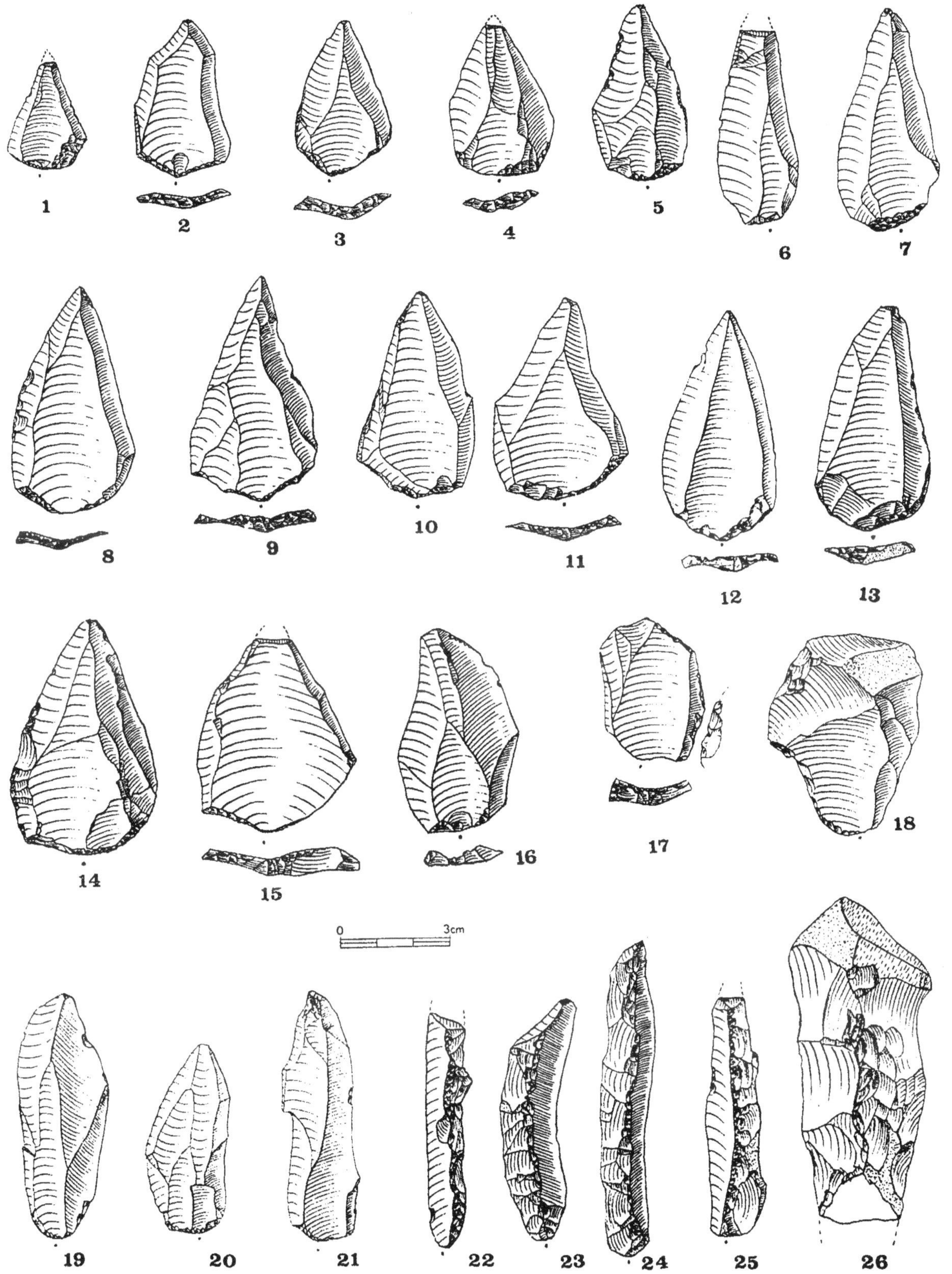

Fig. 3. Bohunice. Bohunicien. 1-16 - pointes Levallois; 17, 18 - éclats Levallois; 19-21 - lames Levallois; 22-26 - lames à crête.

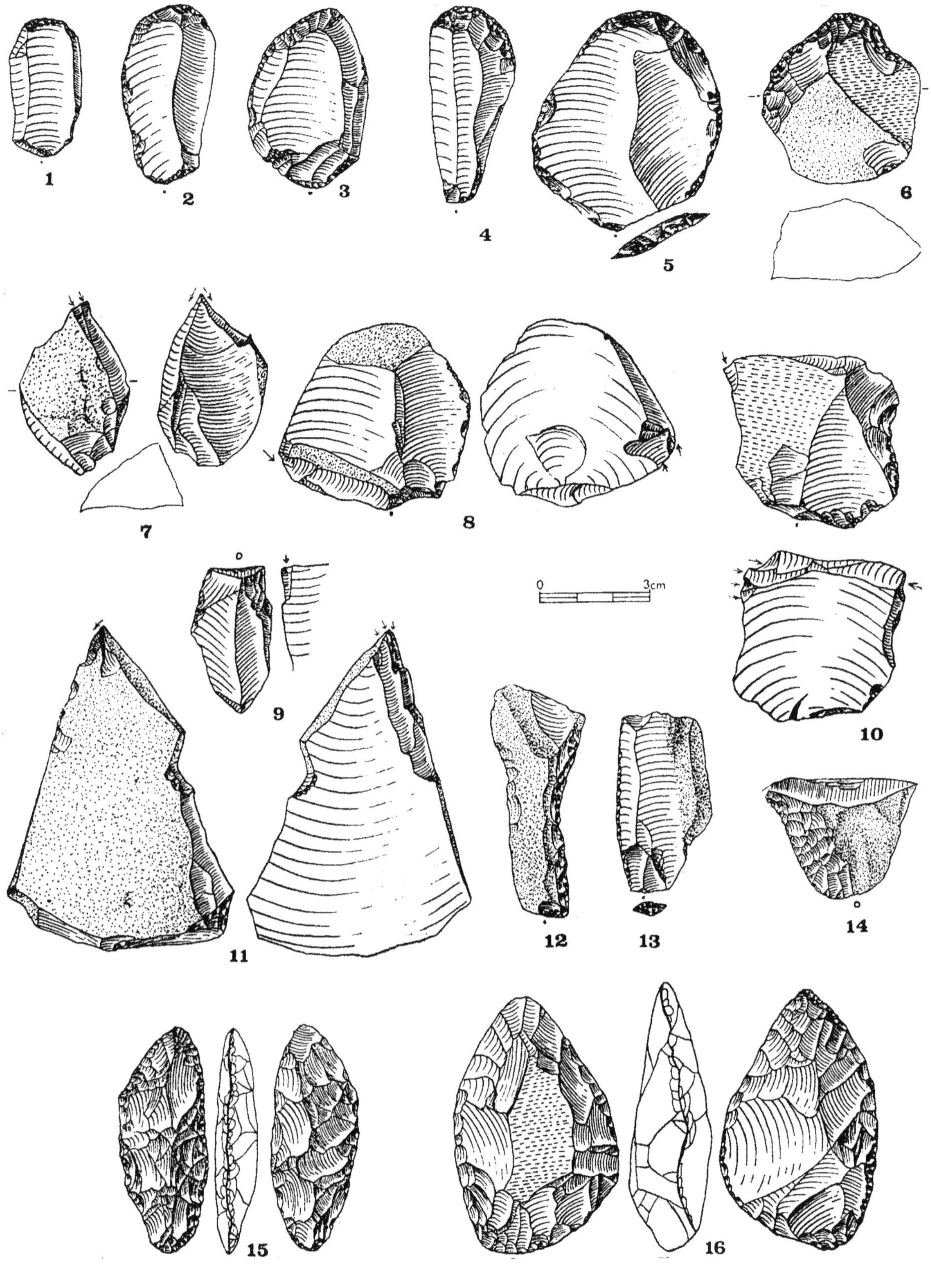

Fig. 4. Bohunice. Bohunicien. 1-5 - grattoir sur lame et sur éclat; 6 - grattoir caréné; 7, 11 - burins dièdre droits; 8 - burin dièdre dèjeté; 9 - burin sur cassure; 10 - burin transversal double; 12, 14 – racloirs; 13 - lame retouchée marginal; 15, 16 - pointes foliacées.

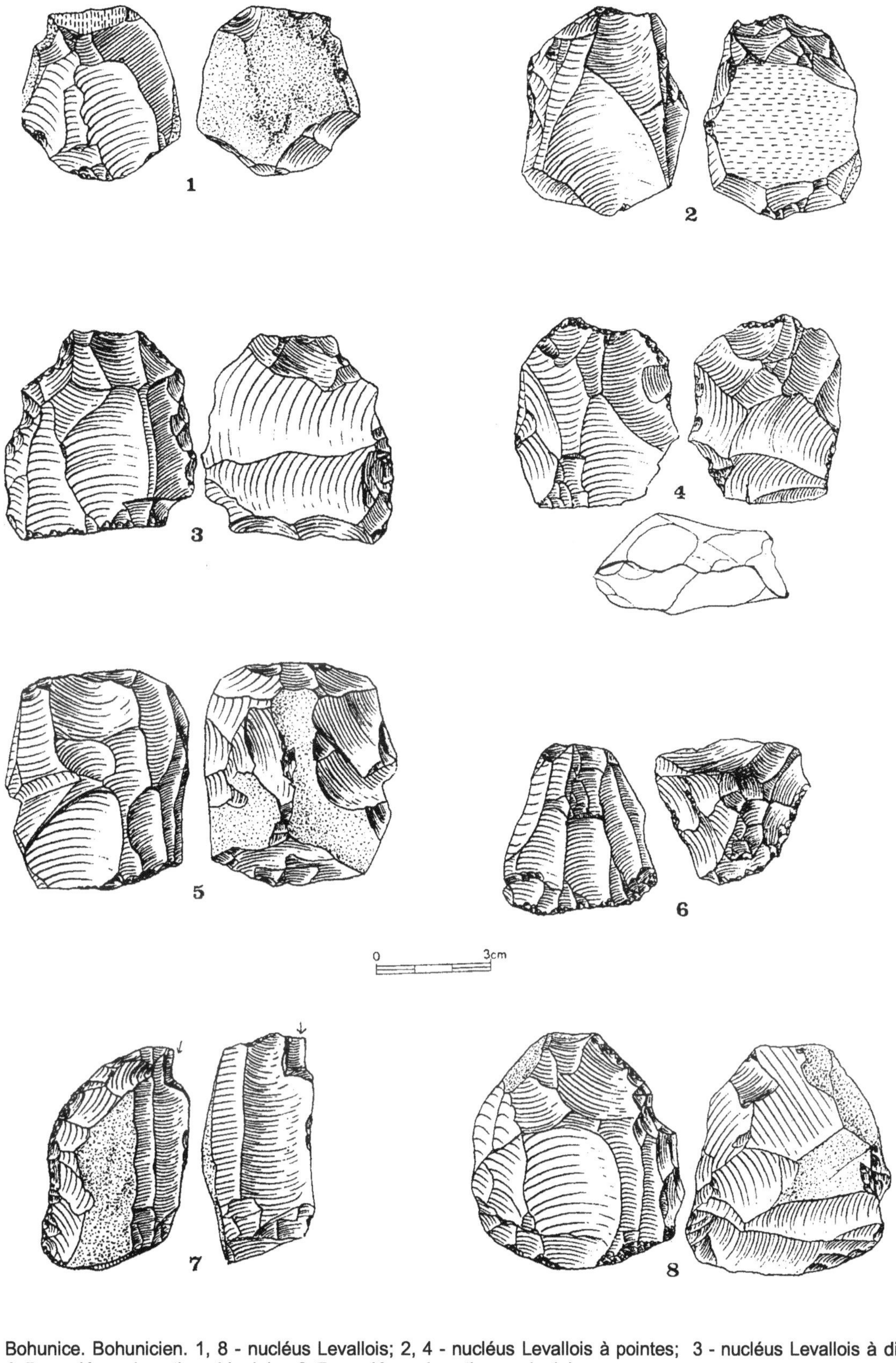

Fig. 5. Bohunice. Bohunicien. 1, 8 - nucléus Levallois; 2, 4 - nucléus Levallois à pointes; 3 - nucléus Levallois à direction changeé; 5 - nucléus prismatique bipolaire; 6, 7 - nucléus prismatique unipolaire.

Adresse de l'auteur: Moravske Muzeum, 65937 Brno, Czech Republic. anthrop@mzm.anet.cz

Bibliographie

Boëda, E., 1995. Caractéristique technique des chaînes operatoires lithiques des niveaux micoquiens de Kůlna. In: *Paléo Suppl.* No.l, p. 57-72. Notes du Colloque du Miskolc 1991.

Bosinski, G., 1967. Die mittelpaläolithischen Funde im westlichen Mitteleuropa. *Fundamenta* A:4, 206 p. Köln-Graz.

Combier, J., 1967. Le Paléolithique de l'Ardèche dans son cadre paléoclimatique. *Publ. Inst. Préh. Univ. Bordeaux, Mém.* 4, 462 p., Bordeaux.

Demidenko, J. E. and Usik, V. L., 1993. On the Levallois technique in the Upper Palaeolithic. In: *Actes du XXe Congrès UISPP Bratislava 1991*, p. 230-242.

Kozłowski, J. K. et Kozłowski, S. K., 1996. *Le Paléolithique en Pologne.* 239 p. Préhistoire d'Europe No.2. Grenoble.

Marks, A., 1983. The Sites of Boker Tachtit and Boker: A Brief Introduction. In: Marks, E. (ed.), *Prehistory and Palaeoenvironmenst in the Central Negev, Israel.* Vol. III. The Avdat/Agev Area, part 3, pp. 15-38. Dallas.

Mook, W. G., 1988. Radiocarbon-Daten aus der Kůlna-Höhle. In: K. Valoch, 1988 (ed.), *Die Erforschung der Kůlna-Höhle 1961-1976,* p.285-286. Brno.

Oliva, M., 1984. Le Bohunicien, un nouveau groupe culturel en Moravie. *L'Anthropologie* 88, 209-220.

Révillion, S., 1994. *Les industries laminaire du Paléolithique moyen en Europe septentrionale: L'example des gisements de Saint-Germain-des-Vaux/Port-Racine (Manche), de Seclin (Nord), et de Riencourt-les-Bapaume (Pas-de-Calais).* Publications du Centre d'Etudes et de Recherches Préhistoriques 5. Villeneuve d'Ascq: CERP.

Rink, W. J., Schwarcz, H. P., Valoch, K., Seitl, L. and Stringer, C. B., 1996. ESR Dating of Micoquian industry and Neanderthal remains at Kůlna Cave, Czech Republic. *Journal of Archaeological Science* 23, 889-901.

Svoboda, J., 1991. Stránská skála. Vysledky výyzkumu v letech 1985-1987. *Pamàtky archeologickè* 82, 5-47.

Valoch, K., 1976. *Die altsteinzeitliche Fundstelle in Brno-Bohunice.* Studie Archeologického ustavu CSAV Brno, 4:1, 120 p., Praha.

Valoch, K., 1982. Die Beingeraete von Predmostí in Maehren (Tschechoslowakei). *Anthropologie* (Brno) 20, 57-69

Valoch, K., 1988. Die Erforschung der Kůlna-Höhle 1961-1976. *Anthropos* 24, N.S. 16, 318 p., Brno: Moravské muzeum.

Valoch, K., Nerudová, Z., Neruda, P., sous press. Stráská skála III - Ateliers des Bohunicien. *Památky archeologické.*

Vencl, S., 1977. Aurignacké osidleni v Hradsku o. Mélnik. *Archeologickd rozhledy* 29, 3-44.

TOWARD MODERN HUMANS: YABRUDIAN AND MICOQUIAN, 400 – 50 kyears ago.
Edited by A. Ronen and M. Weinstein-Evron. BAR S850, 2000.

LE COMPLEXE TECHNO-TYPOLOGIQUE DU BÁBONYIEN-SZÉLÉTIEN DE LA HONGRIE DE NORD-EST ET LE YABROUDIEN DU LEVANT

Árpád Ringer

Introduction

La recherche du Paléolithique de la Hongrie, développée entre 1891 et 1906, a mis au jour, pour la plupart, des industries à pointes foliacées tant au nord-est du pays que dans sa partie occidentale transdanubienne. Toutes ces industries provenaient de gisements en grottes. Les chercheurs hongrois ont interprété ces outillages lithiques comme la manifestation d'une civilisation archéologique à trois ou quatre phases d'évolution, correspondant au Solutréen français. Du point de vue chronologique, ils l'ont placée dans la lignée de développement Moustérien-Aurignacien-Solutréen-Magdalénien, appliquant donc la nomenclature et le système chronologique du Paléolithique française.

En ce qui concerne le rapport chronologique du Solutréen de France et celui de Hongrie, une approche nouvelle a vu le jour dans l'étude d'Henri Breuil, parue en 1923 dans l'Anthropologie, dans laquelle le matériel lithique du Solutréen tardif de l'Abri Puskaporos de la montagne de Bükk a été rapproché du Solutréen inférieur de France (Breuil, 1923).

Le préhistorien hongrois Jenö Hillebrand a fait progresser cette idée, en considérant la majeure partie du Solutréen de Hongrie comme étant plus ancien que le Solutréen de France, et en supposant une migration du territoire de la Hongrie vers celui de la France - sens inverse par rapport à ce qui fut généralement accepté - qui expliquerait l'origine de la civilisation à pointes foliacées (Hillebrand, 1935).

Outre l'opinion de Breuil, la conception de Hillebrand a été considérablement influencée par la pointe en os à base fendue, typiquement aurignacienne, en compagnie d'outils lithiques caractéristiques, qui avait été mise au jour dans la couche du Solutréen ancien de la Szeleta, le plus important gisement en grotte de la Hongrie. Par conséquence, la phase ancienne du Solutréen de Hongrie doit être contemporaine de l'Aurignacien.

Beaucoup plus tard, dans sa monographie intitulée 'Le Paléolithique et le Mésolithique de la Hongrie', parue en 1965, Vértes a pu inventorier déjà trois objets en os venant de la couche 3 de la grotte Szeleta: deux pointes à base fendue et un poinçon à section circulaire (Vértes, 1965, 141).

Cependant, l'interprétation de l'industrie à pièces foliacés du Paléolithique moyen et supérieur de Hongrie a complètement été modifiée jusqu'à cette époque. Le changement fondamentale a eu lieu en 1953, lorsque le chercheur slovaque, František Prošek a proposé d'introduire le terme Szélétien pour éliminer le terme Solutréen, en accord avec l'opinion semblable d'André et de Červinka publiée auparavant (Vértes, 1965, 136).

A partir du début des années 1950, la publication de Freund (1952) intitulée 'Die Blattspitzen des Paläolithikums in Europa' a considérablement influencé la recherche du Paléolithique hongrois. On peut attribuer à cela que nos chercheurs ont remplacé le terme Solutréen par le terme Szélétien et que, comme prédécesseur de celui-ci, les industries moustériennes locales ont été envisagées.

Dans les années 1960, la distinction entre le Szélétien de Transdanubie et celui de Bükk a été de plus en plus acceptée. Vértes (1965) a considéré le premier - sous l'inspiration de Freund - comme la continuation du Moustérien type Tata, tandis que Gábori (1954) a cherché les analogies du même groupe - également sous l'influence de Freund - dans les industries à pièces foliacées du Paléolithique moyen de l'Allemagne du Sud. Beaucoup plus tard, Gábori (1976) et Gábori-Csánk (1985) trouvent l'origine du complexe techno-typologique sur le même territoire de l'Allemagne du Sud.

La position de Gerhard Bosinski sur le Micoquien d'Europe (Bosinski, 1967) a entrainé un changement fondamentale dans le classement des industries à pièces foliacées du Paléolithique moyen de Hongrie et dans l'interprétation de leur origine. Dans leur publications citées ci-dessus, le couple Gábori a classé - parmi les civilisations à pièces foliacées du Paléolithique moyen de Hongrie - le Szélétien de Transdanubie dans le cercle du Micoquien d'Europe centrale. Dès 1973, V. Gábori-Csánk a proposé d'introduire la dénomination 'Jankovichien' pour cette industrie de Transdanubie.

Le Bábonyien

En décrivant le Bábonyien, civilisation de la région du nord-est de la Hongrie, le present auteur l'a également classé dans ce cercle, cependant il a mis l'accent sur le fait que les deux ne peuvent pas être considérés comme identiques à cause de la proportion élevée des pièces foliacées (Ringer, 1983). L'analyse technologique et typologique a permis de déterminer la position chronologique de l'industrie. Nombreuses analogies ont rendu probable une contemporanéité avec la phase plus ancienne du Micoque-Prodnikien, datée du dernier interglaciaire (Ringer, 1983). Plus tard, nous avons réussi à compléter et à confirmer la datation typochronolgique par données paléo-pédostratigraphiques et biostratigraphiques, ainsi que par datation par TL.

Au cours de la décennie suivant la publication du Bábonyien en 1983, l'auteur a élaboré le système chronostratigraphique du Pléistocène supérieur de la Hongrie du nord-est. En 1991, à l'occasion du colloque commémoratif du centenaire de la naissance de la recherche préhistorique hongroise qui a eu lieu à Miskolc, les participants de l'excursion - entre autres, Marcel Otte, Janusz K. Kozłowski, Paul Haesaerts et Avraham Ronen - ont examiné et discuté la stratigraphie de certains sites de loess paléolithiques aux alentours de Miskolc. Ils ont constaté unanimement que l'apparition la plus ancienne du Bábonyien appartenait au sol forestier à couleur brune qui correspondait au complexe paléopédologique, nommé 'Rocourt-Warneton' en Europe occidentale, datant du dernier interglaciaire, et plus précisement la partie inférieure de ce complexe pouvant corréler à l'optimum de l'interglaciaire. Plus tard, cet âge de L'Eemien au sens large a été confirmé par la datation biostratigraphique du matériel archéologique de type de Bábonyien avec *Helicigona banatica* parmi les mollusques, les *Hystrix vinogradovi* entre les petits mammifères et des forêts de charmes et de chênes (*Carpinus-Quercus*) dans la végétation (Ringer, 1994).

Datation

En 1999, nous avons chargé Manfred Frechen d'effectuer la datation par TL du complexe paléopédologique fournissant le Bábonyien sur le site éponyme, Sajóbábony Méhész-tető et sur le site de Sajószentpéter Margit-kapu-dülö. Tous les deux sites se trouvent en Hongrie du nord-est, respectivement à 8 et 15 km vers le nord de la ville de Miskolc, dans la vallée du fleuve Sajó.

A Sajóbábony Méhész-tető, les dates par TL obtenues pour le substrat de loess du sol forestier, à couleur brune, du double complexe paléopédologique sont de 173.0±14.2 ka et de 157.9±23.5 ka BP. Ainsi, le sol forestier de type interglaciaire, qui s'ést développé sur cette couche de loess, peut appartenir au Stade Isotopique 5e. A Sajószentpéter, Manfred Frechen a mesuré des dates par TL de 85.3 ±7.0 ka et de 101.4±9.0 ka BP du loess qui couvre l'unité supérieure du complexe paléopédologique. D'après cela, la formation de l'unité supérieure du complexe paléopédologique peut être placée chronologiquement aux Stades 5c-5a.

Outre les sites de plein air, dont la majorité contourne la partie est de la montagne de Bükk, le Bábonyien se rencontre également dans nos gisements dans des grottes classiques, notamment dans les couches les plus inférieures des grottes Szeleta et Balla qui datent du dernier interglaciaire (Ringer, 1983; 1994).

Dans la première partie de cette étude, nous avons présenté l'histoire de la recherche des industries à pièces foliacées du Paléolithique moyen et supérieur de la Hongrie jusqu'à ce qu'on a séparé le Szélétien de la montagne de Bükk, ainsi qu'on a déterminé la position chronologique du Bábonyien, prédécesseur du Szélétien.

Caractéristiques technologiques du Bábonyien

La matière première fondamentale de cette industrie du Paléolithique moyen est identique à celle du Szélétien de Bükk, pour la plupart des galets de porphyre de quartz hyalin. Parmi les matières premières allochtones se trouvent le silex provenant de la Pologne du Sud et celui tacheté de Swieciechów, également d'origine polonaise.

Le débitage n'est pas Levallois. A partir de matière première de galets, les hommes du Bábonyien ont fait d'abord, à l'aide d'une technique spéciale, des plaquettes planes et allongées qui ont servi de supports pour les pièces bifaciales. Parmi les modes de façonnage bifaciaux se rencontre le 'wechselseitig gleichgerichtete Kantenbearbeitung'. Les types des Keilmesser du Micoquien d'Europe centrale et orientale, les couteaux de Bockstein, de Prodnik et de Volgograd sont présents dans l'industrie. Cependant, les couteaux foliacés, les pointes foliacées et les racloirs foliacés sont vraiment caractéristiques. Ces derniers préfigurent les mêmes types d'outils provenant des niveaux 'ancien' et 'évolué' du Szélétien de Bükk. Parmi les types du Paléolithique moyen caractéristiques sont encore les racloirs transversaux bifaciaux, types spécifiques du Micoquien. A côté de ceux-ci, il y a aussi des racloirs de type charentien. La proportion des types du Paléolithique supérieur - grattoirs, outils sur lames - est aussi surprenante. Souvent, ces pièces sont façonnées sur lames débitées à partir de nucléi de caractère du Paléolithique supérieur.

En ce qui concerne l'origine de l'industrie, l'auteur a supposé auparavant (Ringer, 1983) l'existence d'un Acheuléen local dont le caractère devait être semblable à celui du Jungacheuléen - suivant le modèle élaboré par Gerhard Bosinski. Mais jusqu'à nos jours, nous n'avons pas réussi à démontrer l'existence de cette industrie à bifaces en Hongrie. Quant au Jungacheuléen, il s'est révélé plus récent que la phase la plus ancienne du Micoquien d'Europe (Pastoors, 1998).

En 1997 et 1998, en compagnie de Brian Adams de l'Université de Illinois, l'auteur a entrepris des fouilles à Sajóbábony Méhész-tető et à Mályi Öreg-hegy sur une surface plus grande que les précédentes. A Bábony, une partie du matériel provenant du sol forestier brun Eemien provienne avec certitude d'une fosse contemporaine de la genèse du sol. En ce qui concerne les trouvailles des niveaux plus profonds dans le sol, il n'est pas exclu non plus que, originellement, ils gisaient dans le loess du substrat du sol. A Mályi, nous avons pu incontestablement démontrer le même phénomène. D'après cette observation, l'apparition du Bábonyien en Hongrie du nord-est a eu lieu très probablement dans une période pré-éemienne.

D'aprés l'avis du présent auteur (Ringer, 1995), la tradition des industries à pièces foliacées du Paléolithique moyen d'Europe centrale se distingue du *sensu lato* Micoquien depuis son origine. Selon lui, pendant l'interglaciaire Eem en Europe, la tradition micoquienne, qui provenait d'un centre dont la location n'est pas encore claire ou éventuellement de plusieurs sources, a dû se mêler avec une tradition locale à pièces foliacées tout à fait différente de celle-ci. Cela constituerait le noyau évolutif des industries micoquiennes à pièces foliacées.

Le Bábonyien montre des analogies surprenantes avec certains types d'outil du Yabroudien de la grotte de Taboun. Les figures 1 à 3 n'exigent aucun commentaire. Elles formulent la question suivante: Est-il exclu que des influences culturelles venant de l'Asie du Sud-ouest ont pu atteindre le territoire hongrois de l'Europe centrale au cours des périodes de réchauffements précédant l'Holocène, influences dont l'intensité était semblable à celle des migrations ou diffusions qui ont apportées le Néolithique en Europe pendant l'interglaciaire holocène? Ce sont peut-être les analogies entre le Bábonyien et le Yabroudien, présentées sur les figures, qui répondent à cette question. Il est important d'y ajouter et de mettre en relief que le Yabroudien n'est pas du débitage Levallois, non plus, de même que le Bábonyien, il n'est pas facetté et il contient également des éléments charentiens caractéristiques à côté des bifaces.

Le Yabroudien disparaît du littorale levantin pendant le dernier interglaciaire ou, encore plus probablement, pendant l'avant-dernier interglaciaire (Bar-Yosef, 1993). Ce n'est peut être pas par hazard que l'origine du Bábonyien et de ses composants yabroudiens remonte justement à une époque pré-éemienne.

Se projet est financé par le concours hongrois FKFP/0044/1999 : 'La révision des civilisations Szélétien et Aurignacien de Bükk en Hongrie dans leurs cadre palécologie – chronologique'.

Adresse de l'auteur: Département de Préhistoire et d'Histoire ancienne, Université de Miskolc, H-3515 bolringa@gold.uni.miskolc.hu

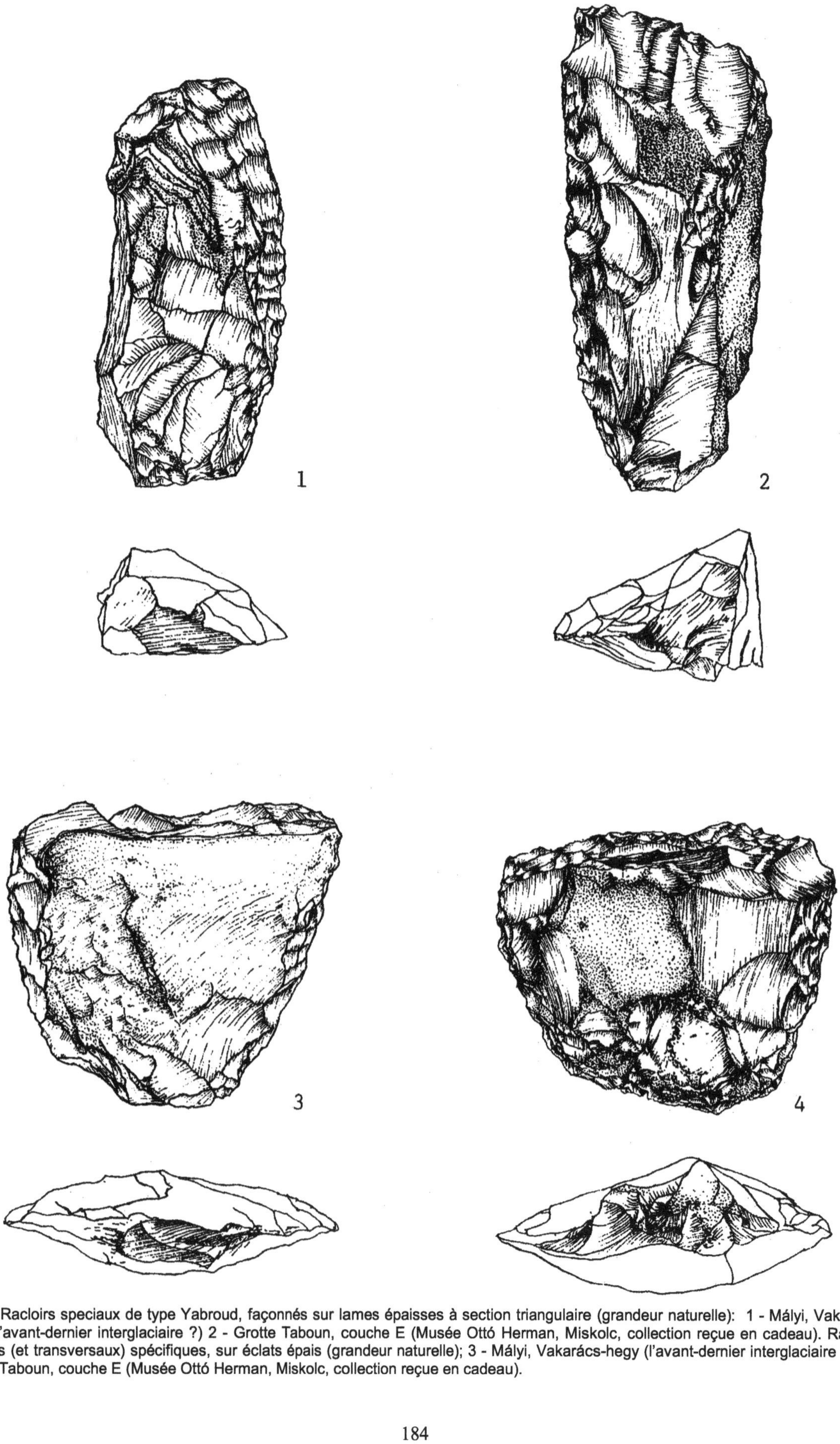

Fig. 1. Racloirs speciaux de type Yabroud, façonnés sur lames épaisses à section triangulaire (grandeur naturelle): 1 - Mályi, Vakarács-hegy (l'avant-dernier interglaciaire ?) 2 - Grotte Taboun, couche E (Musée Ottó Herman, Miskolc, collection reçue en cadeau). Racloirs doubles (et transversaux) spécifiques, sur éclats épais (grandeur naturelle); 3 - Mályi, Vakarács-hegy (l'avant-dernier interglaciaire ?); 4 - Grotte Taboun, couche E (Musée Ottó Herman, Miskolc, collection reçue en cadeau).

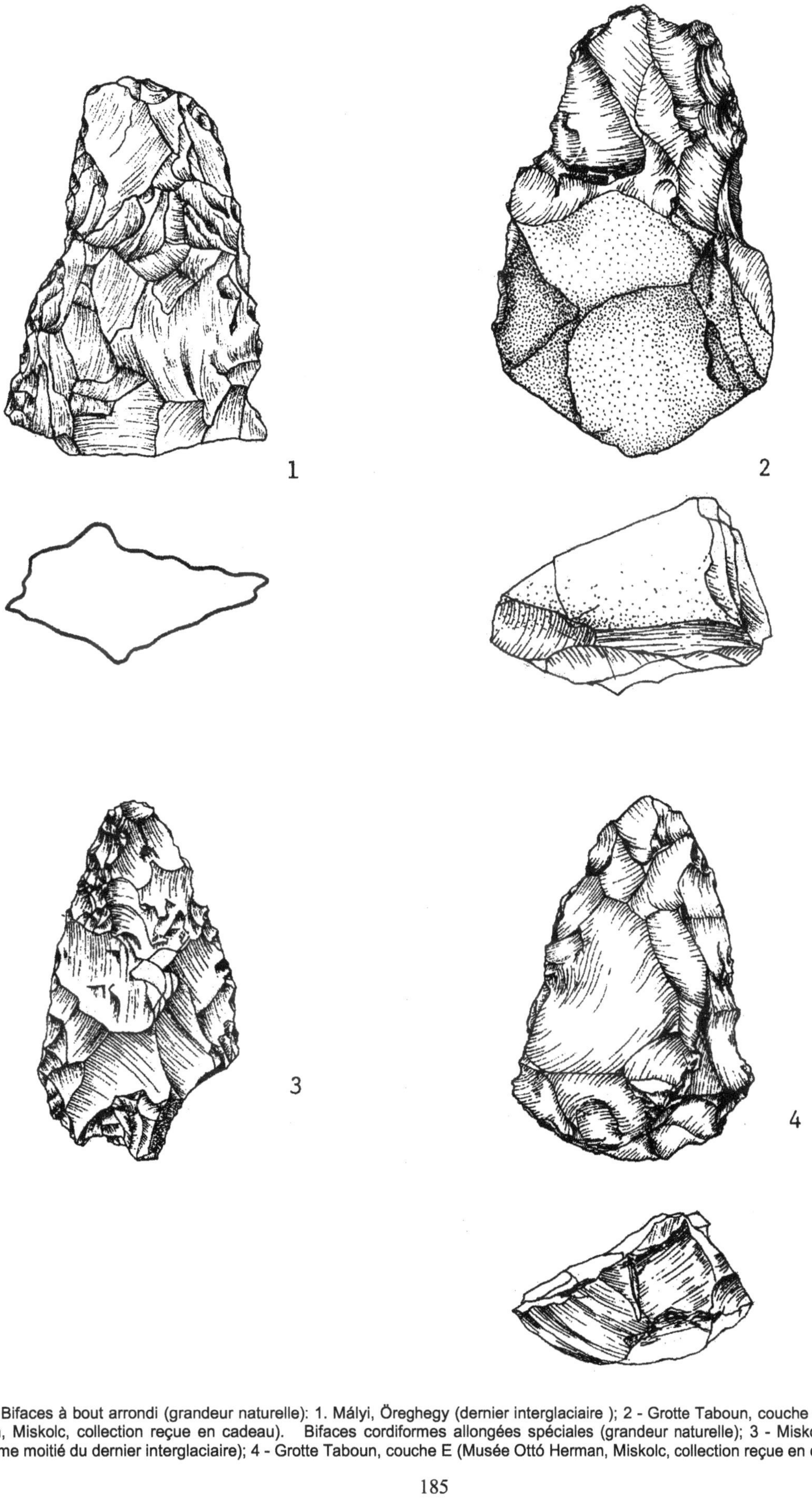

Fig. 2. Bifaces à bout arrondi (grandeur naturelle): 1. Mályi, Öreghegy (dernier interglaciaire); 2 - Grotte Taboun, couche E (Musée Ottó Herman, Miskolc, collection reçue en cadeau). Bifaces cordiformes allongées spéciales (grandeur naturelle); 3 - Miskolc, Kánás-tető (deuxième moitié du dernier interglaciaire); 4 - Grotte Taboun, couche E (Musée Ottó Herman, Miskolc, collection reçue en cadeau).

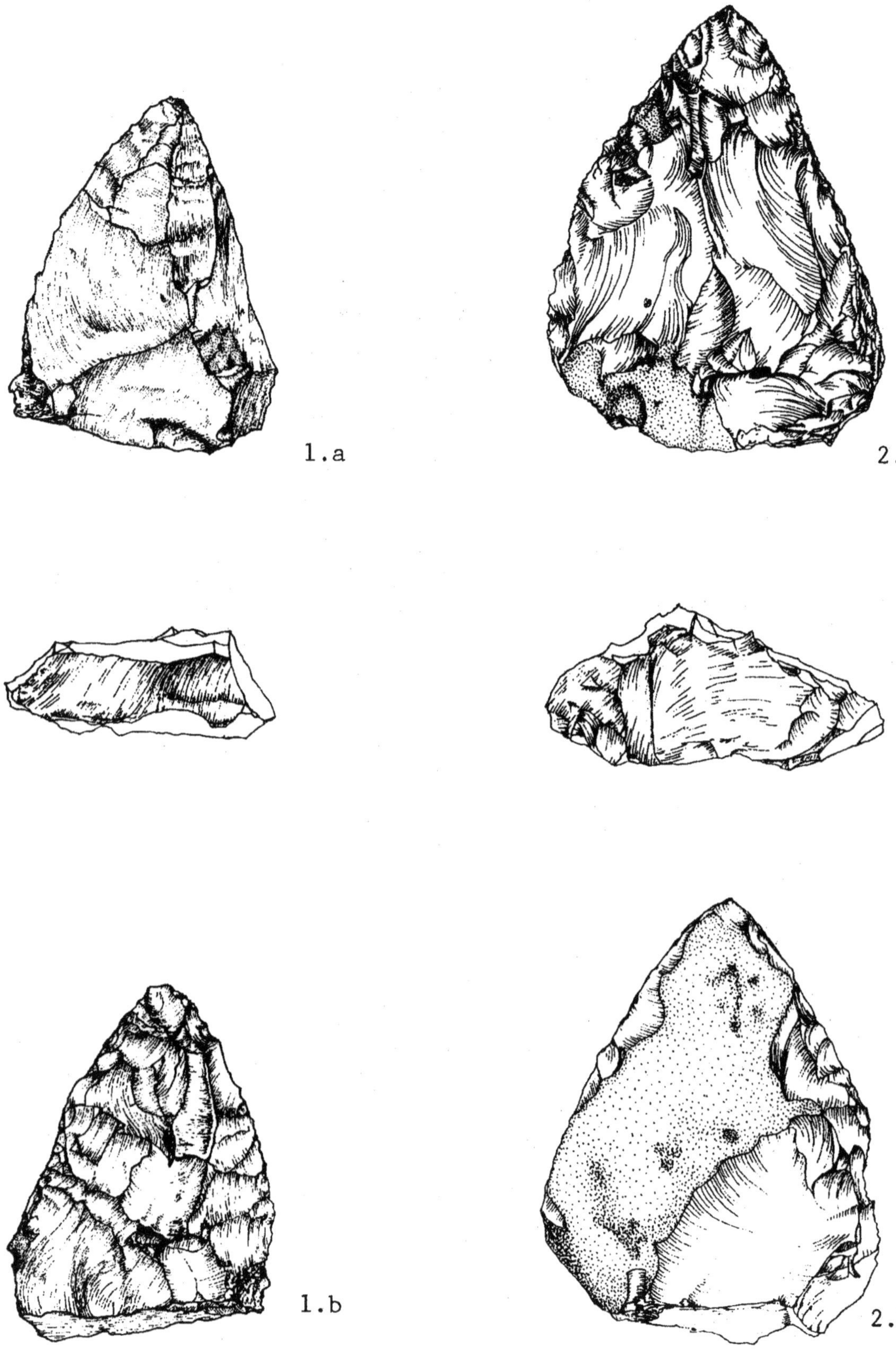

Fig. 3. Halbkeil triangulaires spécifiques, la base garde l'état de débitage, la face ventrale est façonné d'une manière spéciale (grandeur naturelle): 1.a-1.b - Sajóbábony, Méhész-tető (dernier interglaciaire); 2.a-2.b - Grotte Taboun, couche E (Musée Ottó Herman, Miskolc, collection reçue en cadeau).

Bibliographie

Bar-Yosef, O. 1993. The Role of Western Asia in Modern Human Origins. In: Aitken, M. J., Stringer, C. B. and Mellars, P. A. (eds), *The Origin of Modern Humans and the Impact of Chronometric Dating*, pp. 132-147. Pricenton, New Jersey.

Bosinski, G., 1967. *Die mittelpaläolithischen Funde im westlichen Mitteleuropa*. Fundamenta A4. Köln: Böhlau Verlag.

Breuil, M. 1923. Notes de voyage paléolithique en Europe centrales. I. Les industries paléolithique en Hongrie. *L'Anthropologie*, XXXIII.

Freund, G. 1952. Die *Blattspitzen des Paläolithikums in Europa*. Quartär Bibliothek, Band I., Bonn.

Gábori–Csánk V. 1985. *Az öskökori Jankovich kultúra Nyugat-Magyarországon. [La civilisation jankovichienne du Paléolithique de Hongrie de l'Ouest.]*. Thèse de doctorat d'Etat.

Gábori, M. 1976. *Les civilisations du Paléolithique moyen entre les Alpes et l'Oural*. Budapest.

Hillebrand, J. 1935. Die ältere Steinzeit in Ungarn. *Arch. Hung.* 17.

Pastoors, A. 1998. Nouveau regard sur un site paléolithique moyen de plein air. *L'Anthropologie* 102, 523-532.

Ringer, À. 1983. *Bábonyien – eine mittelpaläolithische Blattwerkzeugindustrie in Nordost-Ungarn*. – Diss. Arch. Ser. II. No. 11.

Ringer, À. 1994. *Északkelet-Magyarországi geomorfológiai szintek és régészeti adataik. Felsöpleisztocén folyóteraszok, lőszők és barlangi ?ledékek kronosztratigráfiai rendszere. [Nappes géomorphologiques de la Hongrie du nord-est et leurs données archéologiques. Système chronostratigraphique des terrasses fluviatiles, des loess et des remplissages de grotte du Pleistocène supérieur.]* – Thèse de doctorat PhD.

Ringer, Á., 1995. Les industries à pièces foliacées en Europe centrale: proposition de synthèse. In: *Les industries à pointes foliacées d'Europe centrale. Actes du colloque de Miskolc.* Paléo, Suppl. No. 1, 15-18.

Vértes L. 1965. *Az öskökor és az átmeneti kökor emlékei Magyarországon. A magyar régészet kézikőnyve I. [Les trouvailles du Paléolithique et du Mésolithique en Hongrie. Manuel de l'Archéologie hongroise I.]* Budapest.

THE MIDDLE PALAEOLITHIC KILL-BUTCHERY SITE OF ZWOLEŃ, POLAND

Romuald Schild[1], Andrzej J. Tomaszewski[2], Zofia Sulgostowska[1], Achilles Gautier[3], Andrzej Bluszcz[4], Bodil Bratlund[5], Ariane M. Burke[6], Helle Juel Jensen[7], Halina Królik[1], Adam Nadachowski[8], Ewa Stworzewicz[8], Jan Butrym[9], Henryk Maruszczak[9], and Józef E. Mojski[10]

Introduction

The site of Zwoleń was identified in the fall of 1983 by a local amateur archaeologist, Jan Grudzień, after the operator of a backhoe showed him several large bones of a prehistoric 'monster' excavated from a gravel quarry. The bones turned out to be a mammoth molar, a fragment of a mammoth jaw and scraps of skull. A close examination of the remaining original load of the backhoe scoop by Zofia Sulgostowska yielded a beautiful small biface made of imported chocolate flint. It was obvious that the gravel quarry contained a prehistoric site.

Since then, funds awarded by the Archaeological Inspectorate of the Voivodedom of Radom have financed eight field seasons (1983-1990) of salvage excavations. The fieldwork was jointly carried out by the Institute of Archaeology and Ethnology, the Polish Academy of Sciences and the State Archaeological Museum in Warsaw.

The site of Zwoleń is one of the few Middle Palaeolithic sites of the North European Plain that yielded archaeological material together with bones. The site lies close to the boundary of the maximal extension of the Vistulian ice sheet and is well within its periglacial zone (Fig. 1). The Vistulian ice sheet, as well as the massive land alterations caused by periglacial slope processes, destroyed almost all the Middle Palaeolithic sites located on the Plain.

Synopsis of Geomorphology and Stratigraphy

General Geomorphology and Stratigraphy

The complexity of lithostratigraphic data at the Rifle-Range Site of Zwoleń, exposed in nine cuts, is exceptional. For the purpose of this report, however, we shall concentrate only on the essentials. The site is situated on the southern, right bank of the Zwolenka river, a small western tributary of the Vistula (Fig. 2). The Zwolenka drains the Zwoleń Plateau. The latter is a relatively low plateau in the central section of the Radom Plain which itself forms the southern part of the Mazovia Lowland (Różycki, 1972:276). The topbeds of the Zwoleń Plateau are made up of boulder clays usually associated with the Radomka ice sheet advance, i.e., the maximal stadial (Saale 1) of the Oder Glaciation (Mojski, 1985, 109-115; Lindner, 1992, 525).

The plateau is heavily dissected by the Zwolenka and its tributaries. This dissection often reaches down to the underlying lacustrine sands of the lakes, which were formed during the encroachment of the Radomka ice sheet. The valleys are characterized by flat, relatively wide floodplains and steep slopes lacking clear-cut terrace systems (Różycki, 1972, 277).

The Zwolenka cuts into the Radomka glacial till and underlying lacustrine sands. Along the banks and the slopes of the valley the tills are eroded, and at the Rifle-Range Site the tills are replaced by several suites of intersecting fossil alluvial channels of the Zwolenka. These are covered by slope, sandy sediments of the Vistulian maximum, which grade laterally into the fluvial sands of the overloaded, aggrading Zwolenka (Fig. 3).

Eight TL samples (A. Bluszcz and J. Butrym) place the anaglacial lacustrine sands and overlying glacial tills between ca. 485 and 92 ka BP, with a majority (five) of TL estimates between 436 and 485 ka BP. These measurements are clearly too ancient for the maximum of the Oder Glaciation, which is placed between ca. 273 and 245 ka BP (Lindner 1992, 532). On the other hand, the reliability of the TL method before 250 ka BP is not very good.

The earliest alluvial suite preserved at the site is made up of at least three intersecting, aggrading channels filled with boulders, gravel and sandy silts. Of the five TL estimates obtained from the sands of these channels, three place the TL age between 153 and 207 ka BP (A. Bluszcz), in the Warta Glaciation. The lower Warta channel is overlain by two aggrading, gravelly channels dating slightly over 160 ka BP. The TL measurements indicate an upper middle Warta Glaciation (Saale 3, OIS 6).

Several thermokarst channels with vertical walls and generally flat bottoms either dissect or develop in the lower sections of the alluvial Wartan deposits of the Zwolenka. The upper sections of the channels are filled by redeposited earlier gravel packages of Wartan silts and alluvial sands. Their age is not certain. On one hand, the channels are truncated by the earliest Wartan Zwolenka channels. However, the TL sample for the alluvial sands of the thermokarst gave an age estimate of 95 ka BP (A. Bluszcz). It is most likely that the thermokarst processes associated with the Wartan deposits of the Zwolenka formed in the maximum of OIS 6.

No sediments of the last interglacial are preserved at the site. The Zwolenka was certainly downcutting then. The Vistulian deposition begins with silty channels dated by two TL samples at 98 and 117 ka BP (A. Bluszcz), probably indicating OIS 5d. Again the sediments of the Amersfoort/Brörup complex (OIS 5c) are not present, undoubtedly a sign of a downcutting phase of the river.

The thick gravel bed containing pebbles, cobble and detached blocks of laminated floodplain silts are evidence of an aggradational cold phase and a heavily overloaded river with seasonal runoff. The gravel contains bones and lithics of the Lower Cultural Level, often embedded in the blocks of the floodplain silts, indicating that these silts were the primary sedimentary environment of the archaeological deposits. The sands from this aggradational phase gave two TL age estimates of ca. 80 and 62 ka BP (A. Bluszcz).

The early Vistulian gravely sediments containing the Lower Cultural Level were screened through 0.5 mm mesh. Unfortunately, the remains of rodents were very scarce. Only five molar fragments were identified to the generic level (A. Nadachowski). These belonged to some unidentified species of voles (*Microtus* sp.) and to the collared lemming (*Discrostonyx* sp.). The collared lemming was widely distributed during the Vistulian and is considered an indicator of steppe-tundra, cold-climate communities. Voles are inhabitants of different environments, i.e., they belong to steppe or meadow-plant communities.

The gravelly deposits of the early Vistulian age are dissected by a series of channels filled with fine-grained deposits. Two of these channels contain regularly laminated silts and sands deposited by floodwater in the Zwolenka oxbows. Both channels include archaeological deposits designated as Middle and Upper Cultural Levels. Seven TL estimates range from 85 to 40 ka BP, with a majority clustering around 75 ka BP (A. Bluszcz).

These channels are, in turn, overlain by alluvial deposits of a gravelly, aggradational phase showing several intersecting channels, TL-dated at around 70 ka BP (six estimates; A. Bluszcz). The Zwolenka alluviae of this stage are again deeply dissected (down to 2 m) in certain areas by several thermokarstic channels. These are probably coeval with the Swiecie stadial (Vistulian 2) and the first Vistulian ice wedges in the loess region, which are TL-dated to slightly before 60 ka BP and therefore to the cold maximum of OIS 4 (see Van Andel and Tzedakis (1996) for comparative data).

Above this series are several Zwolenka channels of the Middle Vistulian age of ca. 40 TL ka (A. Bluszcz). They are, in turn, overlain by alluvial and slope deposits dated to the maximum of the Vistulian, at around 29 to 21 TL ka (three estimates; A. Bluszcz). The TL age curve of the Zwolenka sediments at the Rifle-Range Site (Fig. 4) shows a very flat section in the area where the archaeological deposits occur, indicating relatively intense deposition within a short time. It seems most likely that the Lower Cultural Level dates at the very late Odderade Interstadial (late OIS 5a) around 80-75 ka BP. The Middle and Upper Cultural Levels probably represent early OIS 4, slightly before the first cold spell or the following warm episode of Ognon, just around 70 ka BP (compare Van Andel and Tzedakis, 1996).

Geomorphic Environment and Taphonomy of the Cultural Deposits

Clearly the bones and associated artefacts of the Lower Cultural Level had originally been left on a floodplain where they were rapidly covered by flood silts. The lithics and bones of the Middle and Upper Level, on the other hand, lie in the laminated silts and sands of the oxbows. Although many of the bones show heavy chemical damage, the flint artefacts are of mint condition, indicating no transport whatsoever. It is as if they were left on the ice and then submerged with the thaw.

The stratigraphy and geomorphology of the site indicate that at the time of its early use by humans the Zwolenka flowed precisely at its present location in a deep, ravine-like valley. The site is situated in the clear narrowing of today's valley, in proximity to a modern water mill. In the early Vistulian the gorge must have been much narrower, probably measuring a few tens of metres across, since most of the sub-recent widening of the valley is due to Holocene meandering.

Excavations

The goal of the rescue excavations at Zwoleń was to save what had been exposed and endangered in the gravel quarry pit (Fig. 5) where the main cut (Cut I) was placed (Fig. 6). Furthermore, to facilitate further protection it was necessary to find the boundaries of the site by trenching. Eight trenches (II-IX), located to the south and east of Cut I, yielded enough data to place the boundaries of the archaeological locus. To the west of Cut I, the archaeological beds have not been preserved. It has been established that bones and lithics occur over an area of at least 7,500 sq m. The Rifle-Range Site of Zwoleń is, therefore, one of the largest Middle Palaeolithic sites in the world.

In total, the excavations and trenching covered 523 sq m of which Cut I measured 347 sq m. The density of artefacts and bones vary within the site. The highest number of bone inventory entries is in Cut V (7 per sq m) while in Cut I it drops to 2.3 per sq m. Similarly, there is one flint artefact per sq m in Cut V while in Cut I the figure is 0.8 artefacts per sq m.

Bones outnumber lithics in all the cuts except for Cut VIII where there were no bones, but a single flint artefact. In total there are almost one thousand inventory entries for bones while only 327 flint artefacts have been found.

The Lithics

Raw Material (H. Królik)

The lithics are made of four types of flint: chocolate, Cretaceous erratic, Turonian, and Cracowian Jurassic (Fig. 7). The chocolate flint makes up over 80% of the assemblage. It comes from the area of Polany Kolonie II (Fig. 2), some 40 km to the SSW. There it occurs in the Oxfordian or Kimmeridgian limestones as flattish, rather small, rounded, homogeneous nodules ideal for biface production. The cortex of the Zwoleń chocolate artefacts is fresh. This would indicate that the flint was mined, at least in shallow, open pits. The Polany Kolonie II flint was also exploited for bifaces at the turn of the Neolithic and Early Bronze (Schild et al., 1977).

The second in importance is a locally available Cretaceous erratic flint that makes up around 10% of all finds. This flint occurs in Pleistocene glacial deposits and has often been redeposited in more recent alluvial beds. The third is a Cretaceous Turonian flint that occurs in siliceous limestones some 80 km to the south (1.5% of the total). A single piece, a sidescraper, is of special interest. It is made of a special variety of the Jurasic Cracowian flint which occurs in the area of Brzoskwinia, almost 250 km to the SW (identified by Professor B. Ginter).

Technology (J. Tomaszewski)

Cores are rare (Table 1). All are highly exploited and all are for flakes, except for a single blade core. The cores may be classified as discoidal and as displaying changed orientation. A few resemble Levallois flake cores due to a delicate preparation of alternating striking platforms and exploitation of both faces. Some of the cores show a delicate retouch post-dating flake removal, suggesting their secondary use as tools. The cores were exploited with hard hammer. Nine flakes suggest the use of a soft hammer.
The biface technology is quite well represented at the site. Relatively numerous biface trimming flakes of all sizes indicate production and reshaping of the tools on the spot. Over 80% of the biface trimming flakes show characteristics of a soft hammer.

Typology (J. Tomaszewski)

Tools are not very numerous (82 in total). However, compared to the other lithic categories, they appear over-represented. This is most likely due to problems recovering small artefacts in gravelly deposits. Both flake and biface tools (Figs. 8-10) are well represented (Table 2). Among bifaces, symmetric and asymmetric categories occur (Figs. 9 and 10) including backed ones (Fig. 10, 1).

Taxonomic Association

Both the technology and the typology indicate a clear-cut association with the Technocomplex of Backed Bifaces, often also identified as post-Micoquian or Micoquian. Only one other such site (Königsaue, Mania and Toepfer, 1973) occurs in the North European Plain.

Traces of Wear (H. Juel Jensen)

Just six implements show traces of wear, three of which are somewhat better identified. The first is a sidescraper exhibiting traces which indicate cutting of fresh hide. The second implement displays a polish that may be attributed to scraping bone. The third reveals a polish which may have been caused by contact with bone, and possibly represents butchering activity. A series of fairly well-

Lithic Structure	Lower L.	Middle L.	Derived above Middle L.	Upper L.	Derived post occupation L.	Late Alluvium	Derived post Lower L.	Back-dirt	Cuts III-VIII	Total
Cores	2				2			2		**6**
Primary flakes				2	6	2		3	1	**14**
Ordinary flakes	4	4	3	1	17	11	1	2	5	**48**
Pseudo-Levallois points	1	1?			2	1		1	1	**7**
Levallois flakes				2?	1?	1?				**4**
Blades				1		2				**3**
Biface trimming flakes	10	10	7	11	39	22		11	16	**126**
Ordinary chips	4	1	3	2	9	4		2	4	**29**
Retouch chips		1	1	1	2	2				**7**
Tools and fragments	14	6	5	3	21	14	1	7	11	**82**
Hammerstone		1								**1**
Total	**35**	**24**	**19**	**23**	**99**	**59**	**2**	**28**	**38**	**327**

Table 1. Zwoleń, Inventory of Major Lithic Elements.

Tool Types	Lower L.	Middle L.	Derived above Middle L.	Upper L.	Derived post occupation L.	Late Alluvium	Derived post Lower L.	Back-dirt	Cuts III-VIII	Total
6. Mousterian point					1					**1**
9. Side-scraper, simple straight	1					1		1		**3**
10. Side-scraper, simple convex	2		1		1	1			1	**6**
11. Side-scraper, simple concave						1				**1**
12. Side-scraper, double, straight	1				1					**2**
15. Side-scraper, double, biconvex		1			1					**2**
19. Side-scraper, convergent, biconvex, symmetric	1		1		1					**3**
21. Side-scraper, convergent, asymmetric	1	2			2	2				**7**
23. Side-scraper, transverse, convex		1				1				**2**
25. Side-scraper inversely retouched					2	1			2	**5**
28. Side-scraper, bifacial			1	1		3		1	1	**7**
29. Retouched side-scraper, double, alternatingly	1				1					**2**
32. Burin, typical								1		**1**
33. Burin, atypical						1				**1**
39. Raclette	1								1	**2**
42. Notched pieces								2	1	**3**
43. Denticulated pieces						1				**1**
48-50. Retouched flake	1	1	1	2	4	2	1		3	**15**
62a. Bifacial edge piece	1				1					**2**
62b. Undetermined and fragments									2	**2**
Bifaces symmetric	2				4			1		**7**
Bifaces asymmetric	1	1						1		**3**
Unfinished bifaces					1					**1**
Fragments of bifacial tool	1		1		1					**3**
Hammerstone		1								**1**
Total	**14**	**7**	**5**	**3**	**21**	**14**	**1**	**7**	**11**	**83**

Table 2. Zwoleń, Tool Types (Bordes' List).

preserved, unretouched flakes, mostly biface preparation products, was also tested but revealed no traces of human use.

Fauna (A. Gautier)

Of almost 1000 inventory entries, 408 bones have been identified (Table 3). There is a high predominance of horse (MNI 38), over 60% of all MNI. Far behind the horse is the wisent (MNI 6), followed by the mammoth and rhinoceros (each MNI 4) and the reindeer (MNI 3). The other species of fauna probably form an accidental association.

The age and sex distribution of the horses suggest the killing of nursery herds. The remains of younger equids fit into discrete age groups, suggesting kills separated by periods of approximately a year, i.e., seasonal. Indeed, according to the cementum annuli analyses (A. M. Burke), these kills would have occurred in winter, although one specimen could have been a late winter or spring kill. A single tooth indicating an indeterminate warm period dates

Provenance \ Animal Group	Pike	Perciform	Bird	Marmot	Beaver	Watervole	Arctic fox	Mammoth	Woolly rhinoceros	Horse	Medium ruminant, roedeer?	Reindeer	Large cervid, elk?	Steppe wisent	Totals
Cut I, Thermokarstic Channels	-	-	-	-	1	-	-	1	1	5	-	-	-	-	**8**
Later Warta Channel 2	-	-	-	1	-	-	-	-	-	-	-	-	-	-	**1**
Channel 1 of First Early Vistulian Meandering Phase	-	-	-	-	-	-	-	-	-	1?	-	-	-	-	**1?**
Lower Cultural Level	-	-	-	-	2	-	-	23	9	77	-	5	2	16	**134**
Structures preceding second EarlyVistulian Meandering Phase	-	1	-	-	2	2	-	4	1	5	-	-	-	3	**18**
Middle Cultural Level	-	-	-	-	-	-	-	4	8	43	-	-	-	3	**58**
Redeposited between Middle and Upper Cultural Level	-	-	-	-	-	-	-	3	2	18	1	-	-	4	**28**
Upper Cultural Level	-	-	-	-	-	-	-	-	-	11	-	1	-	1	**13**
Redeposited above Upper Cultural Level	-	-	1	-	-	-	-	8	11	48	-	1	-	13	**82**
Late Alluvium	-	-	-	-	-	-	-	1	1	9	-	1	-	1	**13**
Cut III, lower level	-	-	-	-	-	-	-	1	-	-	-	-	-	-	**1**
middle level	-	-	-	-	-	-	-	-	-	1	-	-	-	-	**1**
Late Alluvium	-	-	-	-	-	-	-	-	-	-	-	-	-	1	**1**
Cut IV, lower level	-	-	-	-	-	-	-	-	2	3	-	-	-	-	**5**
middle level	-	-	-	-	-	-	-	1	-	-	-	-	-	-	**1**
Cut V, lower level	-	-	-	-	-	-	-	6	4	13	-	1	1	2	**27**
Provenance uncertain	1	-	-	-	-	-	2	3	3	5	-	-	-	2	**16**
Totals	**1**	**1**	**1**	**1**	**5**	**2**	**2**	**55**	**42**	**239**	**1**	**9**	**3**	**46**	**408**

Table 3. Zwoleń, Faunal Remains.

from the late alluvium (maximum of the Vistulian) and may not be associated with the Middle Palaeolithic site.

The comparison of the interskeletal distribution of some of the Solutré equids (Table 4) with those from Zwoleń indicates that at Zwoleń we are dealing with several degraded assemblages. Elsewhere, Gautier (1984) estimates that in most cases, the ratio of preserved identified bones versus the total bones originally left at a site is less than 1%. On the basis of the foregoing, we can assume that our odd forty horses represent but a small fraction of the thanatocoenoses that have accumulated during the repeated utilisation of the Zwoleń site.

Due to the degradation of the material, it is virtually impossible to study butchering practices and differential removal of meat-bearing body parts. The various elements of the equid skeleton present at the site suggest killing and butchering on the spot, and a limited transport. Complete skulls with articulating mandibles, and with no signs of intentional breakage for the removal of brain or tongue, may indicate that the bagged animals were not used intensively (Fig. 11).

The winter appears not to be the only season during which Middle Palaeolithic animals died in the valley of Zwolenka. Assuming that rhino births took place in late-winter/early-spring, a cutmarked mandible of a woolly rhino calf aged 2-4 months may indicate that, unlike horses, other species were hunted in summer (B. Bratlund).

Cut Marks (B. Bratlund)

The mandible of the juvenile rhino was the only piece that carried indisputable cutmarks of a standard type. Considering the study of numerous Stone Age assemblages, I have proposed elsewhere that a minimum of 100-150 well-preserved bone fragments per species is a reasonable requirement before a lack of cut marks can be considered significant. The horse assemblage from Zwoleń is numerous enough to warrant serious expectation for cutmarked pieces. A great obstacle at Zwoleń is, however,

	Thermokarstic Channels	Lower Cultural Level	Structures preceding second Early Vistulian Meandering Phase	Middle Cultural Level	Redeposited between Middle and Upper Cultural Level	Upper Cultural Level	Redeposited above Upper Cultural Level	Late Alluvium	Cut III, middle level	Cut IV, lower level	Cut V, lower level	Provenance uncertain	Total	Solutré *
cranial remains	1	10	-	5	7	3	9	2	-	-	-	1	**38**	**33**
teeth	-	13	-	5	3	2	9	2	-	1	2	1	**38**	+
tooth fragments	2	12	-	2	5	2	10	1	1	-	9	3	**47**	+
vertebrae	-	3	-	2	-	-	-	1	-	-	-	-	**6**	+
ribs	1	17	2	7	3	-	12	1	-	-	-	1	**44**	+
scapula	-	-	1	1	-	-	-	-	-	-	-	-	**2**	**30**
humerus	-	-	1	1	-	-	1	-	-	2	-	-	**5**	**32**
radius	-	2	-	1	-	-	-	-	-	-	-	-	**3**	**62**
mc	-	4	-	2	-	1	1	-	-	-	-	1	**9**	**50**
pelvis	-	3	-	2	-	-	1	-	-	-	-	-	**6**	**20**
femur	-	2	-	1	-	3	-	-	-	-	-	-	**6**	**34**
patella	-	-	-	-	-	-	-	-	-	-	-	-	**-**	**46**
tibia	-	6	1	2	-	-	2	1	-	-	-	-	**12**	**36**
mt	1	3	-	4	1	-	-	-	-	-	-	-	**9**	**44**
mp	-	1	-	1	1	-	2	-	-	-	-	-	**5**	**-**
carpals/tarsals	-	3	-	2	-	-	1	1	-	-	-	-	**7**	**521**
Ph. 1	-	1	-	2	-	-	1	-	-	-	-	-	**4**	**126**
Ph. 2	-	-	-	-	-	-	-	-	-	-	-	-	**-**	**109**
Ph. 3	-	-	-	1	-	-	-	-	-	-	-	-	**1**	**74**
Totals	**5**	**80**	**5**	**41**	**20**	**11**	**49**	**9**	**1**	**3**	**11**	**7**	**242**	**1217**

* estimated minimum number of fragments or specimens

Table 4. Zwoleń Fauna, Interskeletal Distribution.

that the surface of the bones is generally not well preserved. Of a total of 226 bone fragments examined from a variety of mammal species, 76 were totally unsuitable, 78 of medium quality or only suitable for recognition of deep marks, and only 72 pieces were of the kind likely to yield cutmarked specimens. Of these, only 53 had the nearly undamaged surface required for comparative studies of modification traces.

Conclusions

All the circumstantial evidence from Zwoleń points to the fact that the place was intensively and recurrently used in the Middle Palaeolithic to dispatch horse nursery herds. Because of its geomorphic value it was also opportunistically employed to catch other animals. It seems likely that horses, who, during the winter, grazed on plateaux where the snow is often blown away by wind, were driven into the valley and then corralled at the gorge (Fig. 12). Deep snow in the valley would have impaired their movement and made them much easier to kill. It is noteworthy that a large majority of Palaeoindian bone beds of bison kill sites were formed in late-fall/early-winter. This may be due to the use of frozen meat caches used to supplement the lack of fat in the human diet at that time of the year (Todd, 1991).

Zwoleń is presumably a Neanderthal site. At Kůlna, one of the Micoquian beds (7a) yielded fragments of a Neanderthal jaw and skull (Valoch, 1988). The lowest (9b) Micoquian bed has an ESR age estimate of ca. 70 ka (Rink at al., 1996). If the age estimates of Zwoleń are correct, it may be one of a growing number of open-air Middle Palaeolithic kill-and-butchery sites indicating active hunting of dangerous, and difficult to kill, megafauna. At La Borde the hunters apparently slaughtered nursery herds of wild cattle (Jaubert and Brugal, 1990). Wild cattle were also hunted at Mauran (Girard-Farizy and Leclerc, 1981). In addition, a large proportion of cutmarked rhino bones at Taubach (Bratlund) and Biache-St.Vaast (Auguste, 1995) strongly suggest a planned and repetitive exploitation of megafauna long before the Micoquian (Bratlund).

The recent excavations of open-air sites suggest that the

Neanderthals were skilled and active megafauna hunters. These findings contradict the scavenging-and-small-prey theory presented in numerous papers and books, often written by vocational archaeozoologists. The observed 'specialization' in one or two preys is probably an artefact of local geomorphic and environmental conditions which were skillfully utilized by humans. On the other hand, it is increasingly apparent that the fauna assemblages obtained from most caves can not be used to hypothesize modes of meat procurement by humans. Caves are spatially confined *loci* combining hundreds if not thousands of occupational events telescoped in a single archaeological bed. Each of these events may have had its specific logistics and taphonomic history.

Even with firearms, horses are perhaps one of the most difficult hunts due to their intelligence, social organisation, alertness and speed of these animals (compare Baskin, 1976; Berger, 1986; see M. Grumm-Ghzimailo, 1892 and G. E. Grumm-Ghzimailo, 1948 for *Equus przewalski* hunting in western China). The Zwoleń sample therefore indicates a highly competent social organization and a refined hunting skill among the late Neanderthals.

It is not only the evidence of hunting in the Zwolenka valley that suggests that Neanderthals were much more competent and intelligent than most of our colleagues believe. The raw material economy indicates well-established curating practices and inter-group contacts for barter or traffic in raw materials. In fact, the raw material economy at Zwoleń is much more 'evolved' than that of early Mesolithic. Careful planning and curation are indicated by the fact that the hunters in Zwoleń used flint extracted in warmer seasons. Many of the tools were made on the spot to suit the actual need, from nodules carried along for this purpose.

In recent years it has become increasingly evident that except for their physique, Neanderthals were not essentially different from early *Homo sapiens sapiens*, or at least the archaeological record can not convincingly show the difference. In spite of common belief, the late Middle Palaeolithic stone tool variability in Europe and western Asia was much greater than that of the early Upper Palaeolithic Aurignacian Technocomplex, and Neanderthal social complexity was seemingly at the same level with modern humans. Obviously, no one is able to say how a Neanderthal society would have coped in a modern world.

Authors' addresses:
1 Institute of Archaeology and Ethnology, Polish Academy of Sciences, Warsaw, Poland. RSCHILD@archeolog.iaepan.edu.pl
2 State Archaeological Museum, Warsaw, Poland.
3 Laboratorium voor Paleontologie, Vakgroep Geologie en Bodenkunde, Universiteit Gent, Gent, Belgium.
4 Thermoluminiscence Laboratory, Institute of Physics, Silesian Technical Institute, Gliwice, Poland.
5 Archaeoosteological Research Laboratory, Department of Archaeology, Stockholm University, Royal Palace Ulriksdal, Solna, Sweden.
6 Institut for forhistorisk arkaeologi ved Aarhus Universitet, Moesgård, Højbjerg, Denmark.
7 Archaeological Survey of Canada, Canadian Museum of Civilisation, Québec, Canada.
8 The Institute of Systematics and Evolution of Animals, Polish Academy of Sciences, Cracow, Poland.
9 Department of Physical Geography, Maria Sklodowska University, Lublin, Poland.
10 Department of Maritime Geology, State Geological Institute, Sopot, Poland.

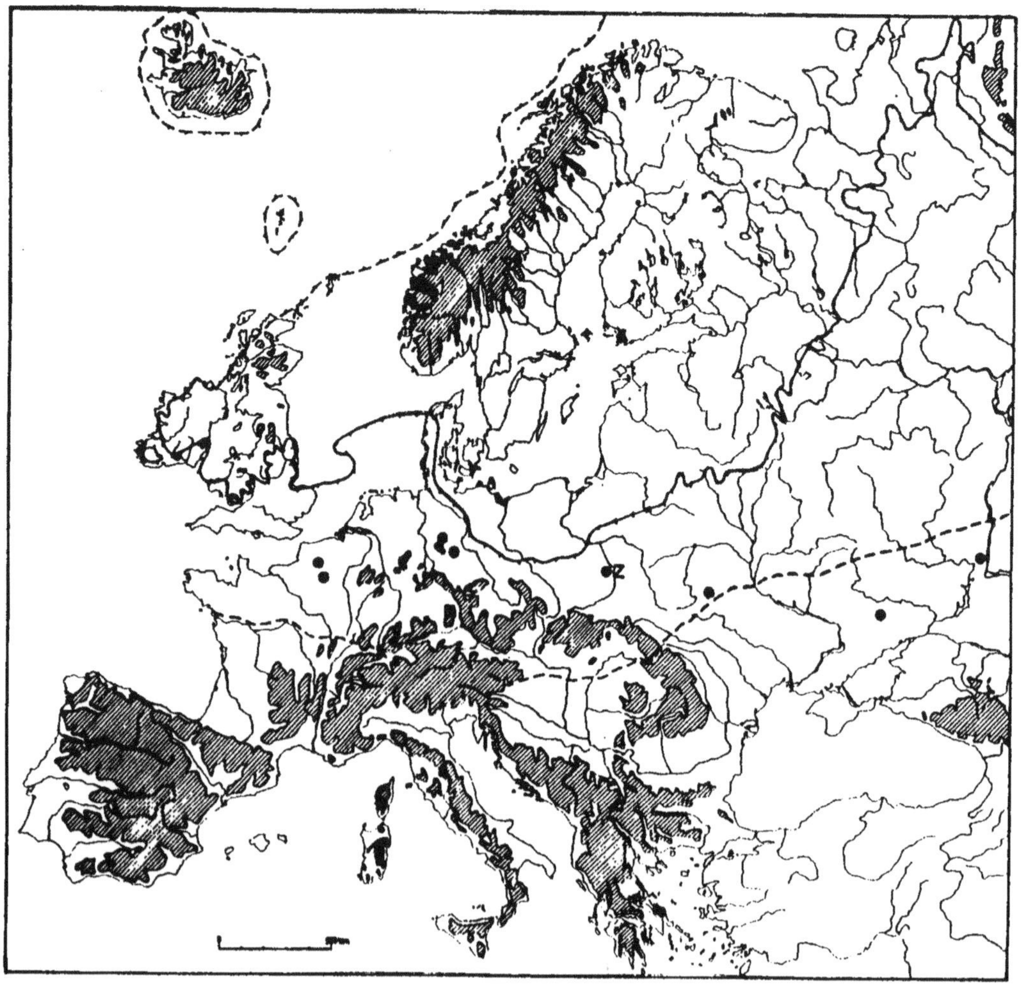

Fig. 1. Map of Europe showing maximal extent of ice sheet in the Vistulian (continuous line) and its periglacial zone (broken line). Sites of the Micoquian affinities on the Plain shown by dots, z - Zwoleń. Sites from the right: Sukhaia Metchetka; Zhitomir; Königsaue; Salzgitter-Lebenstead; Riencourt-les-Bapaume; Le Mont de Bauvry à Bethune (the majority according to Kulakovskaya et al., 1993).

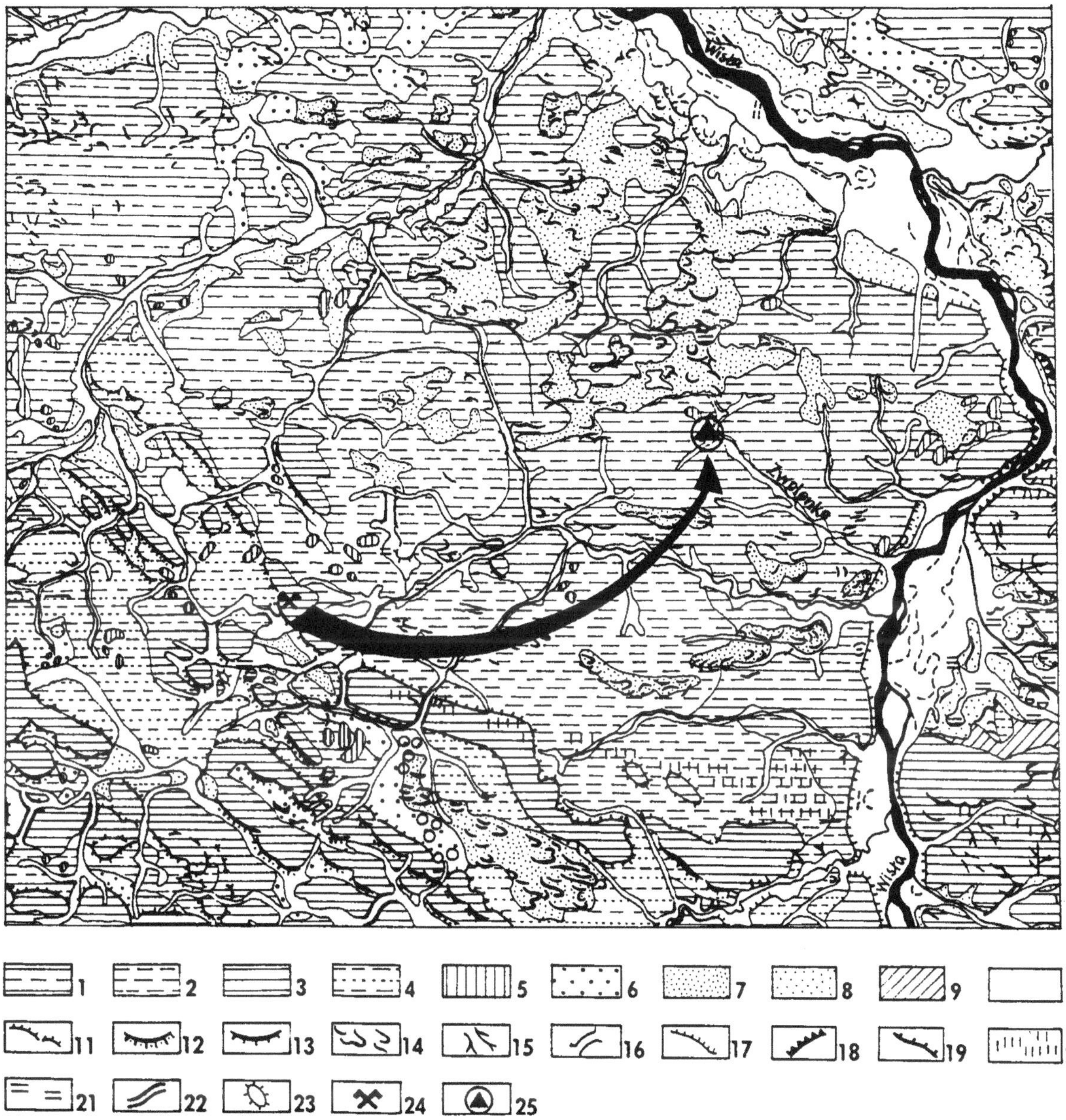

Fig. 2. General Geomorphology of the Radom Plain, according to Geomorphologic Map of Poland, J.E. Mojski (Ed.) 1. Denudate tills and fluvioglacial deposits on plateaux; 2. Plateaux on degraded Miocene deposits overlain by Oder Glaciation sediments; 3. Low plateaux with remnants of Pliocene base-levelling; 4. Low plateaux covered by older glacial deposits; 5. Endmoraines and morainic remnants of the marginal zone of the Radomka stadial; 6. Denudate sandres and kame terraces; 7. Eolian sands; 8. Vistulan river terraces; 9. Low plateaux with remnants of Neogene base-levelling; 10. Floodplains; 11. Low erosional and denudational steps; 12. Sandstone cuestas; 14. Clusters of dunes; 15. Ravines; 16. Dry valleys; 17. Erosional scarps more than 20 m high; 18. Active undercuts; 19. High denudational and/or erosional steps covered by Quaternary deposits; 20. Older relief covered by thick loesses; 21. Plains with riverine and peat accumulation; 22. Gorges; 23. Low limestone relict hills; 24. Chocolate flint outcrops of Polany Kolonie II; 25. Site of Zwoleń.

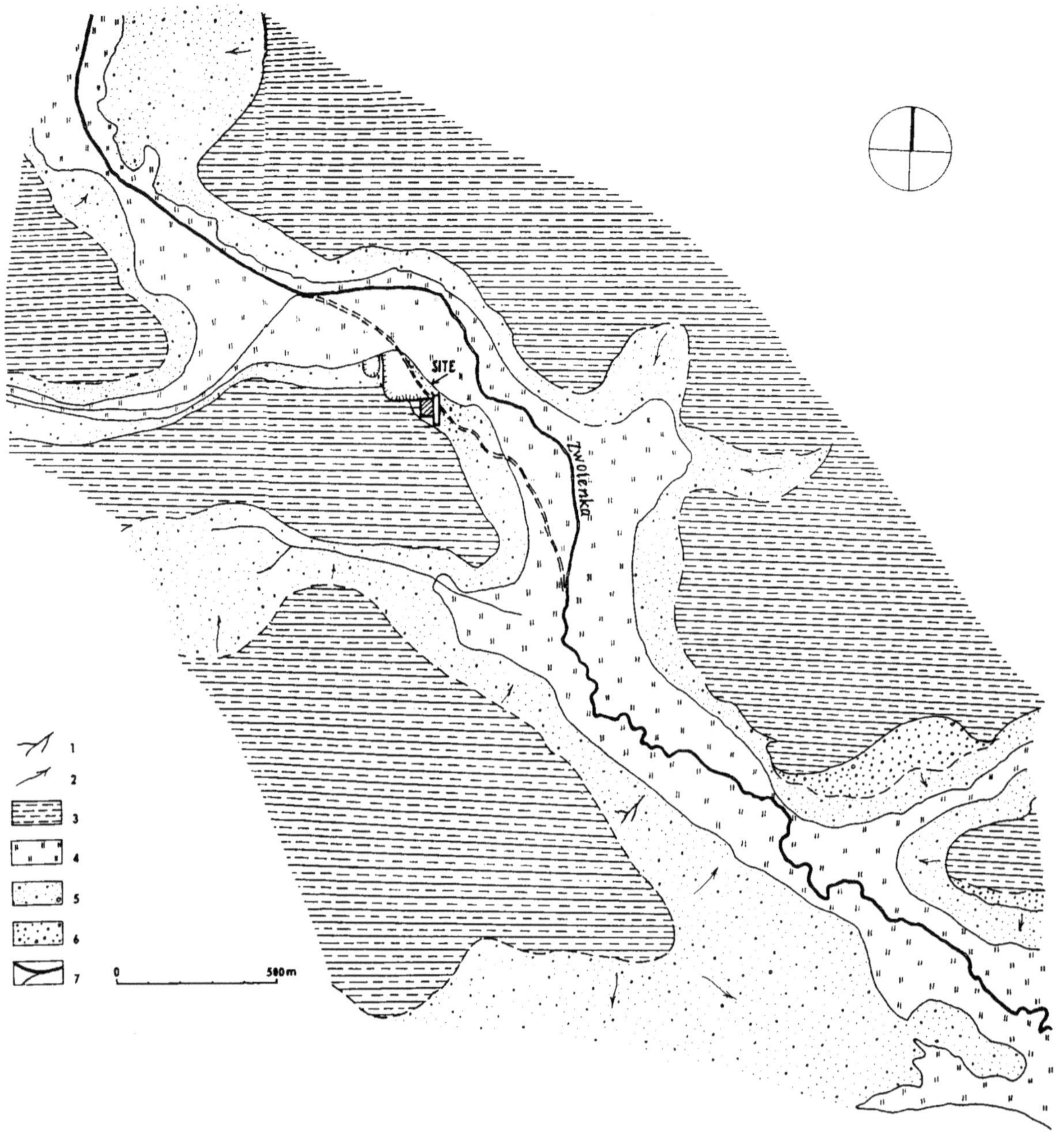

Fig. 3. Basic geomorphology. 1. Fossil Zwolenka channels; 2. Direction of maximal Vistulian slope movement; 3. Tills on plateaux (Radomka stadial); 4. Peats on the floodplain; 5. Slope sands and gravel grading into alluvial deposits; 6. Lacustrine sands; 7. The Zwolenka River and its channels in early Vistulian (dotted). Only Cut I and Rifle-Range shown.

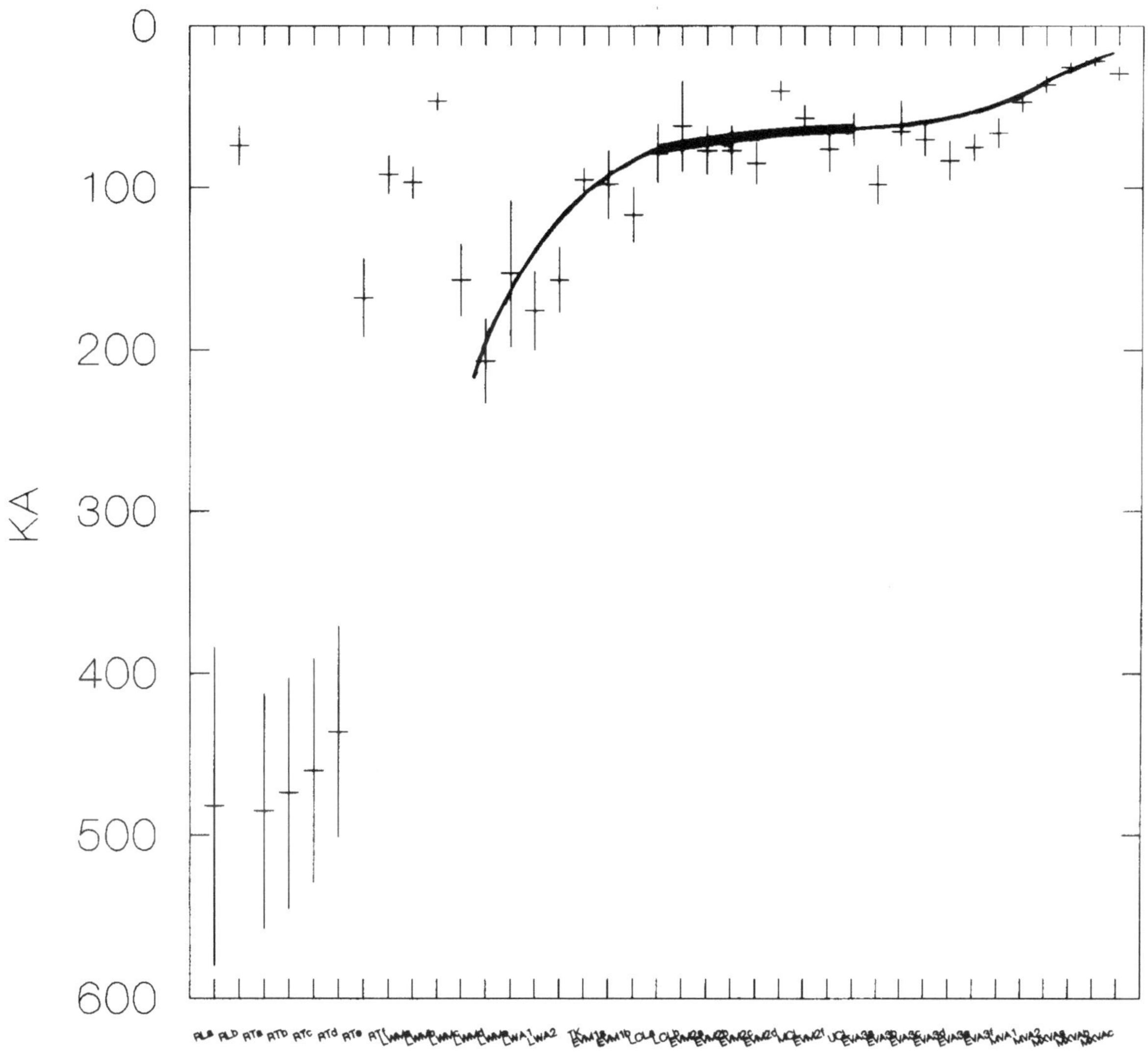

Fig. 4. TL ages from Zwoleń and TL age curve. Thick continuous line shows the TL age for three cultural levels.

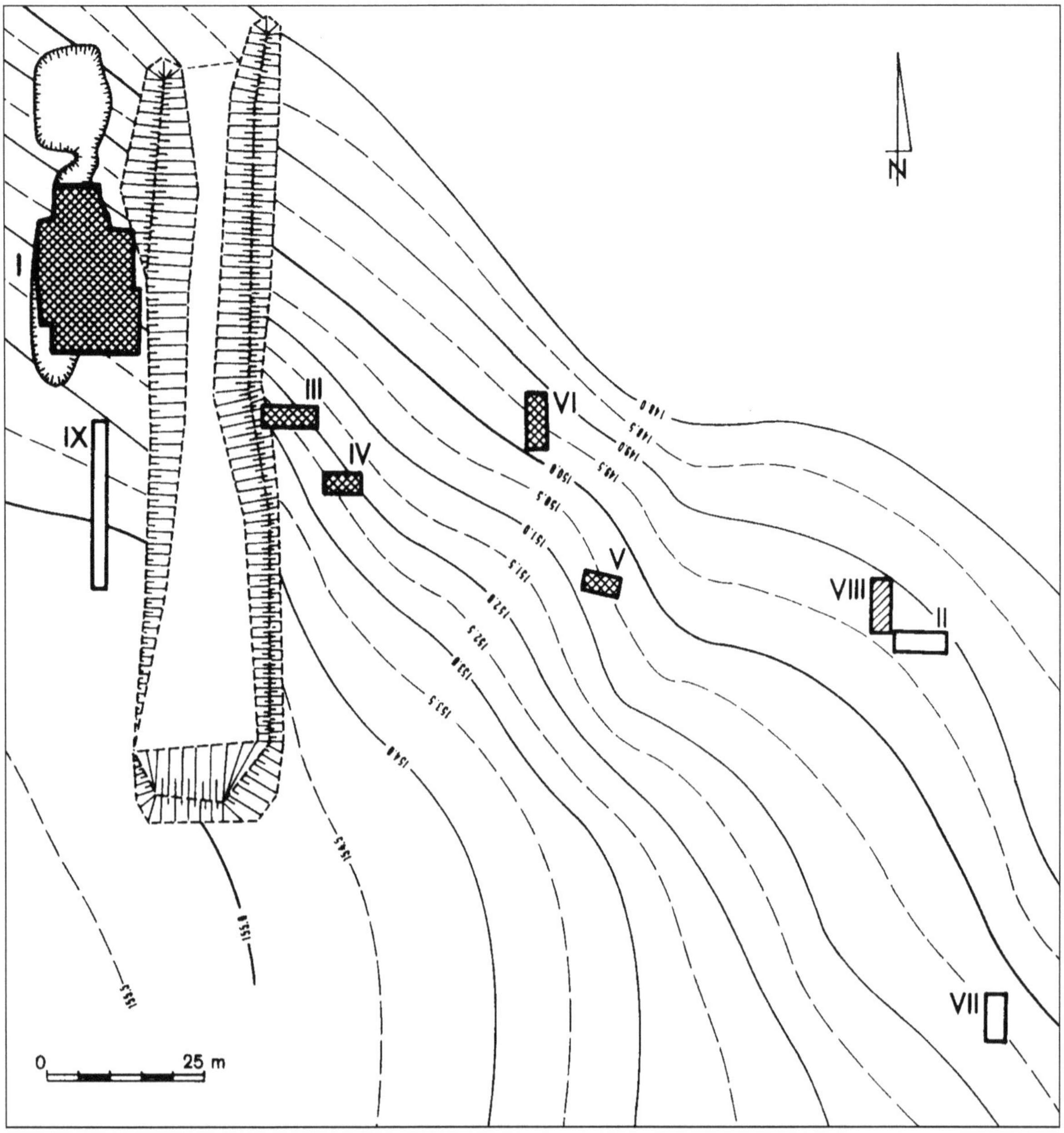

Fig. 5. Contour map and locations of cuts. Cross-hatching - trenches with bones and lithics; diagonal shading - trench that yielded only lithics; blank - trenches without bones or lithics.

Fig. 6. Excavations in the gravel quarry, Cut I.

Fig. 7. Sources of non-local flint material. T = Turonian flint; C = Chocolate flint; J = Jurassic, Cracowian flint (Brzoskwinia).

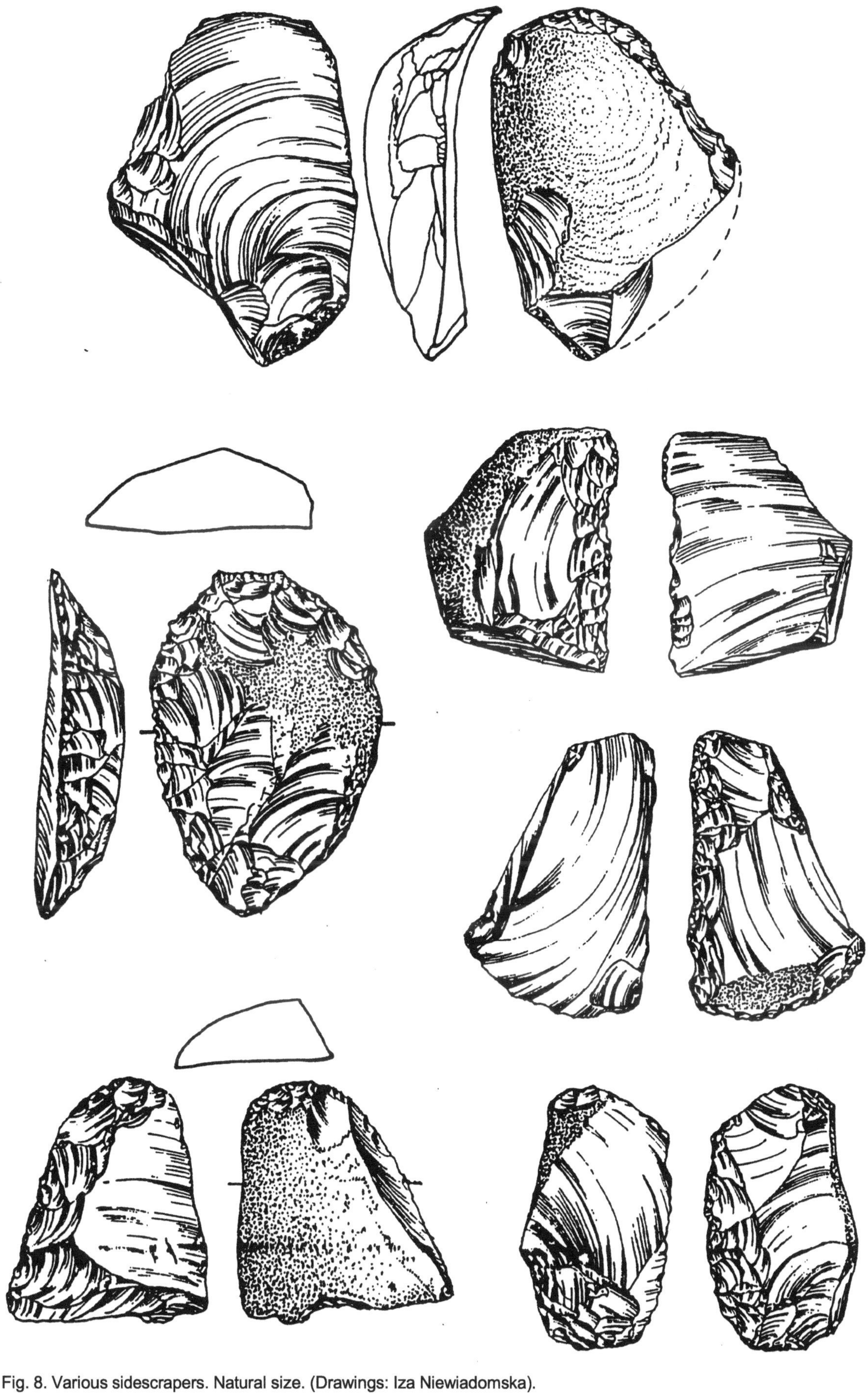

Fig. 8. Various sidescrapers. Natural size. (Drawings: Iza Niewiadomska).

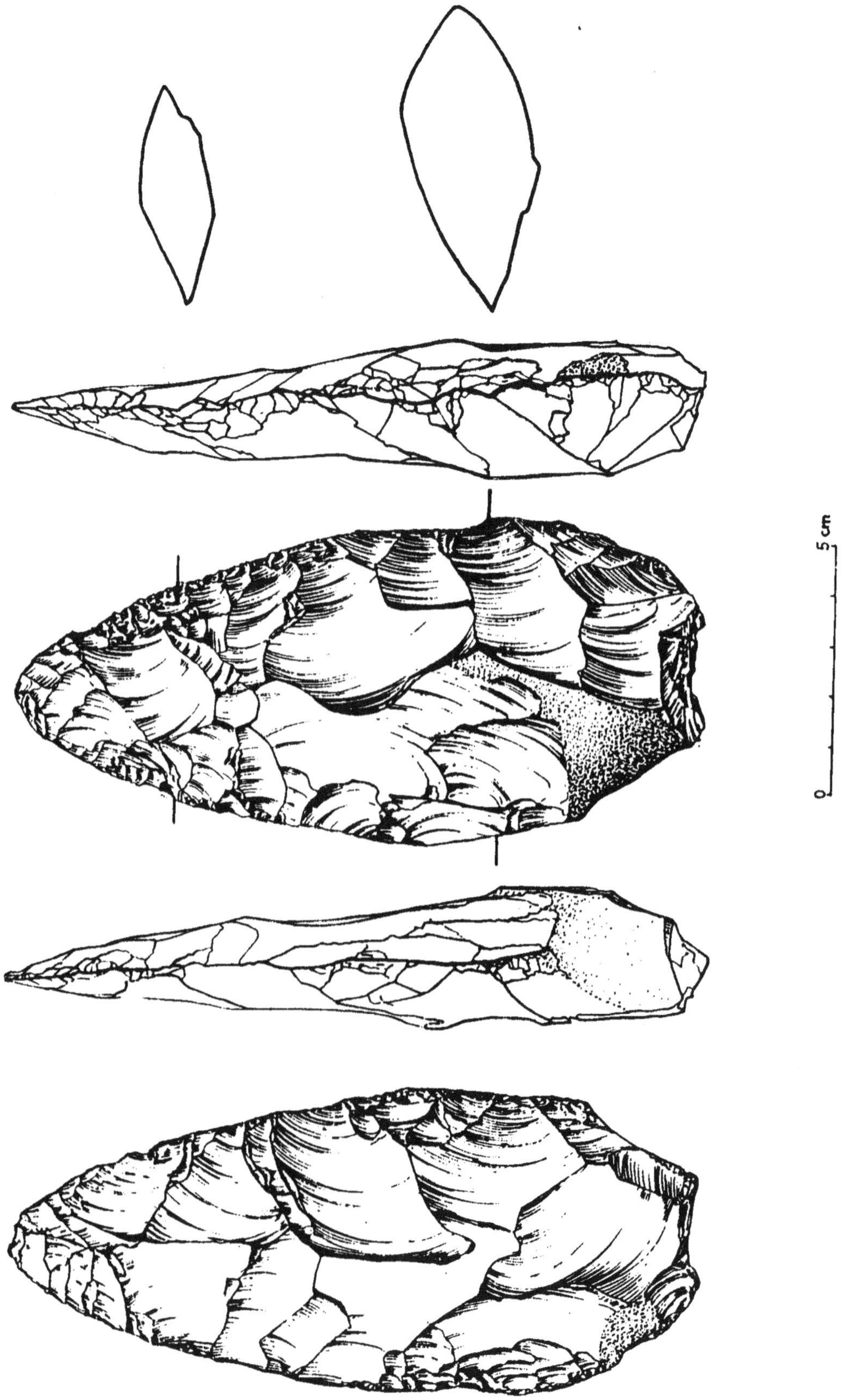

Fig. 9. Elongated amygdaloid biface. (Drawing: Iza Niewiadomska).

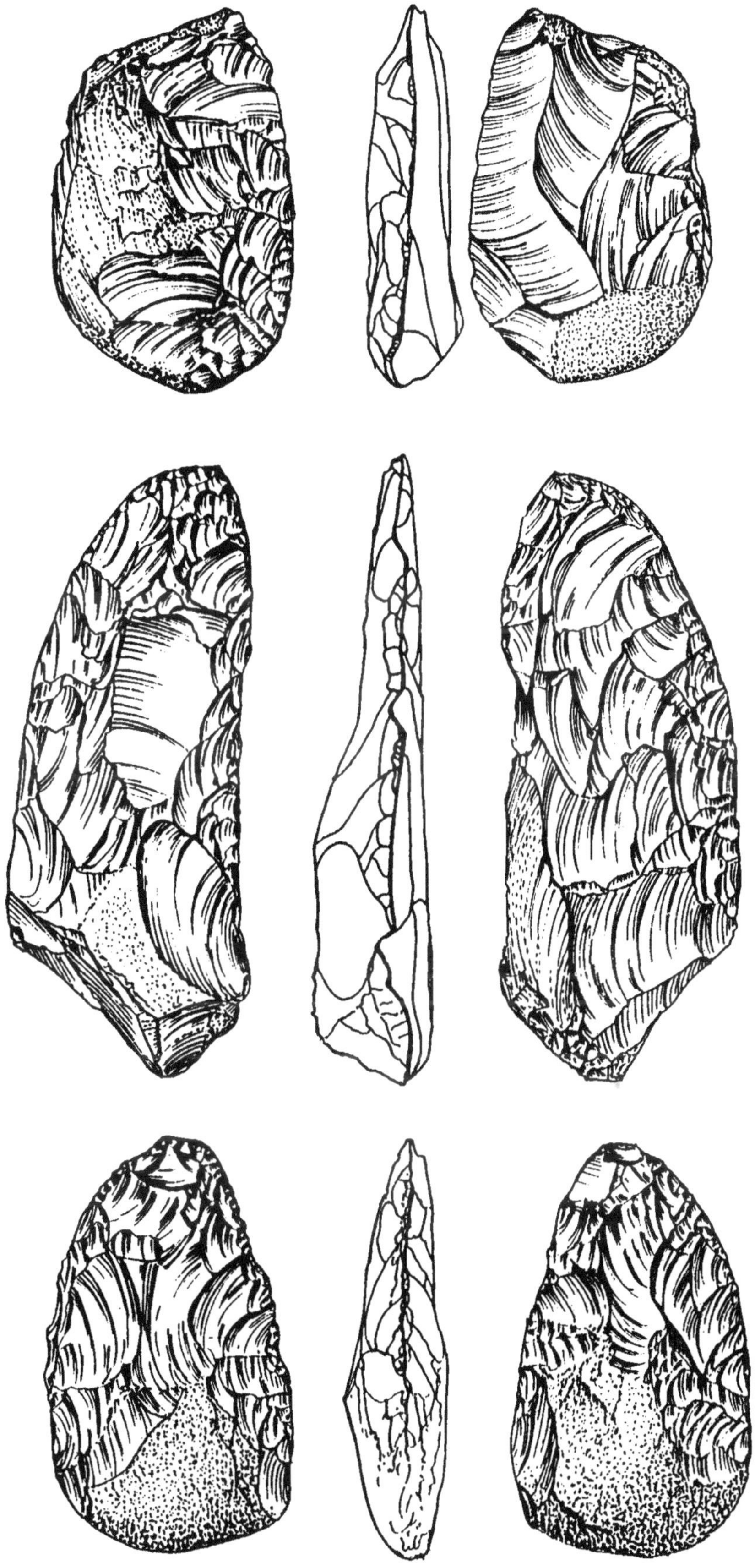

Fig. 10. Top. Prondnik backed knife resembling sidescraper; Middle. Volgograd knife or asymmetric leaf handaxe; Bottom. Asymmetric leaf handaxe with a resharpening scar. Natural size. (Drawing: Iza Niewiadomska).

Fig. 11. Horse skull and upper jaw.

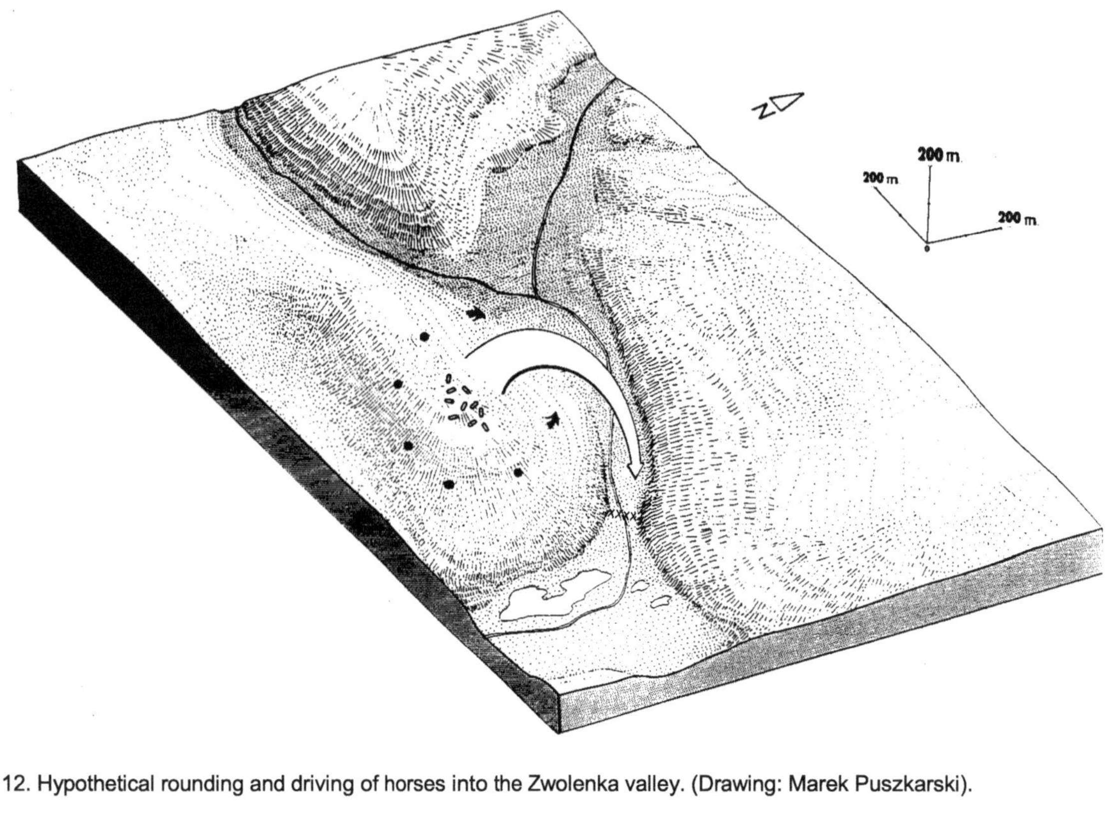

Fig. 12. Hypothetical rounding and driving of horses into the Zwolenka valley. (Drawing: Marek Puszkarski).

Bibliography

Auguste, P., 1995, Chasse et charonage au Paléolithique moyen: L'apport du gisement de Biache-Saint-Vaast (Pas-de-Calais). *Bulletin de la Société Préhistorique Française,* 92 (2), 155-167.

Baskin, L. M., 1976, *Povedenie kopytnykh zhivotnykh (Behavior of ungulates).* Nauka, Moskow.

Berger, J., 1986, *Wild Horses of the Great Basin. Social Competition and Population Size.* The University of Chicago Press, Chicago and London.

Gautier, A., 1984, How do I count you, let me count the ways? Problems of archaeozoological quantification. In: C. Grigson and J. Clutton-Brock eds., *Animals and Archaeology: 4. Husbandry in Europe.* BAR International Series 227, 237-251.

Girard-Farizy, C. and Leclerc, J., 1981, Les grandes chasses de Mauran. *La Recherche* 127, 1294-1295

Grumm-Ghzimailo, G. E., 1948, *Opisanie puteshestvia v zapadnij Kitai.* OGIZ, Gosudarstvennoe Izdatelstvo Geograficheskoi Literatury, Moskva.

Grumm-Ghzimailo, M., 1982, Dikaia Loshad (Equus Przewalski). Iz dnevnika puteshestvia v Kitai 1880-1890. *Niva,* 17, 374-382.

Jaubert, J. and Brugal, J. P., 1990, Contribution à l'étude du mode de vie au Paléolithique moyen: les chasseurs d' Aurochs de la Borde. In: J. Jaubert, M. Lorblanchet, H. Laville, H. Slott-Moller, A. Turq and J. P. Brugal, Les Chasseurs d' Aurochs de La Borde. *Documents d' Archeologie française* 27. Maison des Sciences de l' homme, Paris, pp. 127-145, 152-153.

Koulakovskaya, L., Kozlowski, J. K. and Sobczyk K., 1993, Les couteaux Micoquiens du Würm ancien. *Préhistoire Europeénne,* 4, 9-32.

Lindner, L., 1992, Stratygrafia (klimatostratygrafia) czwartorzędu. In: L. Lindner ed., *Czwartorzęd. Osady, Metody badań, Stratygrafia.* Wydawnictwo PAE, Warszawa, pp.441-613.

Mania, D. and Topfer, V., 1973, *Königsaue. Gliederung, Ökologie und mittelpaläolithische Funde der letzten Eiszeit.* VEB Deutscher Verlag der Wissenschafen, Berlin.

Mojski, J. E., 1985, *Geology of Poland.* Volume I Stratigraphy, Part 3b Cainozoic, Quaternary. Wydawnictwo Geologiczne, Warsaw.

Rink, W. J., Schwarcz, H. P., Valoch, K., Seitl, L. and Stringer, C. B., 1996, ESR Dating of Micoquian Industry and Neanderthal Remains at Kůlna Cave, Czech Republic. *Journal of Archaeological Sciences* 23, 884-901.

Różycki, S. Z., 1972, Nizina Mazowiecka (Mazowia Lowland). In: R. Galon ed., *Geomorfologia Polski.* Państwowe Wydawnictwo Naukowe, Warsaw, pp. 271-317.

Schild. R., Królik H. and Morcibrodzka J., 1977, *Kopalnia krzemienia czekoladowego z przełomu neolitu i epoki brązu w Polanach Koloniach.* Ossolineum, Wrocław-Warszawa-Kraków-Gdańsk.

Todd, L. C., 1991, Seasonality Studies and Paleoindian Subsistence Strategies, In: M. C. Steiner ed., *Human Predators and Prey Mortality,* Westview Special Study in Archaeological Research. Westview Press, Boulder, pp. 217-238.

Valoch K., 1988, *Die Erforschung der Kůlna-Höhle 1961-1976.* Moravske Muzeum, Brno.

Van Andel, T. H., Tzedekis, P. C., 1996, Palaeolithic landscapes of Europe and environs, 250000-25000 years ago: an overview. *Quaternary Science Reviews,* 15, 481-500.

TOWARD MODERN HUMANS: YABRUDIAN AND MICOQUIAN, 400 – 50 kyears ago.
Edited by A. Ronen and M. Weinstein-Evron. BAR S850, 2000.

QUINA-TYPE OCCURRENCES IN EUROPE: SINGLE TOOLMAKING TRADITION OR CONVERGING PATTERNS?

Nicolas Rolland

Introduction

The Quina industry, widespread in western Europe, contains well-defined artefactual characteristics. A standard interpretation regards this industry as a discrete taxonomic entity as well as a distinct toolmaking phylum whose techno-typological properties constitute stylistic signatures by separate hominid social groups.

The alternative presented here argues that Quina assemblages constitute a segment along a classificatory continuum. It is proposed that the components of these assemblages express dynamic reduction processes relating to lithic material constraints and/or a parsimonious exploitation of fine-grain materials, a technological strategy conditioned by various situational factors.

Three examples will be presented to illustrate this model. They are not *case studies* in the sense of specific sites, but rather areas or sources of evidence. The first examines how specific Quina flaking and trimming are better suited for overcoming lithic material constraints relating to texture, shape or size, and is exemplified by localized occurrences. The second and third pertain to region-wide situations, the Aquitaine and Maas basins respectively, where 'classic' Quina characteristics indirectly reflect the repercussions of the last glacial bioclimatic fluctuations on hominid settlement and subsistence strategies.

Following the observation that the 'Micoquian' complex of Central Europe may illustrate analogous lithic material economizing by distinct techniques, it is hypothesized that the apparently homogenous outer structure of Quina-type industry actually results from the converging effects of various circumstances, rather than being a separate tradition.

'Classic' Quina Properties: Taxonomic and Interpretive Issues

Bordes (1953a; 1957; Bordes and Bourgon, 1951) isolated the 'Charentian' group (Ferrassie and Quina) within the 'Mousterian Complex' system for the descriptive classification of Middle Palaeolithic variability. Bordes defined the Quina industry technologically, typologically and quantitatively (i.e., via indices and groups). Other

REDUCTION SEQUENCE	
primary flaking	**trimming**
thick, large clactonian preforms 'salami' slices natural backed knives 'spicchio' pebble-flaking	invasive step-flaking fine edge trimming planoconvex bifacial 'clactonian' rejuvenation trimming
CLASSIFICATORY	
diagnostic typology	**taxonomic: modalities**
convex, transverse racloirs Quina-retouch racloirs bifacial ('tranchoirs') racloirs hachoirs limaces	high racloirs indices high Quina indices high implements frequencies higher blanks/flake tools to cores ratios

Table 1. Quina-type Assemblage Characteristics

researchers (e.g. Lenoir, 1973; Turq, 1989) elaborated on reduction sequence traits.

According to this theory, Quina characteristics are interpreted as flaking and trimming techniques actualizing preferred designs and stylistic idiosyncrasies for finished tools. These are held to be diagnostic of separate social groups (Bordes 1961; Bordes and Sonneville-Bordes, 1970) whose repertoires remain stable, 'closed' systems independent of circumstances (see Table 1 for a summary of Quina characteristics). This view implies that these populations constituted 'nucleated' or autonomous mating networks/ethnic entities (Yellen and Harpending, 1972, 246-248). This model, however, appears incompatible with the prevailing notion that pre-modern hominids were mobile, dispersed, low density, socially 'open' and technologically simple groups (Wobst, 1976).

The taxonomic discrimination between Mousterian Complex industries rests on a detailed artefactual sorting, particularly a polymodal grapho-statistical patterning of racloirs (IR) and Quina (IQ) indices (Bordes, 1953a, Figs. 1-3; Bordes and Sonneville-Bordes, 1970, Fig. 15). Underlying this method is the belief that the quantitative expression of different types of tools was rigidly controlled by a single behavioural mechanism, namely tradition (Bordes and Sonneville-Bordes, 1970, 68. See Greenspan, 1995, 73 for a similar model based on behavioural genetics). This hypothesis and the interpretation that they represent discrete toolmaking traditions were never subjected to stringent testing, however.

The widespread and discontinuous distribution of Quina-type industries and facies across western Eurasia were due to the dispersion of hominids from a single point of origin and homologous developments in different areas. Also, if development was analogous but discontinuous, then time periods also differed. Some facies may be traced as far as Central Asia or the Levant, as suggested by Bordes (1955; 1957). One hypothesis is that certain facies spread into western Europe through a north Mediterranean route, though all facies emerged ultimately in the West during pre-Würmian times, as evidenced in sites such as La Micoque 3 or High Lodge (Bordes, 1950, Table 1; Bordes, 1968, 101). Gábori (1976, 211, and Fig. 65) proposed a different hypothesis, suggesting that the 'Eastern' facies originated in Mediterranean France. Both scenarios are consistent with the *Ferninterpretation* method (establishing a common origin, despite distances) of the Kulturkreis school of ethnology. Taschini (1979, 242-244) attributed a polyphyletic origin to the Charentian industrial complex, whose similarities and facies differences, according to this view, would also reflect traditional idiosyncrasies.

An Alternative Model

Evidence points instead to multiple and independent developments by convergence, e.g., a gradual shift from Ferrassie into Quina in southwestern France (Mellars, 1969; 1993), and Quina-like roots of the Yabrudian within the Acheulean in the Levant (Jelinek, 1982). High Lodge and La Micoque 3 are separated from classic Quina by time gaps too substantial to indicate linear continuity. Furthermore, the Micoque and Quina assemblages display significant techno-typological differences (Bordes and Bourgon, 1951, 6; Rolland, 1986).

Bordes' system of *typologie morphologique* (1967) introduced order in the study of Middle Palaeolithic variability. It represents an etic approach to classification (Hayden, 1984, 80, i.e., objective sorting for the resolution of specific problems). Bordes' behavioural interpretation, however, should have been based independently on an emic approach (Hayden, 1984, 80) or intrinsic properties, rather than relying on the same descriptive system. We consider the toolmaking processes labeled the 'Commont Effect' (Rolland, 1996) to be a concept interchangeable with the 'Frison Effect' (Frison, 1968). The former term seems more appropriate here because, as it was defined earlier and concerns the Middle Palaeolithic directly, it better describes the nature of Quina characteristics.

The Commont Effect: Its Bearing on the Taxonomic Position and Behavioural Significance of Quina

Development of the concept: as the model of the Commont Effect has been detailed elsewhere (Rolland, 1996), only a short description will be presented here. Commont (1913, 138-141, 164, 168) identified the processes involved in lithic material parsimony. These included accelerated blank turnover and transformation of blanks into implements, more exhaustive core flaking and increased tool rejuvenation in sites such as Busigny, Marlés than at Hermies. This model also delineated the various circumstances inducing these changes, such as the properties and availability of raw materials, site locations, seasonal variations, and Pleistocene climatic changes. Commont's pioneering contribution, however, could not benefit from the substantial documentary body or the classification of Mousterian Complex occurrences that subsequently became available.

Bordes (1953b; 1954) elaborated this concept by demonstrating that the classic 'Mousterian/Levalloisian' dichotomy, rather than representing separate toolmaking phyla, amounted to nothing more than Levallois typological indice (ILty) frequency variations. These variations reflected how regional and settlement situations (i.e., open-air loess plain occurrences in northern France) contrasted with enclosed sites in the Dordogne, and conditioned varying degrees of lithic material profligacy and parsimony. Nevertheless, Bordes' study did not exploit the implications and explanatory potential of this model further. Observations were restricted to the Levallois component of the Le Moustier group, the Charentian industries were left out, and these variations were subordinated as lateral facies within the Mousterian Complex.

Leroi-Gourhan (1966; 1969) demonstrated how 'cutting tools' (*sensu lato*) formed a broad kinetic/functional series which ranged from raw flakes or blades to intensively resharpened implements such as Quina racloirs,

Mousterian points, tranchoirs and limaces, when circumstances required exploiting better 'appreciated' fine-grain flint more exhaustively. This tendency predominates among 'Cave Mousterian' assemblages which contain a variety of heavily worn and rejuvenated 'knives' (racloirs, points and limaces) as well as 'scraping' tools (denticulates and raclettes). Leroi-Gourhan's conclusion is in line with Bordes' settlement type opposition, but excludes his industry-type taxonomy.

The Relationship of Quina-Type Industries with Mousterian Complex Industries

The findings depicted in Figs. 1A-F were produced by a binary sorting of artefact classes derived from the Bordes type-list definitions, but according to the Commont Effect which dictates differential transformation of Levallois and non-Levallois preforms/blanks into regularly trimmed implements. This method was applied to a comprehensive sample of Middle Palaeolithic industries of the last glacial in western Europe (Rolland, 1977; 1988; 1996). This data analysis transcends limitations inherent in Bordes' and Leroi-Gourhan's studies by: (1) including all plain and/or utilized flakes and blades, in set I+U (Fig. 1A), incorporating the unretouched non-Levallois flakes/blades, in set I+P (Fig. 1B) which is a category not present in the type-list but is a more or less normally distributed variate (Fig. 1C); (2) covering all Mousterian Complex industries, including the Charentian; and (3) Retaining the open-air versus enclosed sites opposition.

The results indicate that: (1) The Commont Effect unexpectedly covaries closely with each named Mousterian Complex industry type, instead of forming lateral facies contained by them. This finding is in opposition to Bordes' conclusions which were restricted to ILty frequency variations; (2) While differential transformation remains more pronounced in cave and rockshelter sites, this group contains striking contrasts in terms of high implement frequencies among Quina assemblages but low frequencies among Denticulate and most Typical Mousterian. This trend was 'buried' until Mousterian industries could be examined individually with respect to the Commont Effect. Combier (1967, Notes 52, 54) mentioned that Quina occurrences in eastern France contained fewer unretouched flakes and more intensively trimmed tools; (3) Equally significant is the unimodal, asymmetrical (positively skewed) configuration of implement frequencies for the entire sample (Fig. 1D), as well as a markedly skewed racloirs frequency pattern when calculated as part of set I+P (Fig. 1E). Note however, that high implement frequencies are largely controlled by increased racloirs manufacture (Fig. 1F and Rolland, 1988, Fig. 8A); (4) The grapho-statistical pattern of Quina assemblages undergoes a radical modification in

TECHNOLOGICAL

1.	specifically Quina flaking/trimming techniques
2.	lithic material properties variations
3.	'kinetic'/functional
4.	large-sized 'cutting tools'
5.	parsimonious exploitation of fine-grain lithic materials

SITUATIONAL

1.	enclosed sites(caves, rockshelters) settlements (over 92%)
2.	settlement duration and intensity
3.	differential access to lithic resources
4.	home-ranges
5.	ecotone locational situations; mosaic landscapes
6.	subsistence strategies related to higher, more concentrated animal food resources; regional importance of reindeer
7.	ungulate species socio-ecology
8.	bioclimatic conditions colder; steppic, steppe/woodland habitats; more pronounced seasonal variations

PALEOSOCIETAL

1.	dispersed, 'anucleated' mating networks
2.	repertoires (techniques, information) ranges and modalities
3.	low-density demographic patterns, with seasonal semi-sedentarity
4.	frequently associated with Neanderthal hominid finds in western Europe

Table 2. Variables Related to Quina-Type Occurences

both instances, assuming the shape of a tailed segment within an asymmetrical continuum, rather than emerging as a distinct modality.

In conclusion, Quina assemblages represent instances of extreme quantitative expressions of both aspects of the Commont Effect, namely differential transformation into implements and intensive tool reduction/resharpening (Fig. 2). This indicates parsimonious lithic material exploitation, an interpretation corroborated by data on core flaking intensity (Fig. 3). These findings call for a fundamental reappraisal of conventional perspectives concerning the taxonomic and behavioural significance of Middle Palaeolithic variability in western Europe as represented in the Bordes Mousterian Complex. A unimodal continuous pattern suggests the effect of several independent causes (Schreider, 1960, 51) including circumstantial, technological, as well as stylistic sources, rather than a single factor (i.e., tradition). It becomes incumbent to identify the relevant variables (Table 2) liable to condition the lithic parsimony strongly displayed in Quina assemblages.

Explanatory Models

A. Lithic material: Properties and constraints

The case studies depicted in Fig. 4 illustrate various situations where raw material properties or their availability constraints had a direct influence on toolmaking behaviour, producing assemblages dominated by Quina techniques and typology. Material properties include: (1) texture: there is a preference for fine quality flint for the production of sharper cutting tools, as opposed to the use of more expedient coarser materials such as various types of local flints, quartzite, quartz, crystal, lava, and phonolite; (2) size: small, round, flint pebbles requiring the specialized *spicchio* blank release method of the 'Pontinian' facies (e.g., Sant' Agostino: Tozzi, 1970); (3) shape: small or medium-size rounded quartz pebbles (e.g., Érd, Gábori-Csank, 1968) also call for similar reduction processes.

Constraints refer to circumstances in which fine-grain material is locally available but scarce, e.g., Chapelle aux Saints (Bouyssonie and Bardon, 1913), Mas-Viel (Niederlender, et al., 1956), Rond du Barry (Bayle des Hermens, 1987), and La Ermita (Moure Romanillo and Delibes de Castro, 1972), or in which it is imported, e.g., Marillac and Los Aviones, 15-20 km and 30 km respectively (Meignen, 1988, and Montes Bernardez, 1989), thereby requiring economizing by trimming/resharpening.

These Quina occurrences coincide with: (1) isolated situations (Rond du Barry, Massif Central, La Ermita, Old Castile, Érd, Hungary); (2) site clusters (Los Aviones, coastal Murcia, Sant' Agostino, coastal central Italy); and (3) localized situations within a widespread regional concentration (e.g., Aquitaine: Chapelle aux Saints and Marillac, both with classic Quina characteristics).

Fig. 4 reveals that most of the flint component at these sites corresponds with high implement frequencies (primarily 'cutting tools'). This finding indicates economizing behaviour regarding flint, as other materials were exploited casually. The miniature size of the local flint pebbles of Sant' Agostino resulted in high coefficients of unusable flakes and rapid nodule turnover. The hardness of quartzite made flake tools more wear-resistant, a characteristic which would account for the lower frequencies of trimmed/resharpened implements and slower replacement rates at Érd. At both Érd and Sant' Agostino, flint and quartzite were not only locally abundant, but were the only available lithic resources. Use of Quina reduction techniques relates to lithic material properties (texture, shape and size), rather than to availability constraints.

B. Regional case studies

The Aquitaine basin. This region, occupied continuously since the colonization of Europe, contains the largest concentration of Quina occurrences. It is the Charentian type-region, and contains both Ferrassie and Quina type-sites. The morphological unity of the Charentian actually coincides with a region-wide, long-term techno-typological shift from Ferrassie to Quina, which underlies localized variations in lithic material conditions and exploitation strategies (Mellars, 1993). This development implies a self-contained linear trend.

The Charentian succession corresponded with stadial bioclimatic fluctuations correlated with Isotopic Oxygen Stage 4, the onset of more severe, continental conditions, the spread of steppic or steppe/parkland habitats (Bordes et al., 1966; Paquereau, 1974-75), and faunal communities with abundant reindeer (Bouchud, 1966; Guadelli, 1987; Mellars, 1969). These conditions created mosaic vegetational landscapes and optimal animal food biomass (Drury, 1975; Mellars, 1985; Spiess 1979) and impacted directly or indirectly on hominid technology, subsistence, land use and settlement organization. Together they stimulated directional changes in Charentian toolmaking modalities (Rolland, 1996).

The flowchart model in Fig. 5 describes processes leading to the widespread emergence of Quina occurrences in enclosed sites, best represented by localities such as Abri Chadourne, Les Merveilles, Roc de Marsal, Combe-Grenal, and La Quina. This model also accounts for the prevalence of Quina in Charente, which was characterized by a harsher palaeoclimate and greater exposure to westerly and northern winds (Commont, 1913, 138).

The Maas basin. This region lies at higher latitudes than southwest France. Here, repeated interruptions of human occupation occurred during fully stadial episodes, for instance, during the Lower Pleniglacial (Oxygen Isotope Stage 4) when this region was characterized by a cold permafrost climate, sparse tundra vegetation and low-density arctic fauna (Cordy, 1984; Cordy and Simonet, 1990; Haesaerts, 1984). Middle Palaeolithic occupation, represented mostly by Quina occurrences, originated in southern Germany as seen at Sirgenstein, Schulerloch, Vogelherd and Karstein (Bosinski, 1967). Occupation during this period resumed during the milder Interpleniglacial, with enclosed sites concentrated in the Sambre/Maas refugium trough (Ulrix-Closset 1975, 172-174; 1990).

Fig. 6 outlines the processes leading to the analogous Quina horizon in Maas and Aquitine. The Maas region differs in several respects from its Aquitaine counterpart. Maas is characterized by unrelated human populations, a later interpleniglacial occupation episode under habitat conditions more comparable with those of Oxygen Isotope Stage 4 in southwestern France, given latitude differences, higher solar energy input and ecosystem primary productivity, less continentality further South (Fig. 7A-C), more constraining lithic procurement conditions (due to local material of inferior quality and the need to import fine-grain flint back to cave sites home bases) and a predominance of horse over reindeer as an animal food staple.

Discussion and Concluding Remarks

The Quina techno-functional system emphasizes: (1) preference for fine-grain lithic material; (2) optimal cutting tool size; and (3) intensive reduction and parsimonious lithic material exploitation, relating to circumstantial and location constraints. This system seems better suited than either prepared-core or bifacial techniques for obtaining thick, large blanks/preforms from pebble-shaped nodules, and for flaking poor-quality materials. Situational factors compelling a reliance on poorer quality materials would generate higher ratios of unused raw flakes (e.g., Mas-Viel, La Ermita, Los Aviones, Érd) than is the norm in classic Quina occurrences. The recurring need to manufacture large implements (e.g., Les Merveilles, Le Moustier, Roc de Marsal, La Quina) may override a preference for flint. Except in cases such as Romanelli, which evidenced an exclusive use of abundant local limestone (Piperno, 1974), distant access to quality flint may induce economizing exploitation (Marillac, Maas region sites). The similar economizing behaviour characteristic of Aquitaine sites, where quality flint is often locally abundant, is related to other critical factors such as the ecotone situation, animal food procurement, shelter, and residential intensity.

The above regional examples, chronological discrepancies, and other lithic material properties combine to make unlikely the notion of a monophyletic origin, remote in time, for the Charentian group and Quina assemblage type. Instead, they point to several autonomous developments deriving from distinct antecedent repertoires in several regions: The Charentian succession in Aquitaine; the pre-Würmian presence of 'Eastern' Quina (Baume-Bonne) in Mediterranean France; the pre-Würmian emergence (Sedia del Diavolo, Monte delle Gioie) of the Pontinian in central Italy; discrete Early Glacial (Oxygen Isotope Stage 5a-d) occurrences (e.g., Ranc-Pointu and Baume-Bonne in southern France); and the persistence of Quina throughout the Early Würm (Cova Negra).

A polyphyletic derivation, with social traditions playing a dominant role (Gábori 1976; Taschini, 1979, 243) is equally debatable. Traits such as planoconvex or déjétes racloirs, bifacial racloirs or limaces in eastern France, for example, Genay, Vergisson, Figuier and Soyons (Combier, 1967), may indicate both a stylistic drift as well as occurrences within a continuous transition zone (Germolles, Grotte du Docteur) between the Mousterian Complex and the 'Micoquian' of northwestern Europe (Bosinski, 1967). The latter may represent a cluster of distinct techno-typological modalities based on bifacial techniques, also related to lithic material parsimony. As such, this would represent a case of alternative 'isochrestic' stylistic divergence (Rolland, 1990) created by geographic and social distances. Such distances would involve populations occupying a higher latitude in more continental habitats throughout Early Glacial times (Figs. 7A, 7B). Arguing that the Aquitaine Quina horizon merely illustrates an event of two independent intersecting causal sequences, namely the Charentian technological shift and the onset of Lower Pleniglacial stadial, ignores the body of evidence for the several interacting situational variables already discussed and the efficacious role of large-scale environmental changes in modifying Middle Palaeolithic repertoires (Mellars, 1993, 29; and Wengler, 1993 for the Maghreb), as well as the logic of the Commont Effect.

It seems preferable to consider Quina assemblages in descriptive terms, as a segment within the polythetic range of Mousterian Complex toolmaking, the latter constituting a comprehensive reserve of alternative technological options. The dynamic interaction between these repertoires of open, 'anucleated' palaeosocieties and environmental factors, whose intensity, extent or duration varied, would generate Quina occurrences with distribution patterns ranging from random, localized occurrences to region-wide clusters and horizon time-depths.

Author's address: Department of Anthropology, University of Victoria, B.C., 8W 3P5, Canada. nrolland@uvic.ca

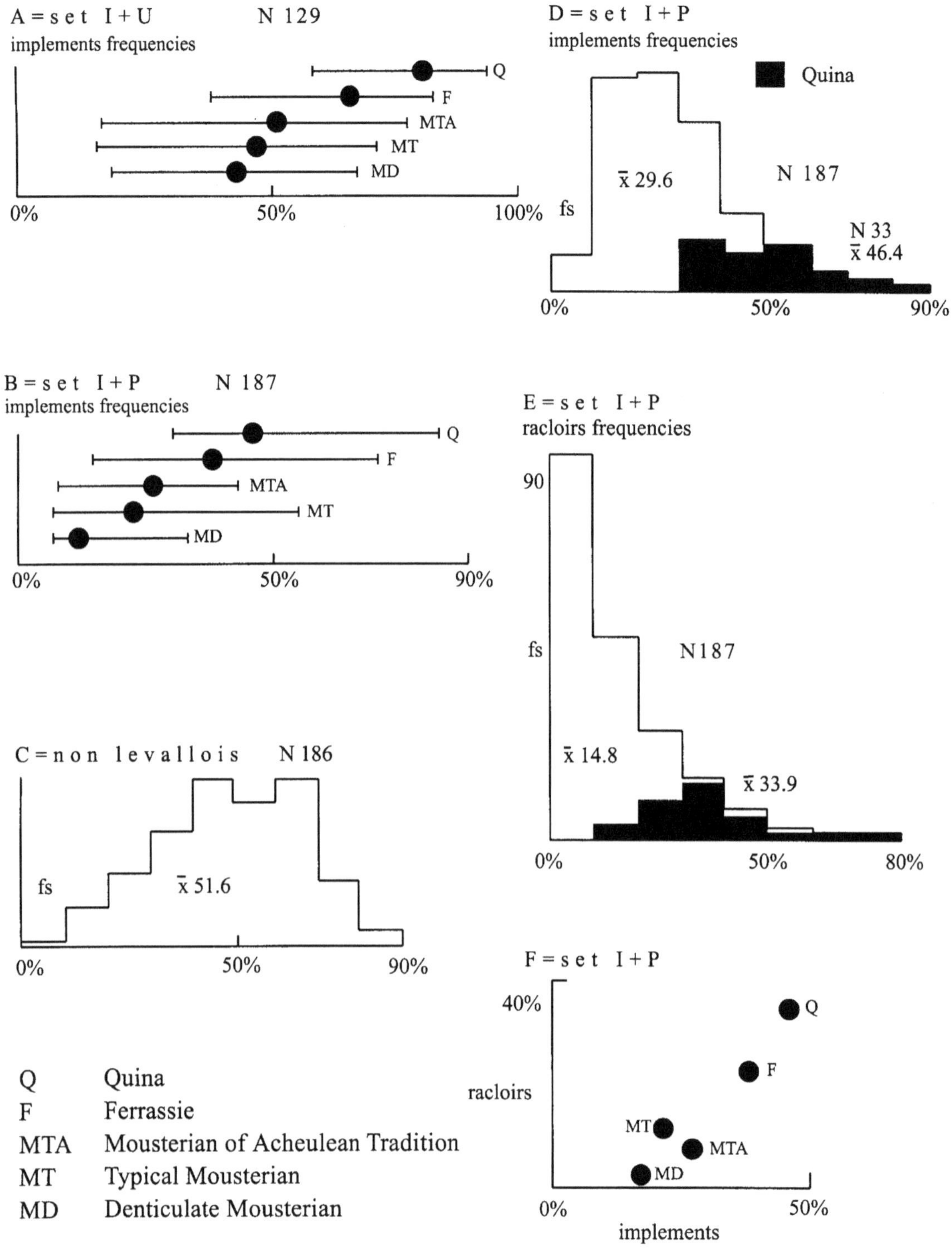

Fig. 1: Frequency variations of major artefacts classes of the Mousterian Complex.

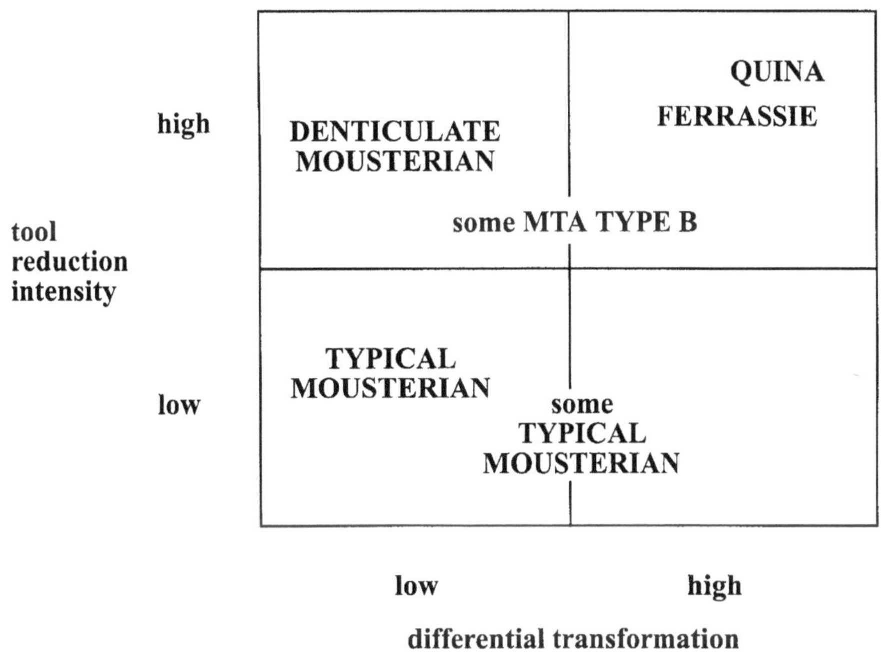

Fig. 2. 'Common effect' and Mousterian complex assemblage types

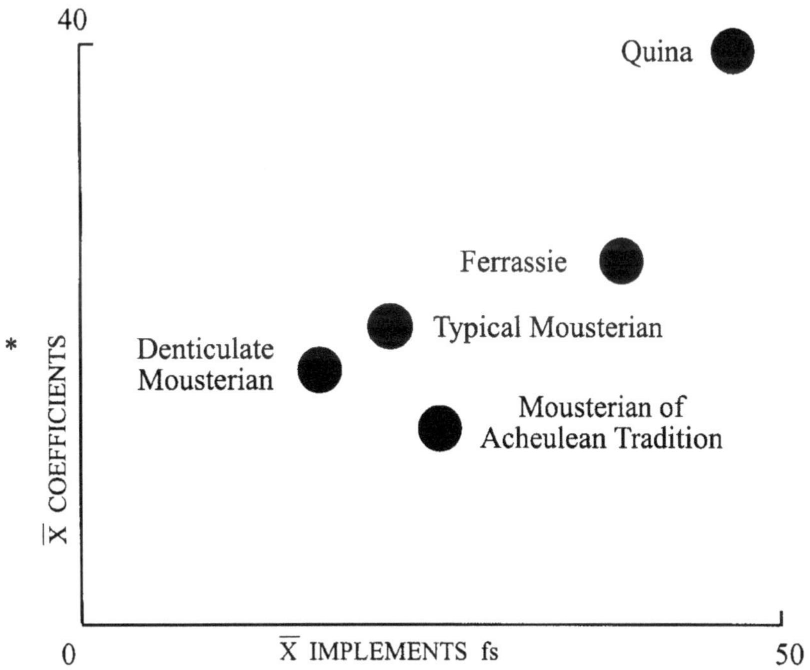

Fig. 3. Lithic material parsimony. * flakes, blades/core coefficients

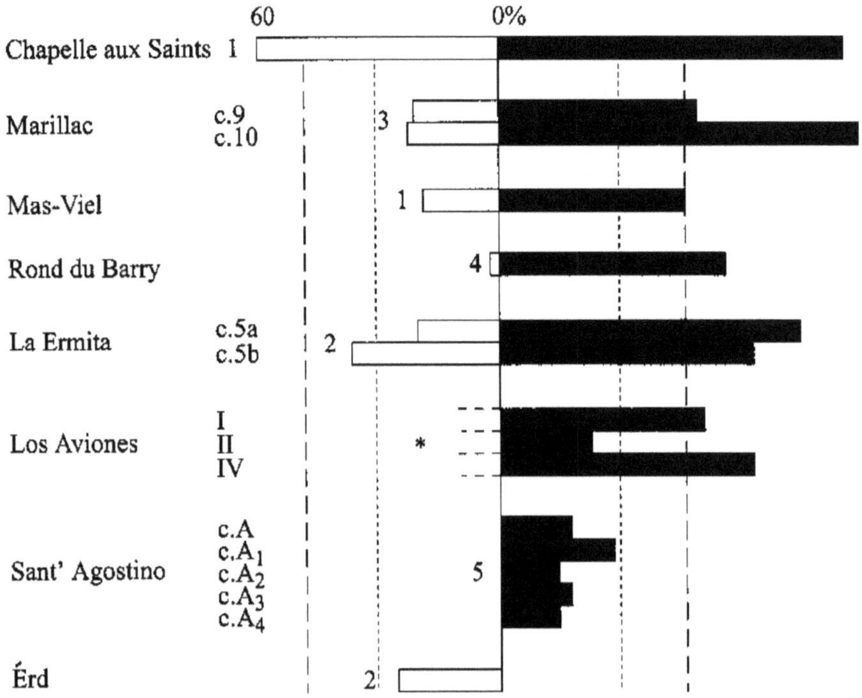

Fig. 4. Relationship between lithic material properties and availability, and Quina assemblages implements frequencies.
1 = quartz, crystal; 2 = quartzite; 3 = coarse local flint; 4 = phonolite, basalt, quartz; 5 = small local flint pebbles; * data counts on quartz and critical component unavailable but very low frequencies of trimmed pieces.
Minimum implements frequencies for classic Quina (30.7%) ....
Mean implements frequencies for classic Quina (46.4%) _ _ _ _ _ _ _ _ _ _ _ _ _ _ _ _

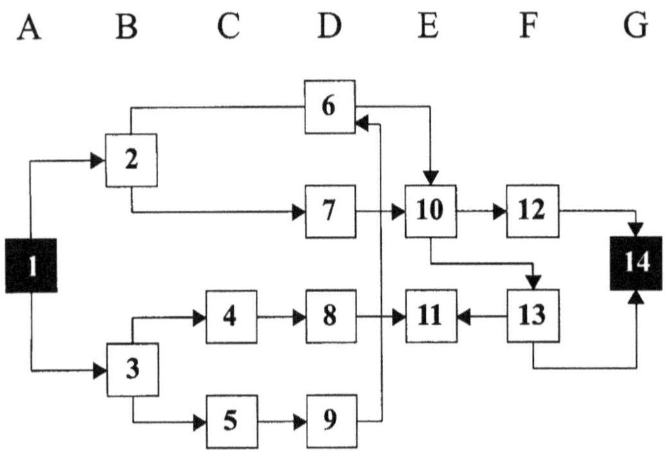

Fig. 5. Aquitaine Basin – flow chart model. A 1: isotopic stage 4, colder, drier phases B 2: severe continental climate, colder, longer winters 3: spread of steppe and open woodland habitats C 4: increased gregarious herbivores biomass 5: reindeer appears, often dominant element of faunal communities D 6: modified seasonal and home range patterns, longer, more intensive cold season residence, increased relative sedentarity 7: lithic resources more difficult to procure and process in winter 8: greater reliance on animal food 9: subsistence based on exploiting concentrated animal prey i.e., reindeer herds aggregating in limestone valley corridors during migratory cycles E 10: pressure on parsimonious use of lithic material exploitation 11: higher fauna to lithic material accumulation F 12: technological shift from Levallois to flaking and trimming techniques more appropriate for anticipated intensive core reduction and repeated resharpening of cutting tools 13: local or exotic lithic material brought into sites at more widely spaced time intervals G 14: Quina type assemblages are the indirect outcome of raw material economizing processes induced by long-term environmental changes in Aquitaine.

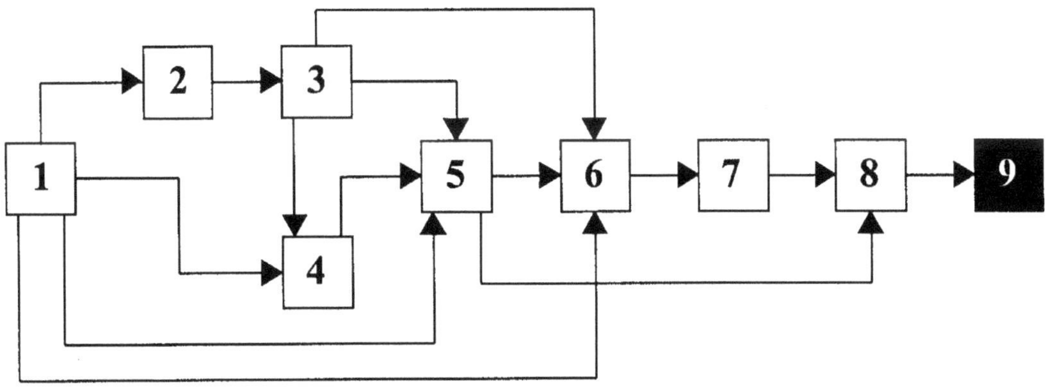

Fig. 6. Maas Basin – flow chart model. 1: bioclimatic transition from Lower Pleniglacial (Early Weichsel stadial) or isotopic stage 4, with intensely cold, dry, continental conditions, permafrost, loess, sparse tundra vegetation, arctic fauna (Cordy's Biozone D), towards climatic amelioration, with cold to moderately cold (permafrost-free) Interpleniglacial (Moershoofd and Hengelo interstadials or earlier stage 3), with July temperatures around 13^0 but still under short latitudinal winter daylight duration. 2: vegetation shifts to a mostly steppic landscape (15% AP), mainly pine, birch. 3: arctic fauna replaced by steppe communities. Some forest species reappear. Horse largely dominant (55-66% at Engihoul), reindeer persistence, presence of red deer, roe deer, boar – Biozone E. 4: human populations gradually reoccupy the region, abandoned during Early Weichsel stadial, now inhabitable due to less severe climate, and availability of gregarious herbivores game herds. Occupants originated presumably from the Southeast (Eiffel, Baden Würtemberg), moving fully into some Maas basin areas during Hengelo/Les Tambourets interstadial, and represented massively by Quina occurrences (Spy, Goyet, Le Walou, Trou du Diable, Trou Magrite, Trou du Sureau, Al 'Wesse, Scladina, Engihoul, Fonds de Foret) 5: Interpleniglacial conditions, with steppe habitats, marked seasonal variations, confined home base settlements almost entirely to sheltered valleys and cave habitats South of the Sambre and Maas trough, which provided protection during winters, game herds movement visibility, and low grade local raw materials (chert, coarse flint, sandstone, quartzite, quartz). 6: year-round foraging requirements, linked with seasonal contrasts and varying food scheduling needs, favoured and extensive home-range system combining semi-sedentary cave occupation during prolonged, cold winters, with warm season semi-nomadic mobility beyond sheltered valleys areas into the region to the North containing a Cretaceous limestone substratum yielding high quality flint, along with nutritious edaphic conditions supporting several game herd species. 7: this extended home-range fulfilled year-round subsistence needs, and offered opportunities for procuring fine-grained flint from Cretaceous outcrops, for on the spot usage, as well as for importing back to winter residence quarters. Flint nodules could be extracted and undergo initial reduction stages more easily under warm season conditions. 8: exploiting imported quality flint in the cave valleys areas during the intensive winter occupation episodes required more parsimonious reduction techniques through several stages (core flaking, blank production, use, trimming, wear rejuvenation to prolong this material's useful time span, contrasting with the casual exploitation of local lithic raw materials. 9: repeated cave residence episodes resulted in the accumulation of exhaustively reduced assemblages, diagnostic of Quina type assemblages, containing tool forms from knives to heavily trimmed racloirs.

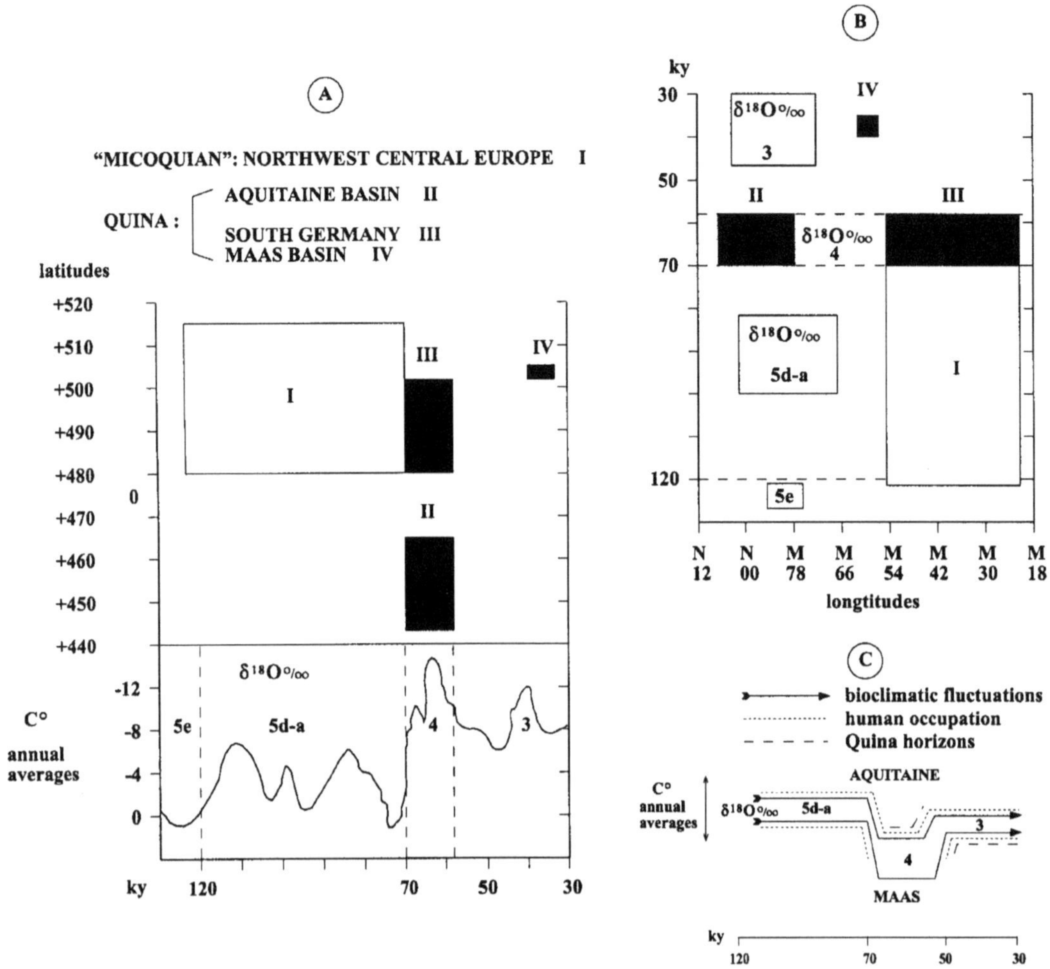

Fig 7. Relationship between bioclimatic phases. (A) latitudes and longitudes gradients. (B) and Quina assemblage type horizons (C).

Bibliography

Bordes, F. 1950, L'évolution buissonnante des industries en Europe occidentale. Considrations théoriques sur le Paléolithique ancien et moyen. *L'Anthropologie* 54, 393-420.

Bordes, F. 1935a, Essai de classification des industries 'moustériennes'. *Bulletin de la Société Préhistorique Française* 50, 457-466.

Bordes, F. 1935b, Levalloisien et Moustérien. *Bulletin de la Société Préhistorique Française* 50, 226-235.

Bordes, F. 1954, *Les limons quaternaires du basin de la Seine*. Archives de l'Institut de Paléontologie Humaine, Mémoire 26. Paris, Masson.

Bordes, F. 1955, L'industrie moustérienne de Techik-Tash. Affinités et âge probable. *L'Anthropologie* 59, 354-356.

Bordes, F. 1957, La classification du Moustérien: état actuel. *Lexique stratigraphique international. Vol I. Europe. Fascicule* 4, 73-77.

Bordes, F. 1961, Mousterian Cultures in France. *Science* 5(3482), 803-810.

Bordes, F. 1967, Considérations sur la Typologie et les techniques dans le Paléolithique. *Quartär* 18, 25-55.

Bordes, F. 1968, *The Old Stone Age*. New York, Mac Graw Hill.

Bordes, F. and Sonnevile-Bordes, D. 1970, The significance of variability in Palaeolithic assemblages. *World Archaeology* 2(1), 61-73.

Bordes, F. and Bourgon, M. 1951, Le Complexe Moustérien: Moustériens, Levallosien at Tayacien. *L'Anthorpologie* 55, 1-23.

Bordes, F., Laville, H. and Paquereau, M.-M. 1966 Observations sur le Pléistocène supérieur du gisement de Combe-Grenal (Dordogne). *Actes de la Société Linnéenne de Bordeaux* 103: 3-19.

Bosinski, G. 1967, *Die Mittelpaläolitischen Funde im westlichen Mitteleuropa*. Fundamenta 4. Köln, Böhlau.

Bouchud, J. 1966, *Essais sur le renne et la climatologie du Paléolithique moyen et supérieur*. Périgueux, Magne.

Bouyssonie, A. and J., and Bardon, L. 1913, La station Moustérienne de la 'Bouffia' Bonneval à la Chapelle-aux-Saints (Corrèze). *L'Anthropologie* 24, 609-634.

Combier, J. 1967, *Le Paléolithique de l'Ardèche dans son cadre paléoclimatique* Bordeaux, Delmas.

Commont, V. 1913, *Les Hommes contemporains du Renne dans la Vallée de la Somme*. Amiens, Yvert et Tellier.

Cordy, J.-M. 1984, Evolution des faunes quaternareis en Belqique. In: D. Cahen and O. Haesaerts (eds.) *Peuples chasseurs de la Belgique préhistorique dans leur cadre nature*. Brussels, Patrimoine de l'Institut royal des Sciences naturelles de Belgique.

Cordy, J.-M. and Simonet, P. 1990, Les faunes moustériennes charentiennes en Europe du Nord occidental. Pre-publication abstracts in *Les Moustriens Charentiens*. International colloquium, Brive – La Chapelle-aux-Saints, 24-30.

Drury, W. H. 1975. The ecology of the human occupation at the Abri Pataud. In H.L. Movius (ed.) *Excavations at the Abri Pataud*. Les Eyzies, 187-193. Cambridge, Peabody Museum.

Guadelli, J.-L. 1987, *Contribution à l'étude des zooécnoses préhistoriques en Aquitaine (Würm ancien et interstade würmien)*. Doctoral dissertation, Inastitut du Quaternaire, Université de Bordeaux I.

Gábori, M. 1976, *Les civilizations du Paléolithique moyen entre les Alpes et l'Oural*. Budapest, Akadémiai Kiadó.

Gábori-Csánk, V. 1968, *La station du Paléolithique moyen d'Érd (Hongrie)*. Budapest, Akadémiai Kiadó.

Greespan, R. J. 1995, Understanding the genetic construction of behaviour. *Scientific American* 272 (4), 72-78.

Haesaerts, P. 1984, Aspects de l'évolution du paysage et de l'environnement en Belgique au Quaternaire. In: D. Cahen and O. Haesaerts (eds.) *Peuples chasseurs de la Belgique préhistorique dans leur cadre nature*. Brussels, Patrimoine de l'Institut royal des Sciences naturelles de Belgique.

Hayden, B. 1984, Are emic types relevant to archaeology? *Ethnohistory* 31 (2), 79-82.

Jelinek, A. 1982, The Middle Palaeolithic in the Southern Levant, with comments on the appearance of modern *Homo sapiens*. In: A. Ronen (ed.) *The Transition from Lower to Middle Palaeolithic and the Origin of Modern Man*, 57-104. Oxford, BAR International Series 151.

Lenoir, M. 1973, Obtention expérimentale de la retouche de type Quina. *Bulletin de la Société Préhistorique Française* 70, 1, 10-11.

Leroi-Gourhan, A. (ed.) 1966, *La Préhistoire* Paris, Presses universitaires de France.

Leroi-Gourhan, A. 1969, Le petit racloir châtelperronien. In: D. Sonneville-Bordes (ed.) *La préhistoire: problèmes et tendances*, 275-282. Paris, C.N.R.S.

Meignen, L. 1988, Un exemple de comportement technique diffrentiel selon les matières premières: Marillac, couches 9 et 10. In: M. Otte (ed.) *L'Homme de Néandertal*. Vol. 4, 71-79. Liège, E.R.A.U.L.

Mellars, P. 1969, The chronology of Mousterian industries in the Perigord region of South-West France. *Proceedings of the Prehistoric Society* 35, 134-171.

Mellars, P. 1985, The ecological basis of social complexity in the Upper Palaeolithic of Southwestern France. In: D. Price and J. A. Brown (eds.) *Prehistoric Hunter-Gatherers. The Emergence of Social Complexity*, 271-197. New York, Academic Press.

Mellars, P. 1993, Technological change in Mousterian of Southwest France. In: H. L. Dibble and P. Mellars (eds.) *The Middle Palaeolithic Adaptation, Behaviour and Variability*, 29-43. Philadelphia, The University Museum.

Montez Bernadez, R. 1989, La grotte de 'Los Aviones' Cartagena (Espagne). *Bulletin de la Société Préhistorique Française* 86, 40-44.

Moure Romanillo, J. A. and Delibes de Castro, G. 1972, El yacimiento musteriense de la cueva de la Ermita (Hortigüela, Burgos). *Noticiario arqueologico hispanico – Prehistoria* 11-44.

Niederlender, A., Lacam R., Cadiergues Dr. and Bordes, F. 1956, Le gisement Moustérien du Mas-Viel (Lot). *L'Anthropologie* 60 (3-4), 209-235.

Paquereau, M.-M., 1974-75, Le Wurm ancien en Périgord: étude palynologique. *Quaternaria* 18, 67-159.

Piperno, M. 1974, L'industria musteriana su calcare di grotta Romanelli (Otranto). *Memorie dell'Istituto Italiano di Paleontologia Umana* 2, 69-90.

Rolland, N. 1977, New aspects of Middle Palaeolithic variability in Western Europe. *Nature* 255, 251-252.

Rolland, N. 1986, Recent findings from la Micoque and other sites in south-western and Mediterranean France: Their bearing on the 'Tayacian' problem and Middle Palaeolithic emergence. In: G. N. Bailey and P. Callow (eds.), *Stone Age Prehistory: Studies in Memory of Charles McBurney*, 121-151, Cambridge University Press.

Rolland, N. 1988, Variabilité et classification: nouvelles données sur le 'complexe moustérien'. In: M. Otte (ed.) *L'Homme de Néandertal*. Vol 4, 169-178. Liège, E.R.A.U.L.

Rolland, N. 1990, Existe-t'il un Moustérien oriental? In: J. Kozłowski (ed.) *Feuilles de Pierre. Les industries à pointes foliacées du Paléolithique supérieur européen*, 95-112. Liège, E.R.A.U.L.

Rolland, N. 1996, The behavioural significance of Quina assemblage type: exploring the 'Charentian' succession case study in Aquitaine. In: E. Carbonell i Roura (ed.) *The Last Neandertals, the First Anatomically Modern Humans*, 129-145. Tarragona, Universita Roviara i Virgili.

Schreider, E. 1960, *La biométrie.* Que sais-je? Presses Universitaires de France.

Spiess, A. 1979, *Reindeer and Cariboo Hunters. An Archaeological Study*. New York, Academic Press.

Taschini, M. 1979, L'industrie lithique de Grotta Guattari au Mont Circé (Latium): definition culturelle, typologique et chronologique de Pontinien. *Quaternaria* 21, 179-246.

Tozzi, C. 1970, La grotta di Sant' Agostino (Gaeta). *Rivista di Szienze Preistoriche* 25, 3-87.

Turq, A. 1989, Approche technologique et économique du faciès Moustérien de type Quina. *Bulletin de la Société Préhistorique Française* 86 (8), 897-899.

Ulrix-Closset, M. 1975, *Le Paléolithique moyen dans le Bassin mosan en Belgique.* Wetteren, Universa.

Ulrix-Closset, M. 1990, Le Charentien en Belgique. Pre-publication abstracts in *Les Moustériens Charentiens.* International Colloquium, Brive – La Chapelle-aux-Saint, 65-67.

Wengler, L. 1993, Relations entre le milieu naturel et les modifications culturelles au Paléolithique moyen. Le cas des Monts d'Oudja (Maroc oriental). *Bulletin et Mémoired de la Société d'Anthropologie de Paris* 5, 355-378.

Wobst, M. 1976, Locational relationships in Palaeolithic Society. *Journal of Human Evolution* 5, 49-58.

Yellen, J. and Harpending, H. 1972, Hunter-gatherer populations and archaeological inferences. *World Archaeology* 4, 244-253.

DISCUSSION

Avraham Ronen: Your remark about Quina is very interesting. Bioclimatic conditions may apply to Quina, but I do not think that they apply to the Yabrudian because, if the dates are accepted , it is a tremendously long time. The Yabrudian in Tabun is 5 - 7 metres thick. It does not seem to be bound to one particular climatic phase. Based on TL and ESR dates, it lasted from about 400 - 350 thousand up to about 250 thousand. And in this interval you have the Pre-Aurignacian/Amudian industries.

Nicolas Rolland: I am talking about the Aquitaine, which is a pretty large region. It is an episode which may have lasted about ten thousand years, it coincides exactly with Isotope Stage 4. Virtually all the occurrences are with reindeer and cold-loving fauna. In the Maas Basin it is something similar except that it is in Isotope Stage 3, but much further north, and conditions were probably almost the same as in Isotope Stage 4 in south-western France. The only suggestion I would think of here in the Near East, is that we are dealing with a totally different environment.
I would like to mention that in Europe, population movement is governed by climate. In the Near East and North Africa we have to think about precipitation. But, what you have just mentioned gives the exception that might disprove that. Or, it might be an exception which may be explained in its own terms.

Nicholas Conard: Concerning the question whether the Keilmesser group is limited to caves. In the north we have open-air sites like Koenigsaue and Lichtenberg, in the south we have almost exclusively caves but this is very much, I would like to think, an artefact of the history of research. In southern Germany, work has almost exclusively been carried out in caves rather than in open-air sites. My guess is that when we eventually find open-air sites in southern Germany we will also have Keilmesser forms.

Avraham Ronen: We can only discuss what we presently know. In spite of very good surveys, we do not have in the Mediterranean climatic zone of Israel any Yabrudian or Aurignacian open-air sites. Furthermore, when there is Aurignacian in a cave, you do not find a remnant of it on the terrace of the same cave.

Gerhard Bosinski: Concerning human behaviour, it does not sound very likely that they were so bound to caves. Is it not possible that this is a question of conservation and shore lines. In the Rheinland we do not have many Upper Palaeolithic open-air sites, but this does not mean that people were not there. It is a question of erosion and other geological and topographical reasons. Here in Israel the sea is so close and perhaps the shore was different so that many Yabrudian sites are not preserved. It is hard for me to accept that Yabrudian people were bound to caves. When caves were available they were used, but normally people lived in open-air sites.

Paul Callow: I would like to add to that. There is also a consideration of intensity of reduction of the tools. We have been discussing this in a number of occasions during the last days. It may well be that in circumstances where you are not reducing the industry to that extent, in an open-air environment, you will not call it Yabrudian.

Vadim Ranov: In central Asia we practically do not have Upper Palaeolithic in caves, only in open-air sites on terraces. It is hard to say why.

Nicolas Rolland: It is an interesting problem. I was very struck by the fact that the Yabrudian is found in caves. I tend to accept that this is an empirical generalization. This is exactly the same situation as in western Europe with the Quina, two industries which resemble each other very much. There are a few open-air Quina sites, but they are the exception that proves the rule. There is Chinchon, Plateau Baillard which is actually the extension of the cave site of Moulin du Milieu, and La Fontaine further north. These are the only ones. You do not find a single site in the entire loess plain of northern France, nor in Brittany. Did they live only in caves? No, I would not say that. It is the same in the Maas Basin. Ninety percent of the caves in the Maas Basin are Quina. I think that what we observe here are behavioural and bio-climatic factors. When it affects a whole region, I think that we have to look at settlement systems in the changing bio-climatic conditions.

PART IV

CONCLUDING REMARKS

GENERAL DISCUSSION

Gerhard Bosinski: We spent a few days together in this meeting and many different points were discussed. I will try to arrange a few points for general discussion, with your help.
Since the title of this conference begins with Yabrudian, I suggest to start our discussion with this industry. There is an open-air site in Syria, excavated by Le Tensorer, so perhaps the Yabrudian is not as bound to caves as Avraham thinks.

Avraham Ronen: I, too, thought that Nadawiyeh el-Askar was an open air site until I saw pictures. It is not a real open-air site, it is a dolina, a twenty metres deep water hole. The stratigraphy, and the layers sloping into that hole, are so much like Tabun that I thought at first that Le Tensorer was showing my slides from Tabun. It seems indeed that the Yabrudian is found only in caves or cavities.
Considering Yabrudian typology, I would argue that Yabrudian scrapers were not transformed from any simple or double scraper, contrary to Dibble's theory. The thick and large Yabrudian blank was made from the onset, as was also a Quina type scraper. We have to assume that function, perhaps also style, determined a tool's form. It could not have been a simple and largely automatic transformation.

Nicolas Rolland: In a very qualified way I agree that blank selection has a major role in a morphology of the end product, the racloir. But, I still believe that these were selected because they were thick for anticipation of prolonged use. In Quina they have different kinds of blanks as well. Many of these can be reduced quite a bit. The Clactonian resharpening flake which Michel Lenoir developed, would go through a lot of use if you kept resharpening it. That is why they were selected large initially and appreciated in a certain way.

Gerhard Bosinski: Of course there is a re-work and re-work. But I can not remember that I often found something transitional between side-scraper and transverse scraper.

Nicolas Rolland: No. I think that these are the result of different blanks. We should not underestimate the variety of strategies. I think that the selection is because they had good properties. We do not know all these properties. I still maintain this has to do with cutting, generally speaking, a bit like the bifaces at Zwoleń, but I think that there was more than one process of reduction and blank selection.

Gerhard Bosinski: So the Yabrudian is limited to cave/cavities, and there is a selection of blanks, and there is a question of style.

Nicolas Rolland: That is less sure, unless we take it in the broad sense - habits which can vary as a technique for resharpening. I think that plano-convex is style, but they are not creating it just for the shape of it. It is not a mental template. You might have difference in style between populations in the sense of James Sacket.

Avraham Ronen: If we try to define Yabrudian, it is a culture or a tradition without blades. When there is a blade, it is almost always a cortical product, a natural backed knife. This is one thing that characterizes the Yabrudian. The other thing is that they do not use the Levallois technique, either not knowing or not wanting it. The small Levallois flakes we do sometimes find are probably handaxe preparation products
At the other end we find the lithic industry of Ruhama, perhaps Evron-Quarry too, so small that you can not further reduce it. It seems to have been small from the outset, because a good part of the cores measure two centimeters in length. We have two strategies in the Lower Palaeolithic, then: one strategy generates large products that can be later reduced, the other strategy generates small products that can not be further reduced.

Jan Burdukiewicz: We find this kind of very small artefacts in central Europe, in Poland and in Italy. They are made of different raw materials, sometimes of a very good quality. People used this strategy intentionally.

Nicolas Rolland: To get back to what I have said in my presentation, I think that as archaeologists, we should not overlook that what people had in mind was how useful the kind of flake was to them. There was a definite selection. They would produce a mass of flakes and select those which are really suitable. They had their own classification, which flakes would be useful for cutting, which would be useful for scraping and which for processing harder material. That gives us a lead.

Naama Goren-Inbar: I do not see any direct connection between whatever is in western Europe sometime around 90 thousand years ago, and the Yabrudian in this part of the world, which is older than 300 thousand years. Looking at the Yabrudian phenomenon within the overall Acheulean framework, there is a selection there which we can trace through the Acheulean. There is evidence that the Levallois technique starts as early as half a million years ago in this part of the world. And if you look at most of the sites which are *in situ*, you will find that you have several distinct core reduction processes, which are aimed at the production of bifaces. It was demonstrated in the past that you can have regular scrapers on Levallois blanks and you can have thicker scrapers one could call Yabrudian, and they are co-associated. Looking at some open-air sites with good collections, you will find thick flakes transformed into transversal and dèjeté scrapers along with thin scrapers made on Levallois or other blanks. Over all, the general composition of the lithics is quite complex in terms of the sophistication of the reduction processes. The question is how many good collections do we have, and how many of these were properly reported. I think we should start from the end of the subject, then go back through time, and we will find that some of the questions will solve themselves, not just the specific cutting edge of this or that tool, but our understanding of the technical, mechanical, typological and stylistic components together within the Acheulean time-frame.

Gerhard Bosinski: You gave us here a rather general contribution, but we are concerned now only with the Yabrudian phenomenon.

Naama Goren-Inbar: The Yabrudian is a component of the Acheulean.

Avraham Ronen: I agree completely. The Yabrudian is a part of the Late Acheulean because it has bifaces and because it predates the Middle Palaeolithic. But it is a non-Levallois facies even though the Levallois technique is practiced during half a million years.

Naama Goren-Inbar: Everything I say comes from what I read on published excavations and some of the material I have looked at. In all the assemblages of Tabun from layer D downwards, which are part of Jelinek's excavations, there is a component of Levallois technique which is reported in all of Jelinek's articles from 1982 onwards. Not only a general Levallois technique but he gives the data concerning the Levallois points, the Levallois blades etc. He has some other specifications of the blades, but the Levallois component is there, as far as I can see in the material.

Gerhard Bosinski: Perhaps the difference between the statements is a different definition of Levallois technique.

Naama Goren-Inbar: Some of us here were students of François Bordes. So, if you wish to go back to the Bordesian time, and I agree to follow, then according to what we have studied in Bordes' lab, some of what we see there is what Bordes would have classified as Levallois technique.

Gerhard Bosinski: Anat claimed that there was no Levallois technique.

Avraham Ronen: I think that the question is how we apply Bordes' method. Many of the 'Levallois' flakes are biface trimming flakes. There isn't a single typical Levallois core, nor a Levallois point. Even Jelinek claims to have found only about 10% of Levallois flakes, which is definitely a non-Levallois industry. Still, I think that the number is exaggerated.

Gerhard Bosinski: In any case, at the time of the Yabrudian the Levallois technique was known, but as in the Quina type, they did not use it.

Daniel Kaufman: I think that there is a number of factors which were not taken into consideration here. The Yabrudian does not look like Acheulean, that is true. It does not look like the following Mousterian industries either. The problem is how could we explain why it looks different, unless we make assumption that it is an integral part of the Acheulean. But how can we explain the variability within the Acheulean? Nicolas Rolland has made an interesting point about anticipation for prolonged use of these very thick flakes. What do we know about how these folks were moving about the countryside? How were they going out and collecting raw material? From where? To what degree was there a need to economize or not in raw materials? All of these things were very important factors in this kind of variability. But we do not have that much information yet in terms of models of settlement patterns during this time. There are very nice models for the Mousterian, Upper Palaeolithic and Epi-Palaeolithic from this part of the world which help explain the lithic variability. But, we have not got to the point yet of applying these kinds of models to what might be going on in the Yabrudian.

Gerhard Bosinski: Not everybody agrees about the Yabrudian as part of the Acheulean. In the European sense, the Yabrudian flake tools are Middle Palaeolithic.

Avraham Ronen: What is your argument, Gerhard, that the Yabrudian is not a part of the Acheulean?

Gerhard Bosinski: It is not only my argument. Think of the work of Alain Tuffreau on the Acheulean. To me the Acheulean is handaxes and cleavers and some flakes. I can not remember that I have seen these in our Yabrud collection.

Naama Goren-Inbar: The extent of variability observed in the Acheulean is enormous, and the number of sites we have is very limited. The fact that some of the Yabrud rock shelters do not have cleavers is not very astonishing because, in Tabun, as you well know, there are over eight thousand bifaces. Just from the cave of Tabun itself. In this collection, the number of cleavers is very small. But recently a new site was discovered in Jordan with 6,000 cleavers in it. Yet it is a small site, as we presently know. This is just an example to show that the extent of variability within the Acheulean is very large. This is not contradictory to the knowledge which was already gained concerning the Acheulean in the Levant. If we go into European comparisons, you may compare our material with Alain Tuffreau's. Tuffreau has demonstrated that by half a million years ago he already had Levallois technique, and by 400 thousand years ago, he already had all the different modes of the Levallois production. By the end of the Mousterian record he had a very complicated set, but all the initiates were already demonstrated in Cagny L'Epinette. But I can not draw any parallels between what we see over there and in this part of the world. Over here we can not differentiate the Yabrudian from the Acheulean on stratigraphical grounds or by any other means. The Lower Palaeolithic started here in Israel about 1.4 million years ago, and it ended sometime before 200 thousand years ago, because then begins the well dated Mousterian sequence in our caves.

Avraham Ronen: We have to think simple, because we know so little about the Acheulean and because it is so remote in time. Why have problems in defining Acheulean? Cleavers do not define our Acheulean. In this part of the world, the Acheulean is defined by bifaces. Cleavers are normally very rare in Israel. Gesher Benot-

Ya'aqov is exceptional for its high number of cleavers. The industries without handaxes are Middle Palaeolithic and the industries with are Acheulean and, until now, always stratigraphically older. Remove the handaxes from an Acheulean assemblage, and you get Middle Palaeolithic. With present knowledge, you will be unable to tell the Acheulean flake tools from a real Middle Palaeolithic. In the same way, if you remove the handaxes from a Yabrudian assemblage, you will get a Quina-type Mousterian, an industry that we do not have in the Levant.

Gerhard Bosinski: Middle Palaeolithic is a question of definition. I do not agree to this definition but I will discuss it later. Handaxes inside Middle Palaeolithic in Europe are a common thing.

Vadim Ranov: I would like to ask Naama, what is the highest chronological boundary for the cleavers in Levant? Are the cleavers a feature of the early Acheulean only, or do they exist in the Middle or Upper Acheulean too?

Naama Goren-Inbar: We do not have cleavers in the very earliest Acheulean at all. The cleavers which were reported in the past from Ubeidiya turned out to be simple big flakes and not cleavers. In this view I contradict some of my colleagues but this is my personal view. The first time we get cleavers in large quantities is at Gesher Benot Ya'aqov, which is very different from other sites and we never had a site like this in terms of cleaver quantities until this recent find in Jordan. Cleavers do continue to appear regularly in many Acheulean sites. The question is, what proportion they have within the assemblages. The number of assemblages with cleavers that were well excavated and well reported is small. At the time of Tabun E there are very few cleavers, but they are present in Ma'ayan Baruch, somewhere in the order of 300 thousand years or more.

Gerhard Bosinski: Perhaps I may add that normally cleavers are made of a coarse material and not of homogeneous silex. Perhaps this is linked with function, and perhaps this may be one of the reasons that in Tabun they are not present.

Naama Goren-Inbar: There are cleavers on very fine-grained flint in Ma'ayan Baruch.

Vadim Ranov: The Indian and Ukrainian colleagues call cleavers flakes with perpendicular edges and without lateral retouch. For me, a cleaver has a lateral steep retouch. The definitions vary at different sites.

Avraham Ronen: I would like to remark on Bosinski's question as to what is the chronological relation between Yabrudian and the Middle Palaeolithic. I think that there is no problem in this part of the world. Stratigrphically, the Yabrudian is everywhere and always below Middle Palaeolithic. In Tabun, in Yabrud, etc.

Gerhard Bosinski: There is a different use of terminology here and in our part of the world. Your Middle Palaeolithic is only the upper part; it corresponds to our Late Middle Palaeolithic from the last glaciation. We have to realize this difference. We got in Europe Middle Palaeolithic sites from 350,000 years and we include the late Acheulean as a part of the Middle Palaeolithic. We have to understand this in order to understand each other.

Avraham Ronen: This is really a different terminology. Our Middle Palaeolithic does not start at the last glaciation. Tabun D, a typical Middle Palaeolithic without handaxes and with Levallois technique, is some 250,000 years old.

Leonid Vishnyatsky: I have a remark concerning the Levallois problem. I think that the Levallois technique began in Yabrud and Tabun very abruptly. In Tabun it began in bed 68 (Jelinek's excavations) and in Yabrud in Layer 10. My question is to Avraham Ronen. You said that the Yabrudians did not produce blades. Does that mean that you consider the Amudian and Pre-Aurignacian as independent industries and not part of the Yabrudian? If so, how, in your opinion, did they appear in the region?

Avraham Ronen: I do not consider the Intra-Yabrudian blade industries as part of the Yabrudian. Hence, I do not accept the term Mugharan Tradition, which came specifically to unify the three occurrences Acheulo-Yabrudian, Yabrudian proper and the Pre-Aurignacian /Amudian. To me the technological differences between the blade and the non-blade industries are tremendous, much more accentuated than between Perigordian and Aurignacian. If you join Pre-Aurignacian and Yabrudian, you may as well join any two cultures in the world. As to your question how did the blade industries appear, I have no answer.

Paul Callow: Avraham referred to the Yabrudian as always staying in caves and suggested that perhaps they never left the caves. The question is why believe that wherever people go, they always take the same tools with them, do the same kind of activities and always leave the same signature. Is there any reason why the Yabrudian should have been making the same kind of things wherever they went?

Avraham Ronen: This, of course, is an interesting and difficult question which I am not sure to be able to answer. Modern ethnography may offer some interesting data in this regard. Take the Middle Palaeolithic as an example. The Middle Palaeolithic open-air sites in the coastal red loam have very similar contents to that found in caves. If so, why expect the Yabrudian - or the Aurignacian - to change their signature when leaving the cave?

Nigel Goring-Morris: As to Avraham's remark that Acheulo-Yabrudians are basically troglodytes. I find difficulties with it. Personally, I would not say that the site of Nadawiyeh is a real dolina. It is rather a water hole. I think that one should take into account that the number of open-air sites of all periods known from the Mediterranean climatic zone in the southern Levant is very low, and hardly any one was systematically excavated. We know practically nothing about what was going on outside the caves.

Gerhard Bosinski: It seems to me important to discuss the controversial use of the terms Lower and Middle

Palaeolithic. The definition of Middle Palaeolithic in my region, and this includes northwestern France as well, is the degree of variability of the retouched flake tools. In this sense, the Middle Palaeolithic comprises the later part of the Acheulean. If we want our meeting to be not only pleasant but also useful, we should realize that either we use different definitions in different parts of the world, or we try to unify them.

Paul Callow: There is no question that we are getting handaxes of Acheulean type occurring very late in the European sequence. They occur with or without Levallois technique, with and to some extent without elaborate flake tools. We are confident about tracking back these industries with substantial quantities of well made flake tools some 300 thousand years at any rate, possibly further. We were discussing High Lodge among other things. Obviously we, at least, have to understand each other's positions. For me, the distinction between the Lower and Middle Palaeolithic can not really be determined on the basis of a single class of objects. I think that what we are seeing here is a set of different behaviours, some of which may or may not be represented in different times and different places. I see it as a rather grey area, in fact. The problems in this area are a matter of definition. However, I am not quite sure how we should resolve this.

Naama Goren-Inbar: Sorry, but I have the feeling that we are going into a time machine backwards. I have all the respect for the European work. We always read whatever is written there and we quote it too, and sometimes we use the same methodology. But I have the feeling that there is some kind of an imposing the northern European data on the Levant. As if no work was ever done in the Levant. This is extremely surprising because the Middle Palaeolithic of the Levant is well published, at least in part. We are familiar with whatever the components of the Middle Palaeolithic are, and I must say that it is extremely different from the European situation. We do not have MTA. We call Middle Palaeolithic here whatever comes under the title of the Mousterian industries. The Lower Palaeolithic in this part of the world - and there are somewhat different opinions - are the industries which culturally come under the term of the Acheulean and the Yabrudian. There are beautiful sequences in the caves, and detailed studies of Kebara, Tabun, Amud and Kafzeh. The Middle Palaeolithic sequence always overlays the Lower Palaeolithic. This is not a question of debate. It is very different from the European picture.

Gerhard Bosinski: We too read and are familiar with the facts as far as they are published. It is only a question of definition. Your Middle Palaeolithic matches the latest part of our Middle Palaeolithic.

Jan Burdukiewicz: There is probably some mixture of two possible explanations. Do we discuss cultural entities or chronology? In prehistory we have many examples where chronology is not in agreement with cultural entities.

Gerhard Bosinski: I personally refer not to cultural, nor chronological aspects, but only to terminology: the use of words in a different way. This is what I wanted to underline.

Romuald Schild: There is a similar situation in North Africa. From 230 thousand years onward you do not have handaxes. Before that there is a lot of handaxes together with so called Middle Palaeolithic tools. That is, mainly side scrapers and Levallois technology. From 230 thousand onwards there are no handaxes. In a way it is similar to the Levant. From 230 thousand on, it is Middle Palaeolithic, whatever you choose to call it.

Nicolas Rolland: Just to add to this confusion. In East Africa you have handaxes and bifacials all the way. Every region seemed to have, at some point or another, a different kind of expression of variability during the Lower Palaeolithic. In the Middle Palaeolithic the variability expands. That is the really important thing. These are very important issues, very basic, and we have to speak a common language, acknowledging that the sequences and content of these regions are different. I am very aware that the Levant is different. But you will find variations like that in central Asia, Africa or India. Southern Europe is different from both northern Europe and the Levant.

Gerhard Bosinski: For me the Yabrudian is Middle Palaeolithic according to the variability in flake tools.

Naama Goren-Inbar: Do you take into consideration that your collection from the Yabrudian is a biased one? That it was heavily selected and only the nice items were transported?

Gerhard Bosinski: In any case, there are many side scrapers, end scrapers and transversal scrapers, and this is the definition of Middle Palaeolithic.

Vadim Ranov: I think that the term Middle Palaeolithic is problematic. For example, the traditional central Asian Middle Palaeolithic we called Mousterian. Mousterian and Middle Palaeolithic were synonym for us. When Joachim Schäfer came to central Asia, he proposed to call it not Mousterian but Middle Palaeolithic. Middle Palaeolithic is the overall cover. Mousterian is only part of the Middle Palaeolithic. We have now an assemblage from the second palaeosol, dated to 200 thousand. It may be Middle Palaeolithic, but not clearly Mousterian. For us, the Middle Palaeolithic has no bifacial tools. For us, Middle Palaeolithic begins from the time we have some elements of prepared cores.

Avraham Ronen: Why are prepared cores a better signal than handaxes? If we took prepared cores in the Levant, the Middle Palaeolithic would have started half a million years ago. Every region may pick its own signal to start the Middle Palaeolithic. But then we should know that we do not talk about the same thing.

Joachim Schäfer: You have prepared cores 500,000 years old. But do you have a Middle Palaeolithic scraper industry at that age?

Avraham Ronen: I do not recall any good scraper industry earlier than what we call Upper Acheulean, like in Berekhat Ram, Holon or Tabun F. If you remove the handaxes from Tabun F, you get a Typical Mousterian.

Gerhard Bosinski: This would be called Middle Palaeolithic elsewhere.

Avraham Ronen: You call Middle Palaeolithic assemblages which we call Upper Acheulean, because of their diversity of flake tools.

Gerhard Bosinski: I would like to sum up this topic, and make clear that we give the same term a different meaning. The title of our conference is not only Yabrudian but also Micoquian. The Micoquian is chronologically a last glacial phenomenon, and I wonder if Yabrudian and Micoquian ever met. I suggest this to be our next topic of discussion.

Avraham Ronen: Yabrudian and Micoquian did meet, at least, in Garrod's report of 1937. She called layer E of Tabun 'Micoquian', and she did not have in mind chronological matter, only typological, technical and stylistic matters. The interesting question is, I think, what makes Yabrudian and Micoquian so closely alike despite this large gap in time. Let us see what Yabrudian and Micoquian have in common, one may perhaps illuminate the other.

Gerhard Bosinski: I personally do not see a great resemblance between the two. To a certain degree, it is my fault that we call these finds Micoquian. At the time, I did not dare to give it a new name. But the proposal of Nicholas Conard, to call these assemblages Keilmessergruppen, is perhaps much better.

Jan Burdukiewicz: Do you mean that there is a genetic connection between Yabrudian and Keilmessergruppen?

Gerhard Bosinski: This depends on the definition of Micoquian. If you take the topmost level of La Micoque, yes. This was the original definition: level 7 of La Micoque. In this sense - yes.

Jan Burdukiewicz: We have the Keilmessergruppen, or backed bifaces, in other countries. I think that this is not the best term for them.

Nicholas Conard: I do not have a problem with very similar kinds of assemblages being separated in time. Flinders Petrie's excavations of workers' barracks in Giza revealed, on the floor of these barracks, Oldowan technology. That does not mean that *Homo habilis* built the pyramids. Similarly, looking at the Ruhama material with Karel Valoch, he remarked that to him it is virtually indistinguishable from Kůlna 11, and quite similar to Bilzingsleben. That does not necessarily mean that there is any cultural connection between these groups whatsoever.

Avraham Ronen: It is not necessarily cultural connections. I meant functional, and perhaps behavioural similarities. If we understand the behaviour related to the Micoquian in Europe, which is much younger, may be it will help us to better understand Yabrudian behaviour, given that they really have some similarities. If the Ruhama material is like Bilzingsleben and Kůlna 11, then it is very interesting to see how these sites may shed light on each other.

Nicolas Rolland: I just want to get back to the question of terminology. I vote strongly in favour of abandoning the term Micoquian. If the term Keilmessergruppen might be a reasonable solution, one thing that strikes me about what we have called Micoquian in central Europe and part of eastern Europe is the bifacial element. Does the term Keilmessergruppen convey that? Messer means a knife, but Keil is more like an axe, is it?

Gerhard Bosinski: As I have defined it earlier, if nobody is against it, Keilmesser is a kind of biface. It includes a bifacial sharp edge and an opposite back of different types. Jan Burdukiewicz gave us a short overview of the different types of Keilmesser. The bifacial component is very characteristic, and again, it is simply not true that the Middle Palaeolithic is without handaxes.

Nicolas Rolland: When François Bordes defined the Mousterian complex his underlying criterion was racloir frequency variations. Could something along those lines be done with the Keilmessergruppen? I know your typology was a major start, but develop it and even quantify it.

Gerhard Bosinski: We had many discussions with the late François Bordes on this issue. There were two main arguments in our discussions. One was the age of the Middle Palaeolithic, because we got Rissian sites. The second was the definition of the Bockstein material. It was only in the end of his life that Bordes had to change a little his model because of new variations of scraper types and points. There are great variations in the Middle Palaeolithic, and I can not imagine that we should exclude any of them.

Romuald Schild: I agree. It is Middle Palaeolithic. However, there are two sealed Late Acheulean sites at Dakhla, certainly before Stage 7, that contain classical Klausenischemesser and Prondniks together with unifacial side scrapers and hundreds of handaxes from amygdaloids through cordiforms, and Levallois technology. I think that they are three hundred thousand years old if not more. They also show the resharpening scars of Prondniks.

Gerhard Bosinski: I agree, if you admit that this is Middle Palaeolithic.

Romuald Schild: No. To me it is Late Acheulean. We published it as Late Acheulean. You can not change it.

Gerhard Bosinski: I can not see any special relations between the Yabrudian and the Keilmessergruppen, not in the type of working edges, perhaps in the type of retouch. But the retouch of the Yabrudian is influenced by the thickness of the pieces. I see no special relations between the Yabrudian and the Keilmessergruppen.

Avraham Ronen: Are Keilmessergruppen racloirs similar

to Yabrudian racloirs, or are they more like the typical Mousterian ones?

Gerhard Bosinski: It depends on the assemblage. There are assemblages with a very low percentage of racloirs, like Bockstein or Buhlen IIIb. There are others, like Schambach and Konigsaue, with numerous racloirs mostly on Levallois flakes. I cannot remember thick side scrapers in these assemblages.

Avraham Ronen: Garrod must have referred, then, to the bifacial component.

Naama Goren-Inbar: The only reason that Garrod called Layer E Micoquian was because of the concave edges of certain bifaces.

Gerhard Bosinski: This statement concerns the history of research.

Nicholas Conard: I was not implying, in my lecture, that we should use the term Keilmessergruppen for all Middle Palaeolithic sites in central Europe. I was using the term for the German material.

Gerhard Bosinski: This was proposed in the past. And I agree, if it concerns only the Bockstein type.

Nicholas Conard: I want to stress that the material is heterogeneous.

Gerhard Bosinski: A large part of the contributions in this conference dealt with the blade industries. Many kinds of blades were mentioned in the lectures. Pre-Aurignacian, Amudian, Seclin, Taubach, Bohunician, Boker-Tachtit, Hummalian, Obirachmat. Is there someone who finds a connection between these blade industries?

Avraham Ronen: To me these blade assemblages are among the most fascinating problems that we presently face in prehistory. They are amazing, and we understand them very little. Each occurrence is in a way unique, with no two alike. Amudian, Pre-Aurignacian and Hummalian are within close distance, but each has a different composition apart from the fact that all have blades. This is a strange phenomenon in a world that normally ignores blades. The Lower Palaeolithic is based on flakes with a few blades here and there, the Middle Palaeolithic is equally based mostly on flakes, and the Yabrudian has almost no blades at all. In the middle of all this you have a sudden introduction of what seems to be a foreign element, as I see it. These blade occurrences appear at a certain moment, towards the end of the Yabrudian and before our Middle Palaeolithic. Jelinek maintained that these assemblages were part of the Yabrudian and came gradually, because, he said, blades were actually present in earlier layers and gradually increased in frequency to form the Amudian/Pre-Aurignacian horizons. Later, the Upper Palaeolithic-like blades have disappeared, equally gradually according to Jelinek, to be entirely forgotten all along the Middle Palaeolithic. I consider this blade intrusion - whether sudden or gradual - as worthy of our attention. For example, why is the Pre-Aurignacian of Yabrud so Aurignacian-like, while the Amudian of Tabun is so non-Aurignacian like?

Naama Goren-Inbar: Amudian, Pre-Aurignacian and Hummalian are different entities. Amudian and Pre-Aurignacian are used alternatively and they mean the same thing in some of the sites in Syria, in Lebanon and here in Israel. But the Hummalian is exactly like Tabun D, which comes into the Middle Palaeolithic in our terminology. Therefore, one should not mix the Hummalian with the Pre-Aurignacian. They are totally different things.

Avraham Ronen: If the Hummalian occurs on top of the Yabrudian, it is like Tabun D. But I think that in El-Kowm, some Hummalian occurs inside the Yabrudian.

Gerhard Bosinski: The blade element in the Amudian and Pre-Aurignacian is overestimated, in my opinion.

Joachim Schäfer: I think that for people who have a broader sense of Middle Palaeolithic, it is much easier to understand this blade industry. For me, it just fits in the Middle Palaeolithic variability. I agree with Avraham when he says that each blade industry is different. Blades are just the result of variability in Middle Palaeolithic production technology.

Leonid Vishnyatsky: It seems to me that there is some confusion regarding the question of blade industries in the Middle or Lower Palaeolithic. In my opinion, one should discriminate between blade industries which are chronologically directly connected with Upper Palaeolithic times and blade industries which were separated from the Upper Palaeolithic by very, very long periods of time, may be even hundreds of thousands of years. The latter occurrences are Amudian, Pre-Aurignacian, Hummalian and others. The former occurrences are Obi-Rahmat, Boker-Tachtit and many others.
On the other hand, one should discriminate between blade industries with lots of blades but only scarce and atypical tools of Upper Palaeolithic appearance, and blade industries which have both blades and typical Upper Palaeolithic tools. Such a discrimination is necessary if we want to escape confusion in discussing problems connected with blade industries.

Daniel Kaufman: Back to the original question. We do have blade industries. The elongated Tabun D type is found mostly in arid regions. Also, Boker Tachtit has been mentioned here. It is the very end of the Middle Palaeolithic, and it is a blade industry. Typologically it is Upper Palaeolithic, except for the Emireh and Levallois points. In Tabun D, Boker Tachtit and Rosh Ein Mor there is a strong Upper Palaeolithic typological component. The typological transition occurs before the technological one.

Karel Valoch: May the Bohunician be compared to Boker Tachtit?

Romuald Schild: Marks says that they are very similar. However, I think that Boker Tachtit is slightly older. I do not think that blade industries that are pre-Last Interglacial are chipping mistakes. They are not chipping mistakes.

There is a site at Piekary, certainly pre-Last Interglacial, which contains opposed platform, non-Levallois cores for blades and backed pieces.

Nicholas Conard: There are a few things that may perhaps be relevant. One thing has always struck me as very unusual in looking at very short lived episodes. The different brief halts may be measured in hours in some cases, or a few days, or at most a few weeks. Assemblages lying right on top of each other are not necessarily of the same group. In fact, they may be very markedly different. It sounds strange to me to use the term Tabun B or D as a homogeneous unit which lasted maybe 50,000 years. In two of my sites, where I had blade assemblages, they are very different, and both have no blades below or above. Both sites look like stations of short hunting episodes.
Just to turn my previous comment into a specific question: do all my colleagues here accept the interpretation that over the period of time of, for example, Tabun D, C or B, may be fifty thousand years, that at this place only one thing was happening over and over again? Is it at least possible, that there are multiple uses over and over again, but taphonomical processes have stirred them up so much that it becomes homogeneous?

Avraham Ronen: It is difficult to answer. Everyone has his own experience and his own ideas. I can tell you, for example, that in the red soils you can sometimes separate horizons of occupation but most of the time you can not. The Middle Palaeolithic of the red soils is always the same. It may be assumed that during the Middle Palaeolithic, one and the same activity was carried out repeatedly in the red soil, and a very similar inventory of tools was repeatedly left there. The same can be said of Tabun C, where the 19 beds distinguished by Jelinek had all similar assemblages.
I am sorry that I was not precise enough when saying why am so amazed at this blade phenomenon. It is not just the blades, but the way they were manufactured. Middle Palaeolithic blades are nothing but elongated flakes, made in the flake technology. But in the Pre-Aurignacian of Yabrud there is a real blade technology. They are struck like Upper Palaeolithic blades. It is enough to read Rust's amazement when he found them in his layer 15, where they could not be anticipated. And it is enough to see the amazement of Skinner when he found blades in his Masloukh excavations. He could not believe that they really belonged there, and he felt obliged to explain them as intrusions through the chimney from an alleged Upper Palaeolithic open-air site. What amazes me is that the Yabrudians, who manufactured thick flakes for thick scrapers, suddenly shifted to these thin, elongated, beautiful blades unlike anything that had been made prior to the Upper Palaeolithic.

Leonid Vishnyatsky: I would like to make a reservation about the Upper Palaeolithic nature of the Amudian blades. You said that they were produced by means of Upper Palaeolithic technology. I want to stress that the preparation of the striking platform is not the same as in the Upper Palaeolithic. That is why Amudian and Pre-Aurignacian blades have thick butts.

Nigel Goring-Morris: You mentioned blades being twice as long as they are wide. Not everybody uses this definition. Jelinek uses anything with parallel sides as blades, and so does Rollefson for the PPNB.

Jan Burdukiewicz: According to Leroi-Gourhan, a blade is five times as long as it is wide.

Romuald Schild: Some say that blades are no more than elongated flakes.

Gerhard Bosinski: Thank you for the precisions. There are now at least four different definitions.

CONCLUDING REMARKS

Avraham Ronen and Mina Weinstein-Evron

Our attention has been repeatedly drawn to the ambiguous term 'Middle Palaeolithic', which apparently has a different meaning when used in the Levant, in west-central Europe, in central Asia or in Africa. Is it a matter of substance or chronology? These two questions will be further addressed below. What may be said, then, on the ties, if any, between the cultural complexes discussed in this volume?

The Yabrudian

The Yabrudian is characterized by bifaces, bifacial backed knives, a racloir-rich flake component and Quina-type typology and retouch. It is a non-Levallois and non-laminar industry. This definition of the Yabrudian is the same as that given by Rust (1950) when identifying this culture and later by Copeland (1983; this volume). The Yabrudian is largely different from the Upper Acheulean, its contemporary. Although still awaiting a detailed study, the Upper Acheulean has apparently different biface forms. Its flake component may be equalled to a 'Typical Mousterian', the only type of the succeeding Middle Palaeolithic known to be present in the Levant. The Upper Acheulean also exhibits a well developed Levallois technique. Blades, rather rare here too, are relatively more numerous and better made than in the Yabrudian. Yabrudian and Upper Acheulean also have a very different geographical repartition: the latter is wide-spread in the entire Levant and adjacent areas (Bar-Yosef, 1994) whereas the former is confined to a restricted area of the central Levant, in northern Israel, Lebanon and Syria (Hours et al., 1973). A final difference to be noted here is that the overwhelming majority of Upper Acheulean assemblages presently known derive from open-air sites (Bar-Yosef, 1994) while almost all the Yabrudian assemblages come from caves. The only exception is the El-Kowm basin in Syria, where open-air Yabrudian sites are encountered (Le Tensorer et al., 1999). Recent dating places the Yabrudian of Tabun between ca. 350 and 250 Kyears (Mercier et al., 1995; nd).

Prehistoric terminology is influenced, in varying degrees, by both substance and chronology. For Garrod there has been no question that the Yabrudian – or the Micoquian, as this entity was first termed by her – was Lower Palaeolithic (Garrod and Bate, 1937; Garrod, 1956). Later, however, given that the Yabrudian was attributed to the Riss-Würm interglacial (Jelinek et al, 1973), and considering that this period had been assigned in Europe to the Middle Palaeolithic, the Yabrudian was declared Middle Palaeolithic and consequently, has been defined as a sort of 'Mousterien de Tradition Acheuléenne' with Quina affinities, non-Levallois (Hours et al. 1973, 246). But a chronological shift does not alter the substance, or context, of a prehistoric industry. In western Europe, MTA is part of the Mousterian culture complex while in the Levant, the Yabrudian clearly predates the Mousterian. By its stratigraphical context and its lithic composition, the Yabrudian should be regarded as one of the terminal industries of the Lower Palaeolithic in the Levant. This assignment can not be altered by chronological considerations.

Early Blade Assemblages

Every Yabrudian cave site had, in the upper part of the Yabrudian sequence, an assemblage characterized by blades with remarkably close affinities to Upper Palaeolithic blade production (Rust, 1950; Garrod, 1970; Lamdan and Ronen, 1989; Ronen, 1992). In spite of occupying a similar stratigraphical position and thus being roughly contemporaneous (their age may be estimated at 250 - 270 kyears), no two of these assemblages have a similar typological composition (Vishnyatsky, this volume). The Yabrud shelter Pre-Aurignacian, for example, has Aurignacian-like tools while the Tabun Amudian has some fine curved-backed blades but is otherwise poor in Upper Palaeolithic tool types. Common to all these occurrences, however, is a mode of blade production which had created slightly thicker products than true Upper Palaeolithic blades, like a very early Upper Palaeolithic (Rust, 1950; Bordes, 1955). These blanks in fact more closely resemble Upper Palaeolithic ones than any succeeding Middle Palaeolithic blades produced in the Levant (Ronen, 1992).

Other Pre-Upper Palaeolithic blade assemblages have been subsequently found in Europe and South Africa. They are assigned to Oxygen Isotope Stages 5 - 4 (Tuffreau et al., 1994; Sitlivy-Escutenaire and Sitlivy, 1996), later than the Levantine occurrences mentioned above. These late occurrences too show a wide diversity of typological compositions: Mousterian in Seclin, Micoquian in Vinneuf and Upper Palaeolithic in Riencourt-les-Bapaume (Tuffreau et al., 1994, 44). Hence, they do not coincide with any well-defined industrial group, or tool-kit tradition, or a specific activity (Tuffreau et al., 1994). There is a growing tendency to view these blade assemblages as a normal part of Middle Palaeolithic technological diversity, especially in its final phase (Tuffreau, 1995; Hovers, 1998; Vishnyatski, this volume; Ranov and Schäfer, this volume). This leaves unexplained the heterogenous Middle Palaeolithic 'blade phenomenon' in a rather narrow time slot, over a wide geographical range with variable environments. The possible seminal role of the much earlier, Intra-Yabrudian counterparts should not be ignored.

The Micoquian

Within its narrow geographical boundaries, the Yabrudian is a homogenous entity. The Micoquian, stretching from east-central France to the Volga, with a considerable environmental variability, is largely heterogenous and does not readily come under a clear definition (Koulakovskaya et al., 1993; Conard, this volume). In its core area, eastern and central Europe, the Micoquian is characterized by a high incidence of bifacial pieces, foliates and, especially, knives. The typical Micoquian backed knives (Bosinski, 1967), or 'Keilmesser' (Conard, this volume) are more generally grouped under 'asymmetrical bifaces' (Koulakovskaya et al., 1993, 10), with as yet insufficient criteria for their definition (Koulakovskaya, 1995; Burdukiewicz, this volume). Technologically, it is a non-Levallois and non-laminar industry (Koulakovskaya et al., 1993), with a disc-core mode of production (Boëda, 1995).

The western Micoquian is of a later date (Richter, 1997) and may represent a westward movement into restricted areas (e.g., Farizy, 1995). Here, the typical Micoquian 'backed bifacial knives' are sometimes mingled with such non-Micoquian elements as Charentian tool types (both Ferrassie and Quina) and Levallois reduction method (e.g., Champlost, Farizy, 1995, and Sesselfelsgrotte, Richter, 1997). If these are indeed genuine, unmixed occurrences, then one may perhaps envisage a border-zone behaviour where foreign influences are more readily absorbed (cf. Hovers, 1998), similar to the intensive linguistic creativity where pidgin and 'creole' languages are born (e.g., Lehman, 1973, 8).

In eastern and central Europe the Micoquian ranges from the Eemian through mid-Würm (Burdukiewicz, this volume; Mania, this volume), roughly to 40 kyears (Valloch, this volume). In western Europe it is contemporaneous with Late Mousterian, dated to the early (Farizy, 1995) or middle Würm (Richter, 1997). Hence the Micoquian seems to be far younger than the Yabrudian. But there are some older, pre-Eemian Micoquian occurrences dated to Oxygen Isotope Stage 7 or even earlier, in Poland (Foltyn et al., this volume) and in Egypt (Schild, discussion, this volume). These considerably narrow the Yabrudian - Micoquian chronological gap.

The Quina Mousterian

The Quina industry is very rich in racloirs, many of which are convex, tranversal or dèjeté, made on thick and braod flakes which allowed for step-retouch in three or even more series of trimmings. The retouch affects a large part - sometimes 1/3 - of the dorsal face, and in the case of retouch on two edges, most of the dorsal face is affected. The technique is non-Levallois.

As repeatedly mentioned during this conference, these are pretty much the characteristics of the Yabrudian racloirs as well. Remove the bifaces from the Yabrudian, and one is left with a Quina type assemblage. The Yabrudian was in fact claimed to be a 'Quina variant' (Rolland, 1996, 130). Quina Mousterian is mostly dated to Oxygen Isotope Stage 4, while the Yabrudian is assigned roughly to Oxygen Isotope Stages 9 - 8. According to the accepted paradigm, this time difference would argue against a direct relationship between these entities.

Time

Cultural connections are evaluated upon the dual measures of substance and time. The degree of similarity is evaluated, then the temporal proximity. Both measures have to satisfy for a positive conclusion to be reached.

Substance and chronology are both difficult to measure. Would archaeologists ever tie ancient Rome with the modern West, were it not for written documents? Although time seems the more easily measurable of the two, this is only an illusion. What most of us know, or concieve, of time is the second half of the 20th century. There is no way to relate our concept of time with Palaeolithic time, 'temps profond' (Olivier, 1998, 19). The huge expanse of Palaeolithic time is staggering (Joulian and Gage, 1997) to the point that it is totally incomprehensible and inconsistently manipulated. Thus, on the one hand, few would question 'continuity' within the 'Mousterian tradition' of 330 - 35 kyears (e.g., Moncel, 1998, 259), but even a shorter time gap could not be tolerated by Bordes (1955) when considering 'continuity'. From the point of view of substance, Bordes had no difficulty to admit that the Pre-Aurignacian could have been a primitive precursor of the Aurignacian; but in view of the time gap, Bordes (1955) resorted to that naive exercise of squeezing Palaeolithic time in order to bring these two entities closer.

At this stage of our knowledge it is not possible to demonstrate connections between Early Blade Assemblages and the Upper Palaeolithic, or between Yabrudian, Micoquian and Quina Mousterian. Similarities clearly exist, and it should be remembered that time gaps are perhaps 'insurmountable' only in our 20th century minds.

Culture-Nature

Bordes' interpretation of Quina as a cultural-ethnic phenomenon (Bordes, 1953; 1981) lost its appeal in the last three decades or so. The trend in vogue favours a particular environmental setting such as adaptation to severe climatic conditions (Rolland, this volume) or the paucity of good-quality raw material (Rolland, 1996; this volume). Some Levallois products which surprisingly occur within an otherwise Quina assemblage in the Belgian cave of Sclayn (Moncel, 1998, 205) were explained as items made on high quality, imported flint, while the Quina industry was made on the local, more mediocre flint. A somewhat similar interpretation was advanced to explain the admixture, in varying doses, of Micoquian, Mousterian and Levalloisian components at Sesselfelsgrotte (Richter, 1997; 1999).

There appears, however, a growing trend to interpret flint tools as a cultural choice. Studying the relations between raw materials and blank production, Guislain (1996, 225) concluded that raw material does not determine flaking or retouch methods. It was similarly suggested that the retouch of a few edges of a blank, and bifacial retouch, are culture-dependent (Marks et al., 1996, 281).

The thickness of Quina, as well as of Yabrudian, racloirs does not appear to reflect an economical use of raw material, nor a production intended for a long use with frequent re-sharpening: the material needed for a single Quina racloir would allow to produce two, if not three 'normal' racloirs, and it is doubtful whether the single Quina implement would have outlived the others. Quina or Yabrudian racloirs, furthermore, are never very small, as would be expected at the end of numerous re-sharpening cycles.

Intensivity of retouch may sometimes be negatively correlated with the distance to raw material source, and not positively, as would be expected by the 'economical' approach. At Combe-Capelle Bas, for example, a richly retouched lithic assemblage is found right on a source of a high quality flint (Dibble and Lenoir, 1995).

It is widely accepted that the activities carried out in caves and in open-air sites differed. Yet two studies independently concluded that there was no difference in the lithic assemblages found in these two settlement types, both in the Mousterian (Ceb and Waiblinger, 1999) and the Micoquian (Koulakovskaya, 1990, 45). Hence, it would seem that rather than economical- or environmental-dependent, lithic technology is above all a social marker, vertically transmitted within the group's boundaries, as recently argued (Hovers, 1998). An observation made by Tuffreau (1995, 423) is relevant here, namely that Middle Palaeolithic blades were used to cut meat, while racloirs were used for wood working. If so, then blades did not come to fulfill any new task which might have become vital in a particular time or place; instead, they were used for a trivial task which, for some 3 million years, has been successfully performed by flakes.

The culture-nature debate reflects on how we view our past. If 'ancient' should be viewed in a natural and biological perspective while 'modern' in a cultural and social perspective, the problem is of course where to place the boundary (Joulian and Gage, 1997; Fiedler, 1999). 'Modern' should be extended, at the very least, back to the appearance of burials (e.g., Kaufman, 1999, contra-Stringer and Gamble, 1993, 212), and probably as far back as the domestication of fire (Ronen, 1998). The variability of lithic technologies during the period covered by this conference, 400-50 kyears; the territoriality of these technologies; the high standardization of production methods and tool forms; and the temporal proximity to modern humans, all point to the strong possibility that the discernible behaviours were determined culturally and socially rather than biologically-environmentally. The Acheuleans may, after all, have been a people (Otte, this volume). Humans do not just live in society, as primates do; humans 'produce society for living' (Godelier, 1998), and are capable of monitoring it and its relations with the world, as required. Hence all human comportment, including tools, is social symbolism, and above all, culturally determined. Cutting meat with blades rather than with flakes is like using chopsticks rather than forks: both transcend the task to be done. It is how a task is done that bears upon the group's identity.

Bibliography

Bar-Yosef, O., 1994. The Lower Paleolithic of the Near East. *J. World Prehistory* 8, 211-265.

Boëda, E., 1995. Steinartefact-Produktionssequenzen im Micoquien der Kůlna-Höhle. *Quartär* 45/46, 75-98.

Bordes, F., 1953. Essai de classification des industries 'moustériennes'. *Bull. Soc. Préh. Française* 50, 457-466.

Bordes, F., 1955. Le Paléolithique inferieur et moyen de Jabrud (Syrie) et la question du preaurignacien. *L'Anthropologie* 59, 486-507.

Bordes, F., 1981. Vingt-cinq ans après: le complexe mousterién revisité. *Bull. Soc. Préh. Française* 78, 77-87.

Bosinski, G., 1967. *Die mittelpaläolithische Funde im westlichen Mitteleuropa.* Fundamenta A/4. Köln.

Cep, B. and Waiblinger, J., 1999. The Use of Cave and Open-Air Sites in Southern Germany. In: *Middle Palaeolithic and Middle Stone Age Settlement Systems.* UISPP Commission 27, p. 9. Tübingen.

Copeland, L., 1983. The stone industries, the Acheulo-Yabrudian of Bezez Cave, Level C. In: Roe, D. A. (ed.), *Adlun in the Stone Age*, pp. 89-194. BAR Internl. Series 159, Oxford.

Dibble, H. L. and Lenoir, M., 1995. *The Middle Palaeolithic Site of Combe-Capelle Bas (France).* University of Pennsylvania.

Farizy, C., 1995. Industries Charentiennes à influences micoquiennes, l'example de l'Est de la France. *Paléo*, Suppl. 1, 173-178.

Fiedler, L., 1999. Repertoire und Gene. *Germania* 77, 1-37.

Garrod, D. A. E., 1956. Acheuleo-jabrudien et preaurignacien de la grotte de Taboun, Mont Carmel. *Quaternaria* 3, 39-59.

Garrod, D. A. E., 1970. Pre-Aurignacian and Amudian: a comparative study of the earliest blade industries of the Near East. *Fundamenta* A/2, 224-229. Köln.

Garrod, D. A. E. and Bate, D. M. A., 1937. *The Stone Age of Mount Carmel,* Vol. I. Clarendon Press, Oxford.

Godelier, M., 1998. Quelles cultures pour quelles primates, définition faible ou définition forte de la culture? In: Ducros, A., Ducros, J. and Joulian, F. (eds.), *La culture est-elle naturelle?,* pp. 217-222. Editions Errance, Paris.

Guislain, S., 1996. Relations entre matières premières lithiques et élaboration technique. *Proceedings XIII UISPP Congress,* Vol. 2, 219-225.

Hours, F., Copeland, L. and Aurenche, O., 1973. Les industries paléolithiques du Proche-Orient, essai de correlation. *L'Anthropologie* 77, 229-280.

Hovers, E., 1998 The lithic assemblages of Amud Cave. In: Akazawa, T., Bar-Yosef, O. and Aoki, T. (eds.), *Neandertals and Modern Humans in Western Asia*, pp. 143-163. Plenum, New York.

Jelinek, A. J., Farrand, W. R., Haas, G., Horowitz, A. and Goldberg, P., 1973. New excavations at the Tabun Cave, Mount Carmel, Israel, 1967-1972: A preliminary report. *Paléorient* 1, 151-183.

Joulian, F. and Gage, J. C., 1997. Hominization and apes: An unnatural kinship. *Diogenes* 180, 73-97.

Kaufman, D., 1999. *Archaeological Perspectives on the Origins of Modern Humans. A View from the Levant.* Bergin and Garvey, Westport.

Koulakovskaya, L., 1990. Leaf-Shaped Implements in 'Eastern Micoquian' of the European Part of USRR. *ERAUL* 42, 43-47.

Koulakovskaya, L., 1995. Aspects typologiques des industries micoquiennes: le site de Korolevo en Ukraine. *Paléo*, Suppl. 1, 207-211.

Koulakovskaya, L., Kozlowski, J. K. and Sobczyk, K., 1993. Les couteaux micoquiens du Würm ancien. *Préhistoire Européenne* 4, 9-32.

Lamdan, M. and Ronen, A., 1989. Middle and Upper Palaeolithic blades in the Levant. In: Hershkovitz, I. (ed.), *People and Culture in Change*, pp. 29-36. BAR Internl. Series 508, Oxford.

Lehmann, W. P., 1973. *Historical Linguistics, an Introduction.* Holt, Rinehart and Winston, New York.

Le Tensorer, J.-M., Jagher, R. and Muhesen, S., 1999. Palaeolithic settlement dynamics in the El Kowm Basin (Central Syria). In: *Middle Palaeolithic and Middle Stone Age Settlement Systems.* UISPP Commission 27, p. 27. Tübingen.

Marks, A. E., Monigal, K. and Demidenko, Y., 1996. The Crimean Mousterian site of Starosele: Industry, dating and fossils. *Proceedings XIII UISPP Congress,* Vol. 2, 279-287.

Mercier, N., Valladas, H., Valladas, G., Jelinek, A., Meignen, L., Joron, J. L., and Reyss, J. L., 1995. TL dates of burnt flints from Jelinek's excavations at Tabun and their implications. *J. Archaeological Science* 22, 495-509.

Mercier, N., Valladas, H., Froget, L., Joron, J. L. and Ronen, A., nd. *Datation par la thermoluminescence de la base du gisement de Taboun (Mont Carmel, Israel).* Compte Rendu Academie des Sciences, Paris.

Moncel, M.-H., 1998. L'industrie lithique de la grotte Scladina (Sclayn), La couche mousterienne éemienne 5. In: Recherches aux grottes de Sclayn, Vol. 2. *ERAUL* 79, 181-249.

Olivier, L., 1998. Le sombre abime du temps. Temps profond et identités collectives aux commencements de l'archeologie du passe pre-classique. In: Ducros, A., Ducros, J. and Joulian, F. (eds.), 1998. *La culture est-elle naturelle?* pp. 99-112. Editions Errance, Paris.

Richter, J., 1997. *Der G-Schichten-Komplex der Sesselfelsgrotte. Sesselfelsgrotte III.* Saarbrucker Druckerei und Verlag.

Richter, J., 1999. In Lack of a Wise Old Man: Late Neanderthal Occupational Patterns in the Altmuhl River Valley, Bavaria. *UISPP Commission* 27, p. 24.

Rolland, N., 1996. The behavioural significance of Quina assemblage type: exploring the 'Charentian' succession case study in Aquitaine. In: Carbonell, E. and Vaquero, M., *The Last Neandertals the First Anatomically Modern Humans*, pp. 129-145. Tarragona, Universitat Roviara i Virgili.

Ronen, A., 1992. The emergence of blade technology: Cultural affinities. In: Akazawa, T., Aoki, A. and Kimura, T. (eds.), *The Evolution and Dispersal of Modern Humans in Asia*, pp. 217-228. Hokusen-Sha, Tokyo.

Ronen, A., 1998. Domestic fire as evidence for language. In: Akazawa, T., Bar-Yosef, O. and Aoki, T., (eds.), *Neandertals and Modern Humans in Western Asia*, pp. 439-447. Plenum, New York.

Rust, A., 1950. *Die Hohlenfunde von Jabrud (Syrien).* Karel Wacholtz, Neumunster.

Sitlivy-Escutenaire, C. and Sitlivy, V., 1996. Variabilité des technologies laminaires avant le Paléolithique Supérieur classique dans la région du Lac Baikal (Siberie, Russie). Étude complète du matériel, analyses comparatives avec l'Europe occidentale. *Préhistoire Européenne* 8, 49-96.

Stringer, C. and Gamble, C., 1993. *In Search of the Neanderthals.* Thames and Hudson, London.

Tuffreau, A., 1995. The variability of Levallois technology in northern France and neighboring areas. In: Dibble, H. L. and Bar-Yosef, O. (eds.), *The Definition and Interpretation of Levallois Technology,* pp. 413-427. Prehistory Press, Madison.

Tuffreau, A., Révillion, S., Somme, J. and Van-Vliet-Lanoe, B., 1994. Le gisement paléolithique moyen de Seclin (Nord). *Bull. Soc. Préh. Française* 91, 23-46.

www.ingramcontent.com/pod-product-compliance
Lightning Source LLC
LaVergne TN
LVHW071206210926
848482LV00015B/733